Introduction to Comets

The return of Halley's Comet in 1986 gave a great boost to interest in comet science. Spacecraft investigations resulted in a significant increase in understanding of the comet, and a greatly increased interest in comet science as a whole. We have now reached an exciting time in cometary research. With several missions launched or about to be launched in the near future, including *Stardust*, *Deep Impact*, and *Rosetta*, cometary research is rapidly being driven forward.

This book describes the wealth of information known prior to the return of Halley's Comet, and the new information discovered since then. It presents material on important background topics including observational techniques, plasma physics, celestial mechanics, the solar wind, and cosmogony. The science of comets is described in order of its discovery, from tail phenomena to coma morphology through to the most recent findings from space missions. The relationship between comets and asteroids is discussed, and future space missions to investigate comets are described. This comprehensive text is a complete and up-to-date treatment of the subject, suitable for advanced undergraduates and graduate students of astronomy and planetary science.

JOHN BRANDT has held positions as a researcher, teacher, and administrator. He served for 20 years as chief of a large NASA laboratory, and was Principal Investigator for the Goddard High Resolution Spectrograph on the *Hubble Space Telescope*. He received the NASA Medal for Exceptional Scientific Achievement in 1978 and 1992, and has had a minor planet, 3503 Brandt, formally named after him for his fundamental contributions to the understanding of solar system astrophysics. Brandt has authored numerous research papers on comets, with an emphasis on plasma tails and the interaction of comets with the solar wind. He was co-author, with Carolyn Collins-Petersen, of *Hubble Vision*, published by Cambridge University Press in 1998. Since 1999, he has been an Adjunct Professor of Physics and Astronomy at the University of New Mexico.

ROBERT CHAPMAN was Professor of Astronomy at the University of California, Los Angeles for 3 years, where he taught various astronomy courses and carried out research on the sun and stars. He then moved to the NASA Goddard Space Flight Center as a research scientist, later becoming a manager and study scientist on several NASA missions. In 1986/7, Chapman was detailed by NASA to the White House, Office of Science and Technology, where he provided support to the

Assistant Director for Space Science and Technology. He has previously written three successful books about comets, including, with John Brandt, *The Comet Book*, which was selected by the *Library Journal* as one of the top 100 science books of 1984. His teachers guide for comet Kohoutek won the America Institute of Physics, United States Steel Foundation Science Writing Award in Physics and Astronomy. Chapman completed his career at Honeywell International, where one of his major duties was supporting HST operations.

Introduction to Comets

Second Edition

JOHN C. BRANDT AND ROBERT D. CHAPMAN

CAMBRIDGE
UNIVERSITY PRESS

PUBLISHED BY THE PRESS SYNDICATE OF THE UNIVERSITY OF CAMBRIDGE
The Pitt Building, Trumpington Street, Cambridge, United Kingdom

CAMBRIDGE UNIVERSITY PRESS
The Edinburgh Building, Cambridge CB2 2RU, UK
40 West 20th Street, New York, NY 10011–4211, USA
477 Williamstown Road, Port Melbourne, VIC 3207, Australia
Ruiz de Alarcón 13, 28014 Madrid, Spain
Dock House, The Waterfront, Cape Town 8001, South Africa

http://www.cambridge.org

First published 2004

Printed in the United Kingdom at the University Press, Cambridge

Typeface Times 11/14 pt   *System* LATEX 2_ε [TB]

A catalog record for this book is available from the British Library

Library of Congress Cataloging in Publication data

Brandt, John C.
Introduction to Comets / John C. Brandt and Robert D. Chapman. – 2nd ed.
p.   cm.
Rev. ed. of: Introduction to comets. 1981.
Includes bibliographical references and index.
ISBN 0 521 80863 4 ISBN 0 521 00466 7 (pbk.)
1. Comets.   I. Chapman, Robert DeWitt, 1937–
II. Brandt, John C. Introduction to comets.   III. Title.
QB721.B798 2003
523.6–dc21   2003043958

ISBN 0 521 80863 4 hardback
ISBN 0 521 00466 7 paperback

Contents

Plate section between pages 248 to 249

Preface

In the preface to our 1981 monograph on comets, we pointed out that comets were not then topics of widespread interest in the astronomical research community. We suggested that the situation might change with the 1986 return of Halley's Comet and the likelihood of *in-situ* studies by spacecraft. As it turns out, we were right on. This is not a self pat on our collective backs; the prediction was not a difficult one to make. We have always been fascinated by these enigmatic objects. A whole armada of spacecraft flew by Halley's Comet in March 1986. The result was twofold: a significant increase in our understanding of that comet; and, a tremendous increase of interest in all comets throughout the astronomical community. Our interest has been sustained through the 1990s by the impact of the fragments of comet Shoemaker–Levy 9 with Jupiter in 1994 and by the extensive studies of the bright and spectacular comets Hyakutake in 1996 and Hale–Bopp in 1997. Recently, another spacecraft has made images of the nucleus of comet Borrelly, and several more are on their way to comets or are soon to be launched. Ground-based observations and theoretical studies have kept pace with the space-based observations. An additional change in our understanding is indicated by the chapter on asteroids. It is now clear that there is much more of a family link between comets and asteroids than we thought 20 years ago. Asteroids and comets are of extra interest as objects that can pose a threat to our civilization through the possibility of catastrophic impacts. We expect the increased interest in comet studies to continue into the foreseeable future.

The purpose of this monograph is to fill in the wealth of details implied in the last paragraph – the historical background of the subject, the observational and theoretical underpinning of cometary science, and prospects for the future – for students, scientists, and interested members of the general public. Cometary science is quite interdisciplinary. You will find in the text, material on observational techniques, plasma physics, celestial mechanics, the solar wind, cosmogony and numerous other background subjects. One of these subjects we found to be particularly interesting. Armand Delsemme, one of the leaders in cometary research, summed up his career in an article, "Recollections of a cometary scientist", that is well worth reading (Delsemme 1998). In the 1950s he was working as an industrial chemist in Belgium, and learned of Whipple's *dirty iceberg* model of cometary nuclei. At the time he had been working on a problem that led him to study what are known as clathrate hydrates. He approached Pol Swings at the University of Liège, and suggested that cometary nuclei may be clathrate hydrates. This important suggestion was the start of his highly productive career as a cometary scientist. As we will show, recent work has suggested that amorphous ice, rather than clathrate hydrates, is the most important constituent of nuclei. But that takes nothing away from Delsemme's illustrious career.

We had to make a decision on the order of topics as we prepared the book. The source of all cometary phenomena begins in the nucleus. Material released from the nucleus flows through the coma and ultimately reaches the tail. Physically, it makes some sense to follow this phenomenological order in the presentation. However, the historical approach also makes sense. When one observes a comet, the most obvious feature is its tail. The morphology of tail phenomena were described early on. The head is also obvious, and the morphology of the coma was described around the same time as tail phenomena. Observations of tail phenomena led to the discovery of the solar wind in the 1950s. The nucleus remained elusive, and definitive observations were not made until during and after the 1986 perihelion passage of comet Halley. In the end, the historical perspective won out in our thinking. We close the book with a summary chapter in which we attempt to show how all the material in the monograph fits together.

References are listed in the "Suggested readings" at the end of the text, organized by chapter. The serious student of comets can find much additional information in the papers, proceedings and review volumes cited there. There is a certain amount of repetition in the "Suggested readings," since a number of the items are relevant to more than one chapter.

We are indebted to many colleagues for illustrations, helpful criticism and assistance in organizing our thoughts as we prepared various sections. The list is extensive, and we cannot thank everyone in these pages. However, three individuals have been helpful at an exceptional level, and we are pleased to single them out for particular thanks: they are Walter Huebner, David Hughes, and Donald Yeomans. These gentlemen reviewed the entire document. We also thank Michael A'Hearn, Michael Belton, Stephen Maran, Lucy McFadden and Malcolm Niedner for reviewing major sections of the manuscript. We hope that the reviews have made the monograph relatively free of errors. Those that remain are entirely our responsibility.

1

Comets in history

In recent years, the science of archeoastronomy has taught us a great deal about prehistoric human's knowledge of the night sky. There can be little doubt that early humans came to know the sky – its diurnal risings and settings, its seasonally shifting patterns of stars, its changing lunar phases – as both a clock and a calendar. One can try to imagine the reaction of these prehistoric people to the mysterious appearance of a bright comet in the sky. Probably other notable events would have occurred at the same time – the death of a loved one, the birth of a child, a killing drought, or an especially successful hunt. If they were at all superstitious, they would have viewed the comet as an omen for whatever important event occurred while it was visible.

The history of people's views of nature, has taught us that comets have always been surrounded by an aura of awe and mystery. Many people shared Aristotle's view that the appearance of a comet signaled disaster or drought. The appearance of a bright comet struck fear in the hearts of its viewers, and with the fear came considerable interest. Even today, the appearance of a bright comet sparks immense public excitement. And, of course, comet researchers are deeply interested. The appearance of comet Halley in 1986 is a case in point. The worldwide scientific community prepared an armada of spacecraft to make close-up studies of the comet, and ground based astronomers mounted an International Halley Watch to track the comet. Our ancestors had one great advantage over us. They did not suffer from the light pollution that plagues large population centers today, and makes it extremely difficult for the general population to see the faint details that are visible in dark skies. Comet Halley was a beautiful sight when viewed in the dark countryside, but was barely visible from within the cities.

1.1 Antiquity to the fifteenth century

A study of the history of ancient people's views about comets is made difficult by the fact that we possess few of their original writings today. Instead, we must rely on secondhand sources such as Aristotle and Seneca. Even so, we do have some very old records of comets. Among the oldest is a Babylonian inscription interpreted as a reference to the comet of 1140 B.C. "A comet arose whose body was bright like the day, while from its luminous body a tail extended, like the sting of a scorpion."

Early philosophers debated whether comets were celestial objects or phenomena in the earth's atmosphere. The Babylonians (or Chaldeans) are credited with the opposing ideas that comets are cosmic bodies, like planets, with orbits, and that comets are fires produced by violently rotating air. The Pythagoreans (sixth century B.C.) and Hippocrates (*c.* 440 B.C.) were reported to have considered comets as planets that appeared infrequently and, like Mercury, did not rise very far above the horizon. Hippocrates and his student Aeschylus also believed that the tail was not an integral part of the comet but rather an illusion caused by reflection. Anaxagoras (499–428 B.C.) and Democritus (*c.* 420 B.C.) thought that comets were conjunctions of planets or wandering stars; Democritus apparently believed that certain stars were left behind when comets dissolved. Ephorus of Cyme (405–330 B.C.) reported that the comet of 371 B.C. split into two stars. Seneca considered this impossible and accused Ephorus of spicing up his tales for public consumption. We know today that some comets have split into two or more pieces, as we will discuss throughout this book.

A contemporary of Aristotle, Apollonius of Myndus, rejected the view that comets were an illusion or fire and asserted that they were distinctively heavenly bodies, orbiting the sun. Some of the early concepts of comets seem quite reasonable to us today. Certainly, many thinkers viewed comets as celestial objects. However, the most influential ideas for almost two millennia were those of Aristotle (384–322 B.C.), as set forth in his *Meteorology* (1952). In this famous work he first discussed the views of others, then presented his own concepts. One of Aristotle's arguments used to rule out the planetary nature of comets is the fact that they can appear anywhere in the sky, and are not limited to the immediate vicinity of the ecliptic, the path that the sun, moon and planets follow around the sky. In addition, comets could not be a conjunction of planets or a coalescence of stars because many comets had been observed to fade away without leaving behind one or more stars.

Aristotle apparently was impressed with the irregular and unpredictable nature of comets, particularly when contrasted to his philosophical concept of the unchanging nature of the heavens. Hence, he considered that they could not be astronomical bodies but were the product of meteorological processes in our atmosphere; specifically, they lay below the moon. He wrote:

We know that the dry and warm exhalation is the outermost part of the terrestrial world which falls below the circular motion. It, and a great part of the air that is continuous with it below, is carried around the Earth by the motion of the circular revolution [the same motion that carries the celestial sphere around the Earth]. In the course of this motion *it* often ignites wherever it may happen to be of the right consistence . . . We may say, then, that a comet is formed when the upper motion introduces into a gathering of this kind a fiery principle not of such excessive strength as to burn up much of the material quickly, nor so weak as soon to be extinguished, but stronger and capable of burning up much material, and when exhalation of the right consistency rises from below and meets it. The kind of comet varies according to the shape which the exhalation happens to take. If it is diffused equally on every side the star is said to be fringed, if it stretches out in one direction it is called bearded.

(Aristotle 1952:450)

This embryonic classification scheme for comets actually survived at least until books written on the comet of A.D. 1577. Aristotle apparently accepted the idea that comets were omens of droughts and high winds. On Aristotle's own ground, this follows somewhat logically because of the "fiery constitution" of the exhalation. Finally, Aristotle thought that the Milky Way was composed of the same material as the comets.

It is easy to be impatient with and critical of Aristotle's views, but this is not fair. His hypothesis, considered in light of the physics of the era, was a good attempt to explain the sudden appearance, unusual movements, and highly irregular shapes of comets. Aristotle himself considered his explanation satisfactory if it was free of impossibilities. Our ire should be reserved for those investigators 2000 years later who could do no better.

Aristotle's ideas gradually grew in importance. Posidonius (135–51 B.C.) synthesized Aristotle's and added some of his own. Although he regarded comets as atmospheric phenomena, he stated that there were more comets than are usually observed because some are lost in the glare when near the sun. This idea came from an observation made by Posidonius himself of a comet near the sun becoming visible during a total solar eclipse. The classification of comets was according to their shapes.

Seneca (4 B.C.–A.D. 65) had a classification scheme similar to that of Apollonius mentioned above. Although he also reviewed previous knowledge, his writings on comets (found in his *Questiones naturales*) are very different. To us, they seem like those of a scientist assaying a situation, and they are filled with apparent flashes of insight. For example: "I cannot think a comet is a sudden fire, but I rank it among Nature's permanent creations" (Hellman 1944:32). We also find:

If it were a wandering star [i.e., a planet], says some one, it would be in the zodiac. Who say I, ever thinks of placing a single bound to the stars? or of cooping up the divine into narrow space? These very stars, which you suppose to be the only ones that move, have, as every one knows, orbits different one from another. Why, then, should there not be some stars that have a separate distinctive orbit far removed from them?

(Hellman 1944:32)

Elsewhere we read:

There are many things whose existence we allow, but whose character we are still in ignorance of . . . why should we be surprised, then, that comets, so rare a sight in the universe, are not embraced under definite laws, or that their return is at long intervals? . . . The day will yet come when the progress of research through long ages will reveal to sight the mysteries of nature that are now concealed . . . The day will yet come when posterity will be amazed that we remained ignorant of things that will to them seem so plain.

(Hellman 1944:33)

Even Seneca was somewhat under the influence of his illustrious predecessor. He classified comets under meteorology, and he discussed weather forecasting from the appearance of comets.

Pliny the Elder discussed comets in his *Natural History*, which appeared about A.D. 77; Seneca is not mentioned as a source. Pliny presented a classification scheme based on appearance (both shape and color) that was used for centuries. However, his discussion of comets included little that was new, and many of his statements were not very specific.

It is curious that comets were not mentioned in Ptolemy's (second century A.D.) *Almagest* and were barely mentioned in his other works, and then only in connection with weather prediction. Ptolemy did argue that events on earth were not inevitably influenced by the stars. Arguments of this nature encouraged the notion (which persisted at least into the sixteenth century) that prayers would help avert the undesirable influences of comets.

There were few cometary studies for many centuries following Ptolemy. Hellman has noted that the years up to the fifteenth century "were not productive of any new cometary theory." Of course, this does not mean that comets were not observed and recorded; appearances of bright comets such as Halley's were recorded, for example, in A.D. 684 in the *Nuremberg Chronicles* and in 1066 in the Bayeux Tapestry (see Fig. 1.1 and Plate 1.1). Men such as Bede (A.D. 673–735), Thomas Aquinas (*c.* 1225–74), and Roger Bacon (*c.* 1214–94) wrote about comets. Despite variations in individual writings, the astrological view of comets was strengthened, particularly the belief that comets were evil omens. The scientific data recorded in Europe and the Middle East were barely sufficient to identify appearances of the periodic comets. Cometary observations by the Chinese have not been mentioned here because they had little or no influence for centuries on the main development of cometary knowledge.

Fig. 1.1. The A.D. 684 apparition of Halley's Comet as recorded in the *Nuremberg Chronicles*. (Yerkes Observatory photograph.)

1.2 Beginning of the fifteenth century to the supernova of 1572

As the fifteenth century opened, European civilization was slowly climbing out of the medieval period, with the beginning of slow progress toward our modern view of comets. Some individuals began to study comets in a systematic manner – gathering facts, probing their nature – rather than exploiting them with the superstitious people. How slowly this change came about can be judged, as noted by Hellman (1944:16), by the fact that Pingré in his famous *Cometographia*, published in the 1780s, "still considered it necessary to refute Aristotle." In this section, we will cite a few examples to illustrate the trends in cometary thought in the fifteenth and sixteenth centuries, up to 1572.

Two bright comets appeared in 1402. Jacobus Angelus of Ulm wrote an extensive treatise on one of these (known today as C/1402 D1[1]). While he took great care in describing that comet, his explanations of what he saw took a decidedly Aristotelian flavor. Aristotelian theory was extended in the 1450s by Matthew of Aquila, who associated comets and earthquakes. He also considered that comets not only signaled evil but could cause evil because, in Hellman's words, "their hot, putrid vapors contaminated the air" (Hellman 1944:73).

Paolo Toscanelli (1397–1482) observed Halley's Comet at its 1456 appearance, as well as several other comets between 1433 and 1472. His positional observations for the comet of 1433 (C/1433 R1) were sufficiently accurate to permit the calculation of an orbit by later workers. Peurbach (1423–61) also observed the comet of 1456. His attempt to measure its distance may have been the first.

A spate of activity was associated with the great comet of 1472 (C/1471 Y1). The principal contribution was made by the legendary but controversial Johannes Muller (1436–76), who is usually known by his adopted Latin name, Regiomontanus. His work on the comet was divided into sections whose titles were a list of problems to be investigated. These include the diameter of the comet, the position of the comet, the length and thickness of the tail, and the distance to the comet by measuring its diurnal parallax. The diurnal parallax is the shift of the comet against the background stars as the earth rotates, and the observer moves as a result. His observations were not sufficiently accurate to permit him to infer a meaningful parallax, and his value of 6° is highly erroneous. Nevertheless, either he or Peurbach was the first to attempt such a measurement. In evaluating Regiomontanus' contribution, we must also bear in mind what he did not do. His writings contain no discussions of the "meaning" of the comet, nor did he issue astrological predictions based on the appearance of the comet.

The concept that comet tails point away from the sun was well known by the mid-sixteenth century. The Italian astronomer Fracastoro (*c.* 1480–1553) wrote *Homocentrica* in 1538 in an attempt to improve upon the Ptolemaic theory of planetary motions by reverting back to concepts of the early Greek thinkers. The effort was not destined to bear fruit, because Copernicus' great *De revolutionibus* was already written and would be published 6 years later. However, Fracastoro did describe his observations of several comets, and he noted that the tails always

---

[1] See section 2.1.2.2 for a description of the comet naming convention adopted by the IAU in 1995.

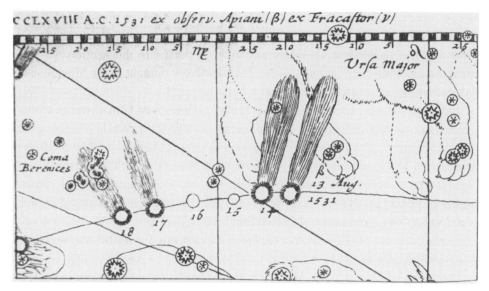

Fig. 1.2. Peter Apian's 1531 observations of Halley's Comet. The observations demonstrate that the comet's tail points away from sun. (*Bulletin de la Société Astronomique de France*, 1910.)

pointed away from the sun. Fracastoro was not alone in this important observation. Peter Apian described observations of several comets that appeared in the 1530s in his *Astronomicum Caesareum* (1540) and remarked that the cometary tails pointed away from the sun. In his *Practica* (1532), Apian described the appearance of Halley's Comet at its 1531 apparition. Figure 1.2 shows Apian's observations of Halley's Comet in 1531.

The close proximity in time of the writings of Fracastoro and Apian has led to some debate over who deserves credit for the discovery. In fact, the exact origin of our understanding of the orientation of comet tails is probably lost in antiquity. Chinese astronomers, in describing their observations of a comet in A.D. 837, said as much, and Seneca in his *Questiones naturales* wrote that "the tails of comets fly from the sun's rays."

In 1550, Jerome Cardan published his *De Subtilitate*, which was a compendium of learning. Cardan's views on cometary science were of considerable interest. He was aware of the parallax method for determining the distance to comets. His views on the origin of comets as summarized by Hellman (1944:93) read: "A comet is a globe formed in the sky and illuminated by the sun, the rays of which, shining through the comet, give the appearance of a beard or tail." Cardan also thought that there were more comets than the ones observed.

The comet of 1533 was observed by Copernicus; unfortunately, the observations have been lost. We may safely conclude that he made no significant contribution to cometary knowledge. A sure sign of at least some maturity of an astronomical subject appeared at the time of the comet of 1556, when some of the earliest catalogs of comets were published. At least three such catalogs were issued within a span of only a few years.

The last key event before the studies of the comet of 1577 was the supernova of 1572, which was visible for a year in the constellation Cassiopeia. Hellman (1944:111) explicitly states that "the influence of the new star of 1572 in molding the astronomical thought of the period cannot be overestimated." The principal observer of the supernova was Tycho Brahe (1546–1601), generally considered to be the greatest observational astronomer of his day. Tycho made and recorded positional observations with the high accuracy for which he is renowned. He failed to detect any parallax, and he placed the new star in the region of the fixed stars. In retrospect, it is curious that Tycho concluded that the new star could not be a comet or meteor because these were formed below the moon.

Tycho's observations were rivaled only by the work of Thomas Digges (*c.* 1546–95) in England, who advocated careful observations of astronomical phenomena in order to ascertain their true nature. The conclusion reached by these authors and several others, not mentioned, was that the new star had no measurable parallax and belonged to the region of the fixed stars. Although there were some dissenting voices concerning the parallax, the basic result was established, as well as an eager and receptive climate for the appearance of the comet of 1577.

1.3 The comet of 1577

Tycho Brahe observed the comet of 1577 (C/1577 V1) from November 13, 1577, to January 26, 1578. His observations and views were contained in a Latin work, *De mundi aetherei recentioribus phaenomenis*, and a German work. Tycho summarized earlier work on comets and ideas concerning the structure of the universe. Aristotle's views on the immutability, that is, the unchangeability, of the heavens were questioned on the basis of the new star of 1572. Tycho also questioned the atmospheric origin of comets because, if that concept were true, he could see no reason for their tails always to point away from the sun. Nevertheless, he felt that the conclusive method for discovering the comet's true place was to measure its parallax.

Tycho measured the position of the comet against the background of stars when the comet was high in the sky and again when it was low in the sky. The shift of the comet relative to the stars between the two observations is related to its parallax. Tycho was aware that he should correct the measurements for atmospheric refraction. If the comet was between the earth and the moon, he should have found a parallax of at least 1°. The average error of Tycho's positions (established by comparison with a modern orbit for the comet) was only 4′. He was cautious and concluded that the comet's parallax was 15′ or less, placing it at least 230 earth radii away. This distance is well beyond the moon, whose distance averages 60 earth radii. Tycho compared his own observations of the comet's position from the island of Hveen (near Copenhagen) with observations made by Hagecius at Prague, and found a difference in position of only one or two minutes of arc. Since Prague and Copenhagen are about 600 km apart, a parallax of 2′ would mean that the comet would be approximately 1 million km from the earth, well beyond the moon's mean distance of about 380 000 km.

Thus, Tycho concluded that the comet was between the moon and Venus. He even attempted to calculate an orbit for the comet. His result was a circular orbit around the sun outside of the orbit of Venus. Of course, he could not represent the observed motion of the comet, assuming it traveled in a circular orbit with uniform motion. He was obliged, therefore, either to assume an irregular motion or to admit that the orbit was not "exactly circular but somewhat oblong, like the figure commonly called oval" (Dreyer 1953:366). Sarton interprets Tycho as having suggested that the comet's orbit was elliptical. Whether an ellipse or just an oval was meant, Dreyer notes: "This is certainly the first time that an astronomer suggested that a celestial body might move in an orbit differing from a circle, without distinctly saying that the curve was the resultant of several circular motions." Tycho's *De mundi aetherei recentioribus phaenomenis* on his studies of the comet of 1577 was first published in 1588. Kepler's conclusion that Mars and the other planets move in elliptical orbits was contained in his book on Mars, published in 1609. The question of the shape of cometary orbits will become more curious when we look at Kepler's views on them below.

Tycho attempted to calculate the linear dimensions of the comet, which, of course, depended on the distance assumed. The measured angular dimensions on November 13 were a head diameter of 8′, a tail length of 22°, and a tail breadth of 2.5°. Tycho realized that the linear dimensions of the comet were tremendous even if the comet looked small to terrestrial observers.

Tycho expressed his opinion that the tail is caused by sunshine passing through the comet: Because comets are not diaphanous, sunlight cannot pass through without effect and because comets are not thick and opaque like the moon, sunlight is not simply reflected. A comet is intermediate and partly holds the sunshine. Because a comet's body is porous, some sunbeams pass through and are seen by us as a tail attached to the head.

Thus, Tycho's studies clearly demonstrated that comets are not atmospheric and are not sublunar. They are supralunar celestial objects that could be studied by scientific methods. The lack of large parallaxes clinched these conclusions. There was no need to take the Aristotelian view seriously thereafter. Tycho's work was confirmed by astronomer Michael Maestlin (1550–1631) as well as by others.

Of course, there were dissenters from Tycho's conclusions about the comet of 1577. Of first importance was Tadeáš Hájek Hájku, known by his latinized name of Hagecius (*c.* 1526–1600); he was considered a leading astronomer of his time. He initially obtained a parallax of 5°, which would have placed the comet well below the moon and, in fact, only about 8 radii from the earth's center. Hagecius' observations were good, and Tycho used them to establish independently the supralunar position of the comet. Hagecius had erred in his interpretation of his imprecise observations. He established himself as a man of considerable scientific character by admitting that he was wrong. He recognized the comet as supralunar in his work on the comet of 1580.

Other observers found a large parallax for the comet, but the quality of their work was not close to Tycho's. The most serious attacks on Tycho would occur after his death.

1.4 Beginning of the seventeenth century to the comets of 1680 and 1682

Johannes Kepler (1571–1630) studied the comets of 1607 (1P/1607 S1, Halley's) and 1618, and published his views on comets in *De cometis* (1619) and in *Hyper-aspistes* (1625). His ideas on the formation of comet tails and the ultimate extinction of comets seem remarkably modern. He wrote:

Gross matter collects under a spherical form; it receives and reflects the light of the sun and is set in motion like a star. The direct rays of the sun strike upon it, penetrate its substance, draw away with them a portion of this matter, and issue thence to form the track of light we call the tail of the comet. This action of the solar rays attenuates the particles which compose the body of the comet. It drives them away; it dissipates them. In this manner the comet is consumed by breathing out, so to speak, its own tail.

(Olivier 1930:9)

Today we know that comets shine by reflected light. Their tails are formed by solar radiation pressure acting on cometary dust and molecules and by the solar wind acting on cometary ions. Kepler also considered that comets were as numerous as fish in the sea.

Kepler's beliefs concerning the orbits of comets, however, are curious. He thought that comets moved along straight lines, but with an irregular speed. Because Tycho had suggested an oval or elliptical orbit for comets, and because Kepler himself found elliptical orbits for the planets, this oversight is all the more puzzling.

Kepler would once again become involved in comet work because of attacks on Tycho by Scipio Claramontius (1565–1652) (sometimes called Chiaramonti) and by Galileo. In Claramontius' work, *Antitycho* (1621), he attempted to prove that comets were sublunar. Among other things, Claramontius may not have been straightforward because, as Drake and O'Malley (1960:374) have noted:

"One of the favorite devices of [Claramontius] was to attack the data of astronomers as inaccurate because they were never in exact agreement. On this pretext, he would throw out all the observations which did not suit his purpose." Claramontius convinced only a few, and his writings did not greatly hinder acceptance of the new approach to comets.

The criticisms by Galileo Galilei (1564–1642) were more serious. Three comets appeared in the year 1618 and an exchange of differing opinions began; the exchange produced Galileo's work dealing with comets, *The Assayer* (1623). Galileo was not in good health, and he did not personally observe the comets extensively. Perhaps he used comets to publicize his scientific methods. In any case, Galileo's antagonism toward Tycho shows clearly in *The Assayer.* Galileo dismissed the parallax observations by noting that the comet could be simply an optical illusion. He suggested that vapors rising from the earth's atmosphere could produce comets seen by reflected sunlight when the vapors had risen outside the cone of the Earth's shadow. A clever argument by Father Horatio Grassi (1583–1654) was that comets were quite distant because they were not magnified by telescopes; Galileo argued that this was not necessarily so if strange optical effects were involved. The curvature of some comet tails caused great difficulties to all concerned.

All in all, the writings concerning the comets of 1618 are not enlightening. For example, luminous gas was supposed to be both opaque and transparent by the opposing sides. It is painful to read that the biblical description of Shadrack, Meshack, and Abednego walking uninjured in the midst of fire after they had been cast into the furnace was used as evidence by both sides.

Kepler responded to the unjust criticisms of Tycho by Claramontius in his book *Tychonis Brahei dani hyperaspistes* or *The Shieldbearer to Tycho Brahe the Dane* (1625). He disposed of Claramontius' views on the parallax by noting the well-established accuracy of Tycho's observations. Kepler responded to Galileo in an appendix to the *Hyperaspistes*. This reply contains some of Kepler's ideas on comets, mentioned above.

1.5 Newton, Halley, and the orbits of comets

The issue of the nature of cometary orbits began to be clarified in 1665 when Giovanni Borelli (1608–79) published the suggestion that the paths of comets are parabolic. The same suggestion was made by Johannes Hevelius (1611–87), who discovered four comets and wrote two books on comets (Fig. 1.3). His ideas on the motion of comets involved an origin in the atmospheres of Jupiter or Saturn and a resisting medium in interplanetary space. A comet began its orbit toward the sun with the flat part of its "disk" oriented perpendicularly to the direction of motion. But when the comet approached the sun, it moved with the edge of the disk

Fig. 1.3. Allegorical figures showing the Aristotelian idea that comets are sublunar (left), the Keplerian notion that comets move in a straight line (right), and the idea of Hevelius (center) that comets arise in the atmospheres of Jupiter or Saturn and move around the sun on a curved trajectory. (Frontispiece from Hevelius's *Cometographia*, published in 1668 in Danzig.)

forward. The reduced effect of the resisting medium caused a departure from the initial rectilinear path; the resulting path could be either a parabola or a hyperbola, although Hevelius did not put the sun at the focus. The comet of 1680 was studied by a student of Hevelius, George Dörffel, who suggested that its orbit could be represented by a parabola with the sun at the focus.

Isaac Newton (1642–1727) wrote the great book *Philosophiae naturalis principia mathematica* (*Mathematical Principles of Natural Philosophy*) or *Principia*, which was published in several editions, the third being published in 1726. The third edition of *The Principia* was translated from Latin into English by Andrew Motte in 1729. Florian Cajori revised the third edition and supplied it with a historical and explanatory appendix. Cajori's version was published together with Newton's *System of the World* in 1934. Comets are an important part of both these works. Edmond Halley, who was instrumental in seeing that the *Principia* was published, played an important role in Newton's interest in comets. Halley's contribution was diplomatic and financial as well as scientific, moving A. De Morgan (1806–71) to write: "but for him, in all human probability, that work would not have been thought of, nor when thought of written, nor when written printed."

Newton's laws of motion and gravitation established the basis for calculating planetary motions. Newton accepted Tycho's parallax measurements of the comet of 1577, and argued that the observed motions of comets – like those of the planets – are explained by the combination of their orbital motion and the orbital motion of the earth. Newton argued that comets move in orbits that are conic sections. If a comet is periodic, its orbit must be an ellipse. He then argued that, for many comets, the orbits are "so near to parabolas, that a parabola can be used for them without sensible error." In Proposition XLI, Problem XXI of the *Principia*, Newton expounds on a method for determining the orbital parameters of a comet moving in a parabolic orbit, given three adequately spaced observations.

Newton used the comet of 1680 (C/1680 V1) as his example. He started with observations made by John Flamsteed, Edmond Halley and himself, made between December 1680 and March 1681 to determine the orbit (Fig. 1.4). He then used the orbit to calculate the path of the comet, as he says, by scale and compass. Halley redid the calculations numerically, to a much greater accuracy than Newton's graphical methods could provide. Halley's calculated positions of the comet differed by less than 5′ from the observations over the time period the comet was observed. As a point of interest, Johann Encke redid Halley's calculations in 1818, and found that the comet came within about 0.006 AU of the sun, making it a sungrazing comet (see section 2.3).

In referring to Halley's orbit and its agreement with observation, Newton wrote in the *Principia*: "The orbit is determined ... by the computation of Dr. Halley, in an ellipse. And it is shown that ... the comet took its course through the nine signs of

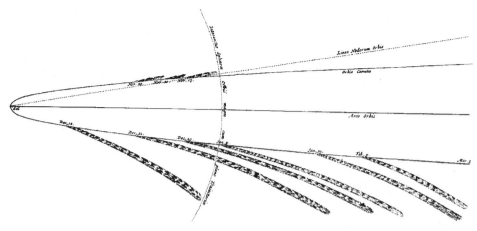

Fig. 1.4. Halley's orbit for the comet of 1680, as presented in Newton's *Principia*.

the heavens, with as much accuracy as the planets move in the elliptic orbits given in astronomy." Thus, Newton concluded "comets are a sort of planet revolved in very eccentric orbits around the sun." These comments were not in the first edition of the *Principia*, since Halley's orbital calculations postdate that edition.

Newton's views on the physical constitution of comets are also very interesting. He points out in the *Principia* that four or five times more comets have been observed in the half of the sky containing the sun than in the other half. He also points out that the brightness of individual comets are related to their distances from the sun and the earth. These two facts strongly support the idea that comets shine by reflected sunlight. Newton reviewed the historical ideas concerning comet tails and concluded that they arose from the atmospheres of the comets. He felt that this would not be difficult because "a very small amount of vapor may be sufficient to explain all the phenomena of the tails of comets."

Newton's views on the orbits of comets were not immediately accepted by everyone. All reasonable doubt would be dispelled by future observations of the comet of 1682. Halley (Fig. 1.5) found the orbits of two dozen comets for which there were sufficient observations and published a catalog of their elements in 1705 (see Table 1.1). The orbital elements for the comet of 1682 showed close correspondence with the comets of 1607 (observed by Kepler and Longomontanus) and the comet of 1531 (observed by Apian). Halley also knew that the great comet observed in the summer of 1456 traveled in a retrograde direction. Although the periods involved in these identifications showed variations (roughly from 75 to 76 years), Halley concluded that all the observations referred to the same comet and wrote: "I may, therefore, with confidence, predict its return in the year 1758. If this prediction is fulfilled, there is no reason to doubt that the other comets will return" (Armitage 1966). He

Fig. 1.5. Edmond Halley. (Yerkes Observatory photograph.)

also wrote that if he were correct, "candid posterity will not refuse to acknowledge that this was first discovered by an Englishman."

Halley knew from Newton's work that the planets Jupiter and Saturn would disturb the orbit of his comet. This was expected from Newton's law of gravitation and was the physical reason for the differences in the revolution period. However, detailed perturbation calculations were not made in England but in France, by the

Table 1.1 *Halley's table of orbital elements of some comets*

The astronomical element of the motions, in a parabolic orbit, of all the comets hitherto obtained.

Passage of perihelion, London time		Longitude of perihelion	Longitude of ascending node	Inclination of orbit	Distance from sun at perihelion
1337 June 2	06:25	37° 59′	84° 21′	32° 11′	0.40666
1472 February 28	22:23	45 34	281 46	5 20	0.54273
[a] 1531 August 24	21:18	301 39	49 25	17 56	0.56700
1532 October 19	22:20	111 07	80 27	32 36	0.50910
1556 April 11	21:28	278 50	175 42	32 06	0.46390
1577 October 26	18:45	129 22	25 52	74 33	0.18342
1580 November 28	15:00	109 06	18 57	64 40	0.59628
1585 September 27	19:20	8 51	37 42	6 04	1.09358
1590 January 29	02:35	216 54	165 31	29 41	0.57661
1596 July 31	19:55	228 15	312 13	55 12	0.51293
[a] 1607 October 16	03:50	302 16	50 21	17 02	0.58680
1618 October 29	12:23	2 14	76 01	37 34	0.37975
1652 November 2	15:40	28 19	88 10	79 28	0.84750
1661 January 16	23:41	115 59	82 30	32 36	0.44851
1664 November 24	11:52	130 41	81 14	21 18	1.02576
1665 April 14	05:16	71 54	228 02	76 05	0.10649
1672 February 20	08:37	47 00	297 30	83 22	0.69739
1677 April 26	00:38	137 37	236 49	79 03	0.28059
1680 December 8	00:06	262 40	272 02	60 56	0.00612
[a] 1682 September 4	07:39	302 53	51 16	17 56	0.58328
1683 July 3	02:50	85 30	173 23	83 11	0.56020
1684 May 29	10:16	238 52	268 15	65 49	0.96015
1686 September 6	14:33	77 00	350 35	31 22	0.32500
1698 October 8	16:57	270 51	267 44	11 46	0.69129

[a] Successive apparitions of Halley's Comet.

astronomer A. Clairaut (1713–65). The job was herculean because all computing had to be done by hand. Clairaut and his associates completed their calculations and reported their results in November 1758. The date of perihelion passage was calculated to be April 15, 1759, with an estimated uncertainty of 1 month. The comet was observed on Christmas night 1758 by an amateur astronomer living near Dresden. It passed perihelion on March 13, 1759.

At its 1759 apparition Halley's Comet was observed by numerous astronomers, including Charles Messier (1730–1817). But none of the observations at that time compare in importance with the fulfillment of Halley's prediction, with its vast philosophical implications for astronomy in general and comets in particular. Comets were shown to be subject to the laws of physics. Their orbits could be calculated and their return predicted years in advance. At least some comets were members of the solar system. Any rational fear of comets as signs of disaster or evil should have vanished. The return of Halley's Comet as predicted was also a triumph for the detailed physics whose development was initiated by Newton and is now called *celestial mechanics*. In other words, Newton's physics clearly applied out to the aphelion distance of Halley's Comet, about 35 AU.

The understanding of cometary orbits was a strong argument against Rene Descartes's (1596–1650) vortices, which were used to explain planetary motion. It is difficult to see how a vortex could pull planets along their orbits, and yet retrograde comets such as Halley's could move in the opposite direction to the planets without any discernible effect. We should also note that any residual support for the concept of solid, crystalline spheres as the means of carrying the planets around the sky in the Ptolemaic system was shattered by knowledge of the high ellipticity of cometary orbits. The first edition (1769) of the *Encyclopaedia Britannica* states that comets "have moved through the ethereal regions and the orbits of the planets without suffering the least sensible resistance in their motions; which plainly proves that the planets do not move in solid orbs."

Halley's data on the orbits of comets showed clearly the contrast between cometary and planetary orbits. The elliptical orbits of comets were very elongated (such that parabolas were a good approximation near the sun), whereas the elliptical orbits of planets showed only nominal elongation (such that the circles were a fairly good approximation of their orbits). Whereas the planes of planetary orbits were closely confined to the plane of the earth's orbit and all went in the same (direct) direction, the planes of cometary orbits could be inclined at all angles to the ecliptic, and their direction of motion could be either direct or retrograde.

The development of the orbit calculation phase of celestial mechanics proceeded relatively rapidly. P. S. Laplace (1749–1827) gave formulas for the calculation of elliptical orbits through the method of successive approximations. Although elegant and rigorously correct, the method was unwieldy and not satisfactory. The practical problem was solved in 1797 by the physician and respected amateur astronomer Wilhelm Olbers (1758–1840). His simple method for determining the five elements needed for a parabolic comet orbit has not been significantly improved to the present day.

The second person to successfully predict the return of a comet was Johann Encke (1791–1865). He studied the available orbital information for comets seen in January 1786 by Mechain, in November 1795 by Caroline Herschel, in October 1805 by J. Pons and others, and by Pons again in November 1818. No parabolic orbit would fit the observations of the comet observed in 1818, and this situation stimulated Encke to undertake a thorough study of the observations. He soon realized that all the observations mentioned above referred to the same comet, which moved in a quite unusual elliptical orbit. The comet's period of revolution was 3.3 years, and its heliocentric distance varied between 0.34 and 4.08 AU. Encke computed the orbit using the method developed by Carl Friedrich Gauss (1777–1865), and published in Gauss's *Theoria motes corporum coelestium* (1809). It permitted calculation of all six orbital elements from three observations of position. Gauss had developed his method as an aid in the search for the asteroid Ceres, which was lost from view five weeks after its discovery. The comet studied by Encke was predicted to return in 1822, and it was observed from Australia. Encke's Comet, as it is now called, has been repeatedly observed to this day.

The 1838 return of Encke's Comet presented the astronomical community with another problem. As Encke had suspected for some time, the comet's period was decreasing at an accelerating rate. On successive returns the comet would arrive at perihelion earlier than predicted by an amount that varied but that was typically about 0.1 day. There were three ways of explaining the observations. (1) The comet could be moving through an essentially uniform resisting medium (proposed by Encke). In such a situation, the comet would spiral in toward the sun while the period decreased. (2) A belt of meteoric particles orbiting the sun would serve the same purpose as Encke's resisting medium. However, variable decreases would occur because comets would be affected only while passing through the belt. In both of these theories, the period of the comet could only decrease. (3) The deviations could be due to the expulsion of material from the comet in a particular direction, that is, a "rocket effect." This was suggested by Bessel after observing Halley's Comet in 1835 and noticing a sunward plume of material that resembled a blazing rocket. Bessel actually computed the orbital consequences of a simple model for a sunward rocket.

Subsequent work would show that the hypotheses of the resisting medium and the meteoric belt both had to be abandoned because comets were found that had increasing periods. Bessel's basic idea has survived and has been incorporated in our current physical picture of comets.

Shortly after the predicted return of Encke's Comet in 1822, another comet was discovered in 1826 that would rival it for historical interest. Investigations on the orbit of the comet discovered by M. Biela indicated that a parabola would not fit the observations, but that an elliptical path with a period of 6.75 years would; the comet's aphelion was near but outside the orbit of Jupiter. The comet was predicted (by several astronomers, including Olbers) to return in November 1832, and it appeared on schedule. Detailed calculations of the orbit had to include the strong effects of Jupiter. The results indicated that the comet had been sighted twice previously, by Montaigne (and also by Messier) in 1772 and by Pons in 1805. The comet's return in 1839 was not observed because the relative geometry always kept the comet so near to the sun on the sky as viewed from earth, that it could not be seen.

Fig. 1.6. The components of comet Biela on February 19, 1846, as drawn by Otto W. Struve. The separation of the two components was approximately 6.5 arc minutes. (Courtesy of Fred L. Whipple, from *The Mystery of Comets*. Smithsonian Institution, Washington, DC, 1985.)

The return of Biela's Comet in 1846 was proceeding comfortably when, in the words of John Herschel (1792–1871), "it was actually seen to separate itself into two distinct comets, which, after thus parting company, continued to journey along amicably through an arc of upwards of 70° of their apparent orbit, keeping all the while within the same field of view of the telescope pointed towards them" (Herschel 1871). At first there was some kind of interaction between the two Bielas, including luminous bridges. Eventually, both parts of Biela developed into complete comets with tails.

Both Bielas returned in 1852, separated by approximately 2 million km (Fig. 1.6). They may have returned in 1858 but, as in 1839, their orbits kept the comets too near the sun to permit observation. The next scheduled appearance for Biela was in 1866, and there was considerable interest in determining the separation of the two parts. However, Biela's Comet was never seen again. Some astronomers expected to see the debris of Biela's Comet in 1872 and, in a sense, they were not disappointed.

The tremendous meteor shower (Leonids) on the night of November 12–13, 1833, (Fig. 1.7) stimulated scientific interest in meteors. Several observers noted that they all seemed to come from the same point in the sky located in the constellation Leo. It was suggested that the original particles were traveling in parallel paths in space. Hence, there was considerable interest in their orbits.

In 1866, Giovanni Schiaparelli (1835–1910) established the connection between the orbits of meteor streams and the orbits of comets. He showed the similarity between the orbits of the Perseids and comet Swift–Tuttle (1862 O1) and between the orbits of the Leonids and comet Tempel–Tuttle (1865 Y1). The natural supposition was that the particles producing the meteor showers were the last remains

Fig. 1.7. Woodcut showing the Leonid meteor shower of 1833. (Courtesy of the American Museum of Natural History.)

of the comets. Biela's orbit was identified by Weiss and d'Arrest in 1867 with the Andromedid meteor shower, now also called the *Bielids*. A shower was predicted for 1872, and a wonderful display occurred. Another excellent display appeared in 1885. The shower was weaker in 1892 and apparently expired in 1899 (although a few possible stragglers are recorded from time to time).

We have seen that a basic understanding of cometary orbits was achieved in the nineteenth century. Most comets travel in very elongated ellipses or perhaps parabolas (i.e., the so-called non-periodic comets), which are oriented at random and are not confined to the ecliptic plane. Many short-period comets had been discovered with aphelia near the orbit of Jupiter. These orbits are compatible with J. L. Lagrange's (1736–1813) suggestion that comets were ejected from Jupiter or with P. S. Laplace's statement that comets could acquire an orbit with a period ~5 years by a close encounter with Jupiter. Further orbital studies would be needed to sort out the various ideas on the origin of comets. The identity of the orbits of some meteor streams with specific cometary orbits was established.

In many respects the determination of the basic facts of cometary orbits was a triumph of Newton's celestial mechanics; predictions such as the return of Halley's Comet in 1759 were spectacular. However, total triumph was short lived, because early in the nineteenth century significant departures from the predictions for some comets were discovered. These comets did not strictly follow the motions predicted by Newton's laws. There seemed to be a type of nongravitational force acting on comets that would not be understood until the middle of the twentieth century.

1.6 Our modern ideas emerge

So far, we have traced the history of cometary physics up to the closing years of the nineteenth century. The late nineteenth and early twentieth centuries were a period of revolution in physics during which many of our modern concepts developed. During the same period, our understanding of cometary physics also increased rapidly. The modern era of cometary physics, which began in roughly 1950, built upon the increased understanding of physics and the growing body of cometary data. Our objective in the remainder of this chapter is to present a summary of our knowledge up to the 1950s. This discussion will provide the groundwork for the details to be presented later.

1.6.1 Forms and tail structure

Both Newton and Olbers made efforts to understand the physical nature of comets. However, this phase of cometary research began most seriously with the research of F. W. Bessel (1784–1846). Bessel observed Halley's Comet in 1835, and noted extensive fine structure near the nucleus consisting of jets, rays, fans, cones, and other forms. Similar structure was also observed by F. G. W. Struve at Dorpat, John Herschel at the Cape of Good Hope, and M. Arago at Paris. In one case,

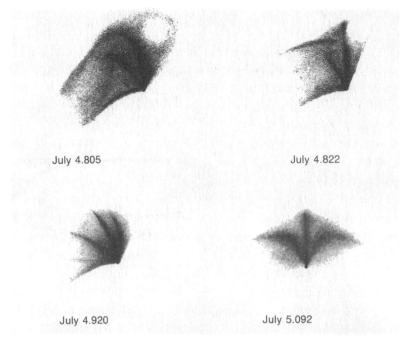

July 4.805   July 4.822

July 4.920   July 5.092

Fig. 1.8. Drawings of comet Tebbutt in 1861, showing spiral structure in the head. (From J. Rahe, B. Donn, and K. Wurm, *Atlas of Cometary Forms*, NASA SP-198, 1969.)

a cone of light was observed extending toward the sun for a short distance, and then to curl backward away from the sun as if propelled by a great force (Fig. 1.8). Bessel, Herschel, and Olbers earlier hypothesized that the force acting on the material in the cone might be electrical, possibly due to a net charge on the sun.

Bessel developed the theory of cometary forms for particles leaving the head of a comet while subjected to different repulsive forces and solar gravity. This approach was followed and extended by F. A. Bredichin (1831–1904). The mechanical treatment of cometary forms is still known as the Bessel–Bredichin theory; see section 4.1.2. Bredichin classified comet tails into three types, depending on the degree of curvature, and assigned each a specific chemical composition, namely, hydrogen to type I, hydrocarbons to type II, and metallic vapors to type III. Type I tails were nearly straight, with a curvature that is hard to measure, and type III tails were strongly curved. The force of repulsion was highest for the type I tails and lowest for the type III tails. Many details of Bredichin's classification scheme are not used today, but the terminology of type I and type II tails remains in use. The Bessel–Bredichin theory of tail forms was extended in the first half of the twentieth century by A. Kopff and by S. V. Orlov, who gave a fairly elaborate classification scheme.

In 1900, the Swedish chemist Svante Arrhenius (1859–1927) suggested a candidate for the repulsive force. His suggestion was the radiation pressure of sunlight. In the first decade of the twentieth century, Karl Schwarzschild (1873–1916) and Peter Debye (1884–1966) made calculations that indicated that the force on a spherical dust particle was at a maximum when its diameter was approximately one-third the wavelength of the incident radiation. In that case, the force of repulsion could be 20 to 30 times the solar attraction. Debye also pointed out that the radiation pressure on a single molecule due to selective absorption and reemission of photons could be even higher.

Solar particulate emission also exert a force on cometary constituents. The origin of the idea that the solar repulsive force could be caused by particles emitted from the sun is hard to pin down, but the idea was certainly extant around the turn of the century. The impetus for much of the work on solar particle emission came from studies of the aurorae and related geophysical phenomena. The solar physicist Richard Carrington (1826–75) observed the first recorded solar flare on September 1, 1859, and noted an intense aurora the following day. Carrington was fully aware of the possible cause-and-effect relationship, but he was wary of jumping to this conclusion. A. H. Becquerel (1852–1908) thought that sunspots might be the source of the auroral particles. In 1892, G. F. FitzGerald (1851–1901) estimated the speed of the particles at "about 300 miles per second" from the time between the central meridian passage of a sunspot and the magnetic storm associated with it (FitzGerald 1892). In a fascinating paper written in 1900 by Oliver Lodge (1851–1940), the aurorae, magnetic storms, and behavior of comet tails were attributed to "a torrent or flying cloud of charged atoms or ions."

Thus, in 1910, A. C. D. Crommelin wrote:

There are at least three theories to explain the repulsion of the tail from the sun: (1) Light-pressure; (2) Electrical repulsion; (3) Mechanical bombardment by electrons, or other tiny particles violently ejected from the sun. It is quite possible that all three act conjointly, as no one of them seems capable of explaining all the facts.

(Proctor and Crommelin 1937:189)

About the same time, some problems and confirmations began to emerge in connection with the simple application of the Bessel–Bredichin (mechanical) theory of cometary forms. The stimulus was cometary photography. A. Pannekoek (1961) wrote in his *A History of Astronomy*:

comet tails, formerly smooth and ghostly, hardly visible phantoms, now appeared on the plates as brilliant torches with rich detail of structure, with bright and faint spots, never before seen or even suspected. Such photos, taken of every succeeding bright comet (like Morehouse's comet in 1908 and Halley's in 1910, often reproduced in scientific and popular reviews) gave a new impulse to the study of comet tails.

(Pannekoek 1961:425)

Comet Morehouse (Fig. 1.9) showed striking parabolic envelopes on the side of its coma (or atmosphere) toward the sun, which were analyzed by A. S. Eddington (1882–1944). Their formation was relatively easy to understand on the basis of the *fountain model*, in which particles were ejected in various directions toward the sun and then repulsed into the tail. The outer envelope of the particle trajectories formed the observed parabola (Fig. 1.10). The problem with the simple interpretation was that it required very large initial ejection velocities and repulsive forces some 800 times solar gravity. Eddington himself was skeptical of these high forces. Also, in comet Morehouse, E. E. Barnard (1857–1923) noted that specific parts of the tail would rapidly brighten where no material was previously visible and with no obvious supply from the nucleus.

Heber D. Curtis (1872–1942) studied knots or condensations in the tail of Halley's Comet (1910). He found that the same feature could be identified on the photographs made on successive nights (Fig. 1.11), and calculated the velocities of the knots. He found speeds in the range from 5 km s^{-1} near the nucleus to about 90 km s^{-1} nearly 0.1 AU from the nucleus. Here the action of the repulsive solar force could be seen directly. The increasing velocities of knots in Halley's Comet, as well as those observed in many other comets, could be used to determine the solar repulsive force. The average value was 200 times solar gravity, and was considered quite high. There was substantial variation around the average, with some values approaching 1000 times solar gravity. In addition, different structures observed at the same time in the same comet tail could show quite different repulsions.

Despite the problems with some aspects of the mechanical theory and despite the different suggestions for the repulsive force, the mechanical theory with radiation pressure as the repulsive force would rule the scene for decades. Russell, Dugan, and Stewart in their classic *Astronomy* (1926) stated:

As the activity increases, the finer particles, repelled by the sun's light pressure, stream away visibly to form the tail, which grows longer and brighter as the comet approaches perihelion. The emissions from the nucleus sometimes take the form of jets, or streams, and sometimes the outflow is more regular, resulting in the formation of envelopes.

(Russell, Dugan, and Stewart 1926:444)

As cometary spectroscopy developed, we learned that the long, straight tails (type I) were composed not of dust but of ionized molecules. The principal emission in the visual wavelength range comes from CO^+. In 1943, Karl Wurm published his results on the repulsive forces from radiation pressure for CO^+ as well as for C, and CN

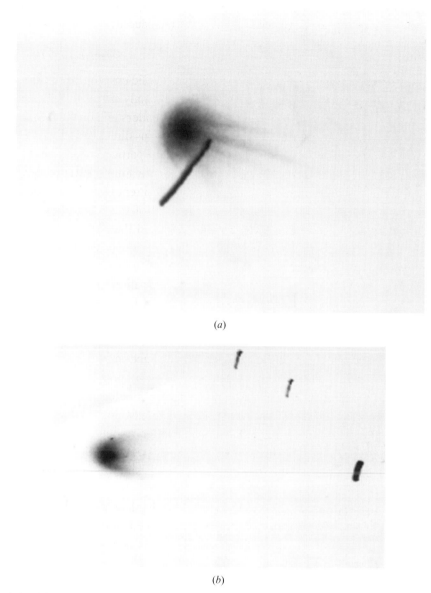

(a)

(b)

Fig. 1.9. Parabolic envelopes in comets. (a) and (b) Photographs of the head region of comet Morehouse (C/1908 R1) showing envelope structures (Greenwich plate, reproduced by K. Wurm, courtesy of J. Rahe); (c) Head region of Donati's Comet (C/1858 L1) in 1858, drawn by G. P. Bond, showing envelopes (Yerkes Observatory photograph).

(c)

Fig. 1.9. (*Cont.*)

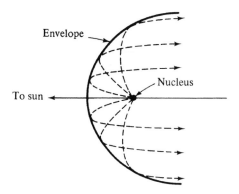

Fig. 1.10. Relationship between particle orbits and cometary structures. In the head, the envelope of particle orbits (dashed lines) form the observed parabolic structures.

(cyanogen), which are found in the coma. This calculation requires knowledge of the solar radiation flux in the relevant wavelength region and the atomic parameters for the molecules involved. Wurm found repulsive forces of 47.0, 1.7, and 0.7 times solar gravity for the molecules CO^+, C_2, and CN, respectively. The low values for C_2, and CN mean that the motions and lifetimes of these molecules are consistent with the observation that they make up the cometary head. The calculated value of 47.0 for CO^+ certainly indicates that this ionized molecule should be driven into the cometary tail, but the value is too low to explain the observed high values of the repulsive force. We will return to this problem when we discuss the modern era of cometary research in later chapters.

The determination of material densities in comet tails depends on spectroscopic and photometric observations. Some idea of the diffuseness of comet tails comes from the fact that the earth has probably passed through comet tails twice in recent history without observable effect. The first time was on or about June 30, 1861 (Great Comet of 1861 or C/1861 J1), and the second time was on the night of May 18–19, 1910 (Halley's Comet; also see section 10.3.1). It is difficult to be precise about the exact times because the comet tails are not exactly straight and because there are no exceptional meteorological phenomena to mark the passage. In 1910 the earth apparently did not pass through the central part of the main tail, but only through its outer edges or a detached streamer. Astronomically, the passage through the tail produced some spectacular views of the tail stretching across the sky, and little else. Unfortunately, the event caused concern and fear because of the belief that poisonous gases from the comet would contaminate the atmosphere.

The very low densities in comet tails are confirmed by the fact that stars generally can be observed through them without diminution or deflection, although apparently there are exceptions. William Herschel (1738–1822) reported that some diminution occurred for stars seen through the tail of the comet of 1807 and that considerable

Fig. 1.11. Comet Halley in 1910, as photographed at Williams Bay, WI. June 6, 15.8^h GMT (top); Honolulu, HI, June 6, 18.5^h GMT (center); and Beirut, June 7, 7.0^h GMT (bottom). (Yerkes Observatory photograph.)

diminution was found for a star seen through the head. By contrast, no effect was observed by either Struve or Bessel for stars seen quite close to the apparent nucleus of Halley's Comet in September 1835. A star field photographed through the head of comet Pons–Winnecke by Georges Van Biesbroeck (1880–1974) in June 1927 showed no shift in position greater than $0.05''$–$0.10''$.

It appears that densities in the head and particularly in the tail are insufficient to dim the light from stars measurably or to shift their position. Exceptions occur, but these would be expected primarily when the star is nearly coincident with the apparent nucleus.

1.6.2 Spectroscopy

As the discussion in the last section showed, progress in the physics of comets depended crucially upon the spectroscopic identification of the chemical constituents involved. Work in cometary spectroscopy began in the 1860s.

On August 5, 1864, Giovanni Donati (1828–73) made the first spectroscopic observations of a comet. When he turned his visual spectroscope toward comet Tempel (C/1864 N1), he saw three faint bands of emission. Sir William Huggins, (1824–1910), one of the pioneering stellar spectroscopists, observed a cometary spectrum in 1866. Not only were the three bands observed by Giovanni Donati visible, but a reflected solar continuum was present, as well.

Cometary spectra were observed and investigated intensively by Huggins, and in 1868 he identified the bands observed by Donati. Swan had seen these same bands in hydrocarbon compounds (such as ethylene vapor, C_2H_4, observed in the laboratory) excited by electrical discharges. The bands also could be observed in the gas obtained by heating meteorites, a result that seemed natural if comets were actually a swarm of meteoroids. Meteoroids are small bodies in space; when they land on earth, they become meteorites. Huggins made the identification by carefully comparing the comet's spectrum directly with the spark spectrum of ethylene vapor; that is, the two spectra were observed side by side in the same eyepiece. Huggins noted:

> ... the apparent identity of the comet's spectrum with that of carbon resides not only in the coincidence of the positions of the bands in the spectra but also in the very remarkable resemblance of the corresponding bands in what concerns their general characters, as well as their relative light. This is well recognized in the middle (green) band where the gradation of intensity is not uniform.
>
> *(Olivier 1930:80)*

Actually, Huggins needed some good fortune to complement his care. The bands, now called the Swan band spectrum of C_2, show a similar intensity distribution in both the laboratory sources and comets. This coincidence does not generally hold for all chemical species or physical conditions.

Photographic records of cometary spectra were first made in 1881. Again, Huggins was the pioneer with his spectrum of the Great Comet of 1881 (C/1881 K1) taken on June 24, 1881. The early photographic results showed (in addition to the Swan bands of C_2) a strong group of emission bands in the violet, which were subsequently identified as CN.

The sodium D lines were first detected in the spectrum of the comet C/1881 K1 and shortly thereafter in three other comets with small perihelion distances. The radial velocity of a comet was measured for the first time in 1882 using doppler shifts of its D lines. Also in 1882, spectrum lines due to iron were observed in the spectrum of a comet that approached very close to the sun.

By the turn of the century, the heads of comets had been shown to exhibit the following typical spectra: (1) the (Swan) emission bands of carbon (C_2); (2) the violet bands of cyanogen; (3) emissions near 4050 Å and 4310 Å, which would be identified as C_3 and CH, respectively; (4) the sodium D lines, usually for comets near the sun; and (5) the continuous solar spectrum with its strongest Fraunhofer lines. Little was known about the tail spectrum. Of course, there were exceptions to the typical spectrum. Comet Holmes (19P/1892 V1) in 1892 showed a continuous spectrum only, whereas comet 5D/Brorsen in 1868 did not show the Swan bands.

The theory of atomic and molecular spectra was still essentially nonexistent by 1900. An understanding of cometary spectra was, of course, dependent on parallel development in the theory. Even the mechanism of emission was not understood. Were comets self luminous or did they shine because of the sun's radiation? The basically correct answer was given by K. Schwarzschild and M. Kron in their 1911 paper on the intensity distribution in the tail of Halley's Comet. They attributed the mechanism to the absorption and reemission of solar radiation, that is, fluorescence. Although there was some debate over the years, Herman Zanstra (1894–1972) showed that fluorescence (and resonance fluorescence) fully accounted for the line and band spectra of comets (Zanstra 1929). In his paper, Zanstra states: "...the theory is proposed that the bright line and band spectra observed in the head of a comet are produced by absorption of sunlight by the gasses and vapours in the comet's head and subsequent re-emission of the same wavelength (resonance) or a longer wavelength (fluorescence)..." He demonstrates that his theory is completely compatible with observations of comet Wells (C/1882 F1).

Information on the composition of comet tails began to accumulate early in the twentieth century with the beginning of observations with objective prisms; that is, large-area, small-angle prisms that are placed in front of the objective lens or mirror

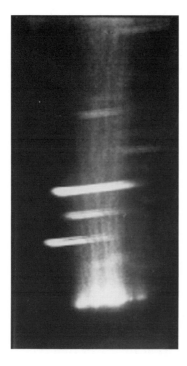

Fig. 1.12. Objective-prism spectrum of comet Morehouse (C/1908 R1) obtained by Frost and Parkhurst. The streaks across the images of the comet are stellar spectra. (Yerkes Observatory photograph.)

of a telescope. The leader in this work was Fernand Baldet. Spectra of comet Daniel (C/1907 L2) and comet Morehouse (C/1908 R1) were obtained, and they showed emission extending into the tails. Comet Morehouse had no detectable continuum in the tail, but the emission bands appeared as doublets (Fig. 1.12). These doublets were duplicated in the laboratory by A. Fowler and subsequently were shown to be due to ionized carbon monoxide, CO^+. In addition, early investigators found N_2^+ in comet tails.

In subsequent years the basic picture would be built up and many new identifications established. The C_2 and CN emissions are confined to the head and are the first (other than the continuous spectrum) to appear as a comet approaches the sun. As the comet nears 1 AU, CO^+ appears in the tail and, if the comet approaches near the sun, the sodium lines appear in the head, whereas the Swan bands (C_2), for example, decrease. If the comet passes within 0.1 AU of the sun, lines of iron and nickel appear. As the comet recedes from the sun, the sequence occurs in reverse order (Fig. 1.13).

We conclude this section on the history of spectroscopy by mentioning an important consequence of the fluorescence mechanism. The intensity distribution within the molecular bands found in comets usually did not resemble the distribution observed in laboratory sources. The problem of understanding these band

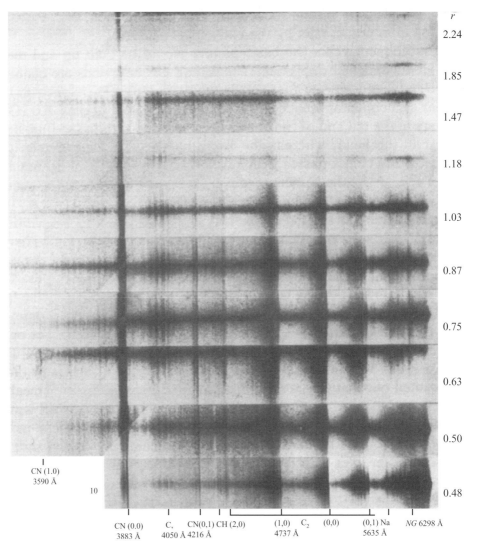

Fig. 1.13. Spectra of comet Cunningham (C/1940 R2) at different heliocentric distances, labeled by *r*. The line labeled NG (nightglow) originates in the earth's atmosphere. (From *Atlas of Representative Cometary Spectra* by P. Swings and L. Haser, Ceuterick Press, Louvain, 1956.)

structures was compounded by the fact that the intensity distribution was significantly different from comet to comet and even varied within the same comet at different times or at different heliocentric distances. The phenomenon, now called the *Swings effect*, was investigated and understood by Pol Swings in 1941. He pointed out that the exciting radiation for the fluorescence mechanism was the solar spectrum, which showed variations with wavelength particularly near strong

Fraunhofer lines. The comet's radial velocity relative to the sun causes a doppler shift of the Fraunhofer lines with respect to the cometary band systems, which change with the position of the comet in its orbit. Thus, a particular line in a cometary band could be bright or faint depending on its position relative to a strong solar absorption line. Swings's hypothesis was confirmed by A. McKellar's observations of comet Cunningham (C/1940 R2). A variant of the Swings effect is the Greenstein effect. The expansion of the coma causes a doppler shift of the Fraunhofer lines with respect to the cometary band systems, which varies with position within the comet. Therefore, cometary line intensities are different at different locations in the comet.

1.6.3 Nuclei and cometary phenomena

The need for a cometary nucleus or nuclei was realized very early in the historical game. The dust and gas observed in the coma and tail had to have a source. It became clear very early that the material in the coma and tail was completely lost to a comet and had to be continually replenished. This conclusion was reinforced by observations of comet Perrine (C/1897 U1) and comet Ensor (C/1925 X1), both of which faded and vanished as they moved in the inner solar system. The source for the coma and tail material could have been a single large body or a swarm of smaller bodies.

Occasional transits of comets across the sun's disk give an opportunity to search for a large nucleus. The Great Comet of 1882 transited the sun's disk and could not be seen. However, it was estimated that if the nucleus were larger than 70 km, it would have been seen. Similarly, Halley's Comet was invisible when it transited the sun on May 18, 1910. Clearly, the nuclei of these comets could not be as large as 100 km.

What is the lower limit for the size of cometary nuclei? The Great Comet of 1882 passed sufficiently close to the sun to vaporize bodies about 1 meter in size. Although the comet showed some disruptions, its brightness did not change appreciably after perihelion. Similarly, the orbit of comet C/1843 D1 took it through the solar corona and within 0.1 solar radius of the photosphere without appreciable effect. The magnitude of the temperatures reached in comets at such close encounters is demonstrated by the fact that vaporization of dust takes place, as evidenced by the appearance of emission lines of metals in the comets' spectra. Thus, the nuclei of these comets could not have been as small as 1 meter, and probably was larger than 10 meters.

In the 1960s, E. Roemer (1963) made brightness measurements of cometary nuclei with the 1-meter astrometric reflector at the U.S. Naval Observatory's Flagstaff Station. These measurements were made when the comets were well away from the sun, and therefore had limited contamination by the coma. Using the comet's distance from the sun and earth, the nucleus' phase function, and an estimate of its albedo, one can infer the size of the nucleus from its magnitude. Roemer deduced sizes in the range 1 to 10 km for several comets, assuming limiting values for the albedo (also see section 7.2.1).

The conclusion of the historical evidence was that a cometary nucleus consisted of a concentrated swarm of meteoroids with sizes in the range 10^{-4} cm to 10 meters. The strongest evidence for this conclusion was the identification of the orbits of several meteor streams with the orbits of specific comets. This concept is called the sand or gravel bank model of a comet.

The mass of this nuclear aggregate was very difficult to measure. Dynamic determinations, if available, would be the most satisfactory. Unfortunately, comets have such low masses that they have never been observed to produce a measurable perturbation on another solar system body. For instance, on July 1, 1770, Lexell's Comet passed within 2.3 million km of earth without producing as much as a 1-second change in the length of the year. In 1805, Laplace demonstrated that the mass of the comet could not exceed $1/5000$ the earth's mass (M_e) or 4.6×10^{23} g. Calculations based on mutual perturbations of the two pieces of Biela's Comet after it split in 1846 give a rough mass of $4.2 \times 10^{-7}\ M_e$ or 2.5×10^{21} g.

Studies of the Perseid meteor stream led B. Vorontsov-Velyaminov in 1946 to estimate a mass for comet Swift–Tuttle of 5×10^{16} g. He also estimated a mass for Halley's Comet in 1910 of 3×10^{19} g from the brightness of the reflected Fraunhofer spectrum; similar estimates were made by several others. If we assume a density of 1 as a reasonable value for most comets (50% silicates + 50% water ice), then the mass estimates, which depend on the cube of the radius of the nucleus, are sensitive to nuclear size estimates. Estimates made around 1984 placed the nucleus of Halley's Comet at about 2.5 km, yielding a mass of about 10^{17} g. Given the magnitude of the uncertainties, we conclude that the mass of Halley's Comet is $10^{18\pm1}$ g and that of an average-size comet is $10^{17\pm1}$ g. (Incidentally, you will find other values quoted in this book for the nuclear density. This is one of the more uncertain parameters for comets, and probably varies from comet to comet.)

Now how does this relatively low-mass nuclear aggregate of rocks or swarm of meteoroids produce all the cometary phenomena usually observed? The historical answer was based on the variation of cometary brightness. The law of cometary brightness can be written, ignoring phase effects, as

$$J = \frac{J_0}{\Delta^2} F(r), \tag{1.1}$$

where the brightness is in units of ergs per centimeter squared-second frequency interval, and Δ is the earth–comet distance, in AU. The geocentric variations was set equal to Δ^{-2}, because all evidence indicates that the only effect of the earth–comet distance is the usual inverse square law of brightness versus distance.

The function $F(r)$ represents the heliocentric distance variation of cometary brightness, a quantity that could well be governed by many factors. Hence, the discussion proceeded empirically by assuming $F(r) \propto r^{-n}$. Cometary brightnesses were usually expressed in magnitudes, in terms of which (1.1) became

$$H = H_0 + 5 \log \Delta + 2.5n \log r, \tag{1.2}$$

where r and Δ are in AU. We can see that H_0 is the magnitude of the comet when $r = \Delta = 1$. This equation was simplified further by defining $(H - 5 \log \Delta) = H_\Delta$, the magnitude if the comet were 1 AU from earth. At length, one found

$$H_\Delta = H_0 + 2.5n \log r. \tag{1.3}$$

According to (1.3), a plot of total cometary magnitude versus $\log r$ should have produced a straight line with slope $2.5n$. This simple formula provided (and still provides) a reasonable representation of the data, with n being variable from comet to comet. Common individual values have been found to range between 2 and 6, with rare cases in the range -1 to 11 (see section 3.3). If the amount of reflecting or scattering material in the comet (rocks, dust, and gas) remained constant as it changed heliocentric distance, n would be 2. However, the average value was more like 4, which indicated a generation of material proportional to the absorbed solar radiation. The material generated was assumed to be principally gas. In the 1940s, B. U. Levin attempted to develop a physical picture of the relevant processes. To understand Levin's work, we must look at the physics of adsorption.

A solid material is known to cover itself with a monomolecular layer of gas or liquid that can be strongly attached in a process called adsorption. The outside surface that is covered includes all irregularities and cracks. The layer attached to the solid body is called the adsorbate, and the solid itself is called the adsorbent. The process whereby molecules are released from the surface of the solid body is called desorption. The processes of adsorption and desorption form the basis of Levin's physical picture. The processes are complex and depend on the nature of both the adsorbate and the adsorbent. The rate of desorption from a completely covered surface can be written as

$$\nu = s/\tau, \tag{1.4}$$

where s is the monolayer capacity of the adsorbent, which is defined as the surface concentration at complete covering, and τ is the average lifetime for a molecule to remain in the adsorbate. The lifetime is given by the empirically tested relationship known as Frenkel's equation, namely

$$\tau = \tau_0 e^{+L/RT}, \tag{1.5}$$

where L is the heat of adsorption, T is the temperature (K), τ_0 is the vibration period of a molecule attached to the surface, and R is the gas constant. Because the principal temperature variation in (1.5) is in the exponential term, the desorption rate is roughly

$$\nu \propto e^{-L/RT}. \tag{1.6}$$

If one assumes the brightness of the comet's head to be proportional to the desorption rate, then

$$J \propto e^{-L/RT}. \tag{1.7}$$

To complete the derivation, we need to know the temperature of the adsorbent at different distances from the sun. This was supplied in 1929 by Herman Zanstra, who first derived the temperature of solid cometary particles assuming an equilibrium between absorbed and reemitted solar radiation. Zanstra's equation, derived ignoring rotation of the nucleus, gives

$$T = \frac{T_0}{r^{1/2}} = \frac{298}{r^{1/2}}(\text{K}). \tag{1.8}$$

This equation might not hold for particles larger than 1 cm if they keep the same side toward the sun. With this caveat, the brightness of the head becomes, in magnitudes,

$$H_\Delta = 2.5 \log J + \text{const} = \frac{L r^{1/2}}{R T_0} + \text{const.} \tag{1.9}$$

At $r = 1$, $H_\Delta = H_0$, and we rewrite the equation as

$$H_\Delta = H_0 + \frac{L}{R T_0}(r^{1/2} - 1). \tag{1.10}$$

Equation (1.10) can be compared with the empirical relation (1.3),

$$H_\Delta = H_0 + 2.5n \log r. \tag{1.11}$$

At first sight (1.10) and (1.11) might appear to be incompatible, but they are not. Over the range of r usually associated with cometary observations, $\log r \sim 0.87(r^{1/2} - 1)$. Hence, Levin's picture was found to be compatible with the observations. In addition, the empirically determined value of n is approximately related to the heat of desorption L by

$$n \approx 0.46 \frac{L}{R T_0}. \tag{1.12}$$

The observations determined a large range of values for L with an average value of 6000 cal mol^{-1}. Typical values measured in the laboratory were near 10^4 cal mol^{-1} and, hence, the cometary adsorbate was rather volatile by terrestrial standards.

Levin's theory had at least one fatal flaw. Laboratory studies showed that 1 g of meteoritic material was capable of supplying a total of 10^{19} molecules, a number that included both adsorbed molecules (discussed above) and molecules that had penetrated well into the solid and, accordingly, required more energy for extraction than the surface-layer molecules. For a comet with a typical nuclear mass of 10^{18} g, the total gas supply available could not exceed about 10^{37} molecules. However, Karl Wurm has shown that the typical comet contains approximately 10^{35} or 10^{36} molecules in its coma, and these molecules leave through the cometary tail in 1 day or less. Thus, the total meteoritic gas supply would maintain the supply of molecules observed in comas of comets for a period of time ranging from 10 days to at most a few months. Of course, we know comets can have comas visible for many months, and some of the periodic comets have been seen on numerous returns. Thus, although the adsorption–desorption picture is compatible with the relative rate at which gas is released, it falls far short of explaining the total amount of gas released. As we shall see later, the solution to the problem is not to produce the gases from a minor impurity, but from the material in the nucleus itself. Ices provide a suitable material, and they can generate an ample supply of gases to maintain comets: 10^{22}–10^{23} molecules g^{-1}.

1.6.4 The model of the nucleus after 1950

We will say that the modern era of cometary astronomy began in 1950. In that year the first of Whipple's papers on the icy-conglomerate model of the cometary nucleus and Oort's paper proving the existence of a vast reservoir of new comets were published; hence it is a natural boundary in time. Though, as with all historical boundaries, it is not rigidly defined. We will presently establish many other lines of investigation that are important to the understanding of cometary phenomena began at approximately the same time. The description of some of these items will be brief in this chapter. Many items will be discussed in more detail in subsequent chapters.

1.6.4.1 The icy-conglomerate model

The icy-conglomerate model of the cometary nucleus was originally put forward by Whipple in three papers beginning in 1950 (Whipple 1950a, 1950b, 1955). In this model the nucleus is presented as a rather large mass of ices such as H_2O, NH_3, CH_4, CO_2, and C_2N_2, with particles with a range of sizes from ~ 1 μm to meteoroids (say a few tens of centimeters), embedded throughout. The structure was assumed

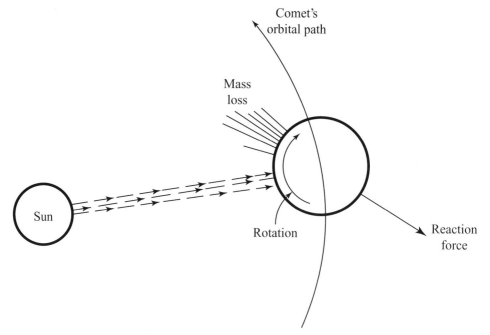

Fig. 1.14. Schematic of the rocket effect.

to have enough tensile strength to resist tidal disruption due to a relatively close encounter with a planet or the sun. The ices postulated by Whipple also provided an adequate supply of gas for the comet because each gram of ices yields 10^{22}–10^{23} molecules. Thus, for equal masses, ices are a better source of gas than are meteoritic grains or stones by a factor of 10^3 or 10^4. However, it is essential that the ices exist in a single lump or perhaps a small number of lumps; very small isolated ice grains would have a short lifetime. Ices are generally poor conductors of heat and, thus, the incident solar radiation heats only a surface layer that is slowly eroded away by sublimation. The dust ($\sim$1 μm size) observed in cometary comas and type II tails is released when the ices sublimate. Some of the dust could remain attached to the nucleus to form an insulating layer.

   If, as is highly likely, the nucleus spins on an axis, the *rocket effect* can operate to produce the nongravitational forces required to explain the motions of some comets, such as Encke. Heat is supplied to the nucleus by solar radiation, and it takes time for the heat to penetrate to the icy layers and to cause sublimation. Thus, the sublimation of the gas molecules (which form the coma) takes place in a preferential direction on a spinning nucleus that makes an angle with the sun–comet line (Fig. 1.14). The recoil due to this preferential, non-isotropic mass loss produces the nongravitational forces on comets. The direction of rotation of the nucleus determines whether the nongravitational force decelerates or accelerates the comet's motion. More recent studies suggest that nongravitational effects are due to a radial outgassing that is asymmetrical with respect to perihelion (see section 2.4.2).

Whipple's icy-conglomerate model explains the source of the gases known to come from comets at many apparitions, explains the nongravitational forces, supplies the dust observed in type II comet tails, and has tensile strength to resist tidal disruption. Eventually, all of the icy material is sublimated, and the remaining dust and stones are spread out by perturbations along the comet's orbit to form meteoroid streams. The model has roughly three-fourths of the mass in ices and the remainder in meteoritic dust and stones. Some of the dust particles emitted by active comets would be spread throughout the solar system by planetary perturbations and the Poynting–Robertson effect. Whipple concluded that the cometary source was sufficient to supply the zodiacal light indefinitely.

The icy-conglomerate model even held hope of explaining the jet-like phenomena so frequently observed near cometary nuclei. The nucleus could be rather inhomogeneous, and sub-surface packets of gas might develop. Donn and Urey (1956) extended the icy-conglomerate model to include explosive chemical reactions involving possible free radicals or unstable molecules. The explosions would contribute to the activity observed in comets, including jets and outbursts.

Donn and Urey also commented on the chemistry of Whipple's model. The presence of methane (CH_4) and carbon dioxide (CO_2) in the same object seemed implausible because in methane, carbon is in its most reduced state, whereas in carbon dioxide it is in its most oxidized state. Because carbon dioxide (in the form of CO_2^+) was observed in comet tails, there was the possibility that methane was not an important constituent of the cometary nucleus.

1.6.4.2 The clathrate hydrate model

The most significant problem with Whipple's original icy-conglomerate model was the rate of production of gas at different heliocentric distances, particularly the usual onset of gaseous activity at approximately 3 AU. Very distant comets usually show only a continuous spectrum. Near 3 AU the CN spectrum appears, and near 2 AU the spectra of C_3 and NH_2 appear. As the comet continues inward and approaches 1.5 AU, the spectra of C_2 (Swan bands), CH, OH, and NH appear. These spectra grow stronger as the comet moves inside 1.5 AU and the spectra of CO^+, OH, N_2^+, and CH^+ appear in the tail. The lines of Na appear near 0.8 AU and, if the comet's perihelion distance is very small ($\sim$0.1 AU), lines of Fe, Cr, and Ni appear. This sequence is an averaged idealization that may not apply to a particular comet. Originally, it was thought that the spectroscopic sequence could be explained by the more volatile molecules sublimating at the larger heliocentric distances, coupled with the subsequent dissociation of various relatively complex parent molecules.

However, when examined in detail, this picture could not satisfactorily account for the onset of gaseous activity near 3 AU.

The onset of activity appeared compatible with the sublimation of water ice; that is, a hypothetical frozen H_2O comet approaching the sun would begin gaseous activity near 3 AU. Water is not easy to detect from the ground and, hence, the spectra observed (CN) could result from other constituents released at the same time. Delsemme and Swings (1952) suggested that the nucleus was composed of solid hydrates of the species thought to be needed, such as $CH_4 + 6H_2O$, and so on. The picture based on solid or clathrate hydrates has been presented by Delsemme and his colleagues in a series of papers (Delsemme and Wenger 1970; Delsemme and Miller 1970, 1971a,b). Some substances form crystal structures that have cavities large enough to permit occupancy by noble gas atoms or molecules. Water molecules can form a hydrogen-bonded framework with cavities that are polyhedra with pentagonal or hexagonal faces. Each unit cube contains 8 xenon atoms and 46 water molecules; thus, the clathrate is essentially a hydrate, Xe-5 3/4 H_2O. Similarly, methane forms a clathrate hydrate, CH_4-5 3/4 H_2O.

Delsemme's introduction of clathrate hydrates into Whipple's icy-conglomerate model appeared to have solved the problem of the onset of gaseous activity in comets near 3 AU. The essential point is not the existence of the exact crystal structure of the clathrates but the existence of cavities in which other molecules can be trapped. The rest of the spectroscopic sequence results from excitation phenomena. The sodium and metals lines arise from the vaporization of dust as the comet nears the sun.

The origin of clathrate hydrates requires high pressures (Lewis 1997), and there is no indication that comets formed under conditions of high pressure or that they experienced high pressures since their formation. As we said above, clathrate hydrates explain the detection of gas emission at heliocentric distances of about 3 AU. As we will see, several comets, including comet Hale–Bopp showed continuous gas emissions of several species at distances as large as 7 AU. An alternate explanation is the possibility of amorphous water ice, which can trap other molecules, which are subsequently released at temperatures above about 135 K. Amorphous water ice is formed at $T < 135$ K, by fast condensation of the gas into the solid form. The fast condensation does not give the water molecules time to reorient themselves into a crystalline lattice and the low temperatures prevent reorientation of the molecules. This condition prevails in dark interstellar clouds (Kouchi *et al.* 1994), and at a distance greater than about 7 AU in the protoplanetary nebula (Mekler and Podolak 1994). Thus, if amorphous water ice does exist in comets, it was either formed before the material arrived in the solar nebula or at a distance beyond Jupiter. This issue is not yet fully resolved, but currently amorphous water ice is favored (section 7.3.2).

Delsemme's model is compatible with other cometary phenomena. For example, the sublimation rates are sufficient to produce the nongravitational forces required to explain orbital data in detail. Delsemme's model also produces an icy-grain halo (caused by the stripping of ice grains from a solid as the ice sublimates) that may reflect enough sunlight to produce the so-called false or photometric nuclei of comets. Delsemme has been the champion of a hypothesis that comets bombarding the earth in its first 600 million years of existence account for the water in our oceans (see section 10.5).

The view of water ices as the principal constituent of the nucleus received an additional boost in 1970 with the observation of huge clouds of atomic hydrogen around comets. Comet Tago–Sato–Kosaka (C/1961 T1) was observed in Lyman α (λ 1216 Å) from the Second Orbiting Astronomical Observatory (OAO-2) by Code, Houck, and Lillie (1972). In April 1970, Lyman-α radiation was measured from comet Bennett (C/1969 Y1) by Bertaux and Blamont using photometers on the Fifth Orbiting Geophysical Observatory (OGO-5). Comet Bennett was also detected in Lyman α by the OAO-2 observers and by Thomas using the other Lyman-α photometer on OGO-5. The impressive aspects of these discoveries were these: (1) they had been predicted in some detail by Biermann (1968) and (2) the clouds of hydrogen had huge dimensions, some 1.5×10^7 km or about 0.1 AU. The clouds around these two comets were rather similar. In December 1970 the hydrogen cloud around comet Encke was detected; this cloud had a dimension of around 10^6 km.

The spectrometer on OAO-2 obtained spectral scans of comet Bennett over the range $\lambda\lambda$1000 to 4000 Å. The strong emission near λ3090 Å is due to the OH radical. Its strength leads us to the conclusion that OH has an abundance comparable to that of H. The fact that both H and OH are present, and are roughly equally abundant, is strong evidence that they originated from the dissociation of water molecules. Furthermore, the large abundances of H and OH lead us to the conclusion that water is the principal constituent of the cometary nucleus. In 1962 Swings and Greenstein discovered the forbidden red oxygen line near λ6300 Å in the spectrum of comets. The presence of this line, with its very low transition probability, indicates that large quantities of O exist in cometary atmospheres. One possible source of the O could be the dissociation of the OH radical; however, the dissociation of H_2O directly into $H_2 + O$ and the dissociation of CO_2 into $CO + O$ probably also contribute to the presence of O. The existence of O does point to a large abundance of OH in the cometary atmosphere, and the evidence is entirely consistent with the idea that H_2O is the major parent molecule for H, OH, and O.

This picture was strengthened and refined by observations of comet Kohoutek. The H_2O^+ ion was identified by Herzberg and Lew (1974) by comparing labora-tory spectra with cometary spectra taken by Herbig and by Benvenuti and Wurm. Evidence has also accumulated for large amounts of CO or CO_2 in cometary atmo-spheres. If the nucleus contained roughly 10–15% CO or CO_2 by number trapped in the clathrate-like ice cavities, the presence of CO in the halo and the predominance of CO^+ as the observed emission in type I tails would be explained. It seems likely, then, that the predominance of CO^+ emission in plasma tails is due simply to the large amount of CO present in comets.

1.7 Comets and the solar wind after 1950

1.7.1 Ion acceleration

An important area of cometary physics was discovered by Biermann (1951). He studied changes in the position of features in type I tails; knots or clouds of gas showed repulsive accelerations of $1 - \mu = 10^2$ or more times solar gravity. As has been noted, such accelerations could not be produced by radiation pressure. Biermann attributed the accelerations to momentum transferred to the tail ions (largely CO^+) by a continuous outflow of ionized material from the solar corona, which he called the *solar corpuscular radiation*. Biermann developed a simple quantitative model for the momentum transfer – between a proton–electron stream and an ion–electron gas. The acceleration on the ions was

$$\frac{dV_i}{dt} \approx \frac{e^2 N_e V_e m_e}{\sigma m_i} \approx 10^{-4.3} \frac{m_e}{m_i} N_e V_e. \tag{1.13}$$

Here N_e and V_e are the electron density and the bulk velocity in the beam, respectively, m_e and m_i are the masses of the electrons and ions, respectively, and σ is the electrical conductivity that has been evaluated approximately.

For $V_e = 1000$ km s^{-1} (from geomagnetic delay times) and an assumed $N_e = 600$ cm^{-3}, the acceleration is 100 cm s^{-2}. At the orbit of earth, the solar gravity is 0.59 cm s^{-2} and, hence, 100 cm s^{-2} corresponds to a $1 - \mu$ of roughly 170, a value adequate to explain the observations. The problem with this solution was the assumed electron density of 600 cm^{-3}. Although this value was in agreement with other evidence at the time, it is now known to be about two orders of magnitude too high.

Biermann's picture was supported by Hoffmeister's (1943) work on the orientations of plasma tails. As viewed in the plane of the comet's orbit, the plasma tails were not strictly radial but lagged behind the radius (in the sense opposite the comet's motion) by an aberration angle of about 5°.

This aberration effect is probably the most durable early cometary evidence for the existence of the solar corpuscular radiation, or the *solar wind* as it is now called. These studies emphasized the facts that the solar wind shaped parts of comets and that a proper understanding of the solar wind interaction was necessary for a sound physical picture of comets (Chapter 6). The study of plasma tail orientations remains a valuable method for inferring solar wind velocities.

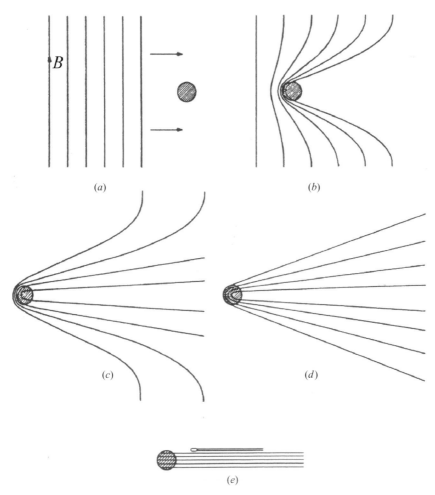

(a)   (b)

(c)   (d)

(e)

Fig. 1.15. Interaction of the solar wind magnetic field, *B*, with a cometary ionosphere. Parts (a) through (d) are a time sequence as viewed perpendicular to the
magnetic field. Part (e) is the view parallel to the field. (After H. Alfvén, 1957, On
the theory of comet tails, *Tellus* 9: 92–6)

1.7.2 Cometary magnetic fields

The interplanetary or solar wind magnetic field was introduced into the study of
comet tails by Alfvén in 1957. If a plasma beam (protons plus electrons) with a
frozen-in magnetic field encounters a comet, molecules that are ionized (perhaps
by the encounter) find themselves attached to the field lines. This effect "loads" the
field lines near the head of the comet, but not those far away. Thus, the field lines
are hung up on the head of the comet, as illustrated in Fig. 1.15, which shows the
view perpendicular to the field of the beam; the view parallel to the beam is also
shown in Fig. 1.15. The two views are different unless there is sufficient turbulence.

   The introduction of the magnetic field by Alfvén naturally explained the narrow, straight streamers that make up the type I tail. The magnetic field might enhance the coupling between the interplanetary and cometary plasmas and perhaps produce the accelerations of features observed in type I tails without the need for the high densities required in Biermann's essentially viscous picture. Hoyle and Harwit (1962) investigated the possibility that the interaction in the field-free case might be enhanced by collective effects; they concluded that plasma instabilities were not effective. A companion study by Harwit and Hoyle also in 1962 showed that capture of an interplanetary magnetic field by a comet could lead to a coherent picture. The compression of field lines in the head could squeeze plasma out of a magnetic tube of force, much like toothpaste out of a tube, and thus account for the initial velocity of tail material of roughly $20 \, \mathrm{km \, s^{-1}}$ near the head. After the material enters the tail proper, it essentially expands into a medium with zero pressure at large distances from the head. Plausible pressure differences could produce the accelerations observed for the tail knots and clouds.

1.8 The origins of comets

1.8.1 Early twentieth-century ideas

The nineteenth century saw considerable speculation on the origin of comets. Orbital information rather than the physical state of the comet was the basis of these efforts.

   As Chambers would write in 1910:

Two provisional answers suggest themselves: either (1) comets are chance visitors wandering through space and now and again caught up by the Sun, or by some of the major planets . . . and compelled to attach themselves to the Sun and by taking elliptic orbits to become permanent members of the solar system; or (2) they are aggregations of primeval matter not formed by the Creator into substantial planets, but left lying around in space to be picked up and gathered into entities as circumstances permit.

(Olivier 1930:208)

Also in 1910, A. C. D. Crommelin wrote concerning the origin of comets:

(1) that they are the products of eruptions from the sun; (2) that they are the product of eruptions from the larger planets in a sunlike state; (3) that they are stray fragments of the nebula which is supposed to have been the parent of our system, and that they remained unattached to any of the large masses that were formed from the nebula.

(Proctor and Crommelin 1937:179)

   Theories based on an interstellar origin and subsequent capture within the solar system have been advocated by Laplace (in 1813), Heis, Schiaparelli, von Seeliger, Fabry, Bobrovnikoff, and Nolke. A theory of purely interstellar origin encountered insuperable difficulties basically because we did not observe comets with hyperbolic orbits. In 1860 the renowned solar physicist R. C. Carrington noted that comets originating outside the solar system would as a rule have hyperbolic orbits. He also noted that as the solar system moved through the supposed interstellar cloud of comets, we should observe more comets coming from the direction of the solar motion than from the opposite direction. Any doubt concerning the lack of hyperbolic orbits was removed by Ellis Strömgren in 1914 when, by considering planetary perturbations, he showed that (within the errors of observation) there was no evidence from observations that any comet entered the solar system with a hyperbolic orbit. In addition, Henry Norris Russell in 1920 showed that planets were not efficient in "capturing" comets. Only Jupiter has a legitimate comet family. Saturn might have a comet family, but Uranus and Neptune certainly do not. The capture process, even if the comets with hyperbolic orbits came long ago, is basically inefficient and cannot account for the observed orbits of comets. Subsequent research by E. Strömgren confirmed Russell's conclusions.

The capture theory might be revived in two ways. One way is the theory proposed by R. A. Lyttleton (1953) in which the condensations in the interstellar medium are caused by the passage of the sun. These condensations travel in hyperbolic orbits with respect to the sun and intersect each other on the line representing the solar motion in the antiapex direction. Collisions of these condensations were thought to produce, according to Lyttleton, the sand or gravel bank model of comets. Because the collisions would be inelastic, the hyperbolic motion could be changed into the elliptical and parabolic motion observed. There are many difficulties with Lyttleton's theory, and it has very few advocates at present.

The second way to save the capture hypothesis would be to have comets originate in an interstellar cloud that traveled at the same speed and direction as the solar motion. Such a hypothesis is highly artificial and has probably never been seriously proposed. A perfectly natural way to achieve the desired result is to have both the solar system (sun and planets) and the comets formed out of the same interstellar cloud. This is the same basic idea proposed above by Chambers (second suggestion) and by Crommelin (third suggestion).

However, the fact that comets had to be solar system objects gave support to other theories of their origin. Crommelin seemed inclined toward a solar origin at least for some comets, such as the sungrazing group of comets that included comet 1882 II. Material was certainly observed to be ejected from the sun at great velocities, and this fact supported the idea. However, it seemed impossible to explain the origin of comets with perihelion distances well away from the sun. The solar hypothesis also suffers from another problem: the sun is too hot. Fortunately, most other theories of cometary origin are not encumbered with this objection.

The final historical theory of cometary origin within the solar system ascribes comets to ejections from the major planets, principally Jupiter. Through the years Lagrange, Proctor, Tisserand, and Vsekhsvyatskij have advocated a planetary origin via ejection. S. K. Vsekhsvyatskij envisioned a volcano-like process in his earlier work (cf., Vsekhsvyatskij 1977), but the large escape speed from Jupiter (67 km s^{-1}) seemed too high to be attained by known physical processes. In his later work the site of origin was transferred to the satellites and rings of the major planets, where the escape speed is lower by a factor of 10. The near parabolic orbits were supposed to have been generated by ejection from the outer planets, starting with Uranus and including hypothetical trans-Plutonian planets.

An alternate proposal places the site of ejection of comets on the satellites of the major planets. The satellites thought by Vsekhsvyatskij to be responsible are Europa and Callisto (Jupiter), Titan (Saturn), Titania (Uranus), and Triton (Neptune). The Galilean satellites of Jupiter were examined closely in 1979 by the *Voyager* space-craft. Intense volcanic activity has been found on Io; however, its density leads to a model of mostly silicate material. Europa and Callisto probably have icy crusts; however, no sign of volcanic activity was found. At least for these three Jovian satel-lites, the volcanic-type ejection of compact cometary bodies seems very unlikely. The most telling argument against the major planets as a source of comets is the ob-served distribution of $1/a$ values, with a peak between 10^{-4} and 10^{-5}. Perturbations acting on comets arising in these outer planets cannot produce such a distribution.

1.8.2 The Oort cloud and Kuiper belt concepts

The question of cometary orbits and the origin of comets was also under active study during our modern era. In fact, some of the speculation about the origin of comets has at times outstripped the necessary parallel progress in the physics of comets. This chapter of our story also begins in 1950 with a study by Oort. The idea of a very large cloud of comets surrounding the inner solar system has been known for some time. Öpik had discussed the idea in 1932, and Young's *General Astronomy* ascribed the idea of a comet-dropping envelope a very large distance from the sun to Peirce in the nineteenth century. But in 1950 Oort developed the idea that has become, with modifications, the most widely accepted view of the origin of comets. Oort's investigation was stimulated by van Woerkom's (1948) study on the origin of comets. The paper by van Woerkom contains many items of interest, in particular, the idea that comets diffuse inward from a field of parabolic comets with Jupiter's perturbations causing the diffusion.

The basic data of Oort's paper were the distribution of original values of $1/a$ (*a* being the semi-major axis in astronomical units); the "original" values refer to

Table 1.2 *Distribution of original values of* $1/a$

$1/a$ (AU^{-1})	N (Strömgren)
≤ 0.00005	10
0.000 05–0.000 10	4
0.000 10–0.000 15	1
0.000 15–0.000 20	1
0.000 20–0.000 25	1
0.000 25–0.000 50	1
0.000 50–0.000 75	1
>0.00075	0

the time when the comets were still well outside the orbits of the major planets. The 19 values for which the mean error in $1/a$ is less than 0.0001 came from the work of E. Strömgren. Oort interpreted the frequency curve implied by Table 1.2 indicating a steep maximum for very small values of $1/a$ and concluded that a substantial fraction of these comets originated in a region of space extending from 20 000 AU to at least 150 000 AU from the sun. Although this distance extended halfway to the nearest star, Oort noted that the comets were not interstellar because of the absence of hyperbolic original orbits; hence, he inferred that they were always members of the solar system. But how could these comets be sent into the inner solar system where we observe them? Oort considered the only viable mechanism to be perturbations caused by passing stars, and he studied that mechanism in detail. He concluded that the cloud and stellar perturbations could account generally for the observed distribution curve of $1/a$ as well as for the random distribution of orbital planes and perihelia, and for the preponderance of nearly parabolic orbits. Oort estimated that the cloud contained roughly 10^{11} comets with a total mass between 1/10 and 1/100 the mass of the earth.

It was thought at the time of the Oort cloud hypothesis that the short-period comets could be accounted for by the diffusing action of Jupiter; van Woerkom's calculations indicated an average change in $1/a$ caused by a comet's passage through the inner solar system of 0.0005, a change that could be either positive or negative. This provides the input to the short-period comets. They are lost by disintegration in the inner solar system and by ejection due to planetary perturbations.

Oort also speculated that comets were formed at an early stage of the planetary system from the asteroid ring and "brought into large, stable orbits through the perturbing actions of Jupiter and the stars" (Oort 1950). Meteoroids were thought to have a common origin.

Finally, Oort offered an explanation for the fact that about 20 times more comets with very small values of $1/a$ were observed than he would predict. This excess of "new comets" was "taken to indicate that comets coming for the first time near the sun develop more extensive luminous envelopes than older comets" and were thus more likely to be discovered. He noted that this phenomenon might warrant additional study. Oort and Schmidt in 1951 studied the differences between old and new comets. Comets entering the inner solar system for the first time show a stronger continuous spectrum, presumably because they are "dustier." The dust could be liberated as the volatiles near the surface evaporate. On subsequent passes through the inner solar system, the volatiles near the surface are not available, and there is less evaporation and dust release. The large dust release for new comets makes them anomalously bright and accounts for their enhanced rate of discovery.

Other orbit-related processes (besides the origin of comets) have been under investigation during this period. In a series of papers Marsden and associates have investigated the nongravitational forces on comets. Initially, they assumed reasonable forms for the necessary functions (because the relevant physics was uncertain), but their papers from the 1980s use functions based on Delsemme's vaporization temperature of an icy conglomerate controlled by the latent heat of water snow. Reasonable values of variables allow the construction of orbits based on both nongravitational and gravitational forces; these give a good representation when compared with the observed positions of comets. The radial nongravitational force is usually about 10 times the transverse component in the plane of the comet's orbit. Thus, the net force makes an angle of about $6°$ with radial direction. However, it is the transverse force that is cumulative and produces most of the observed effects. This force can act either in the direction of the comet's motion or in the opposite direction. The nongravitational force also changes with time. Results for comet Encke show a maximum in the transverse nongravitational force near 1820, a fact that may have contributed to the discovery of nongravitational forces at that time. Marsden and associates have also considered the influence of nongravitational forces on the calculation of original orbits. The nongravitational forces introduce substantial errors into the original values of $1/a$ and, in addition, a systematic effect that reduces the radius of the outer part of the Oort cloud to approximately 50 000 AU.

A statistical approach to the capture of short-period comets by Jupiter was undertaken by Everhart (1972). His basically numerical approach allows him to follow the orbits of tens of thousands of hypothetical comets for up to 2000 returns each. He found that the cumulative effects of Jupiter's perturbations can produce comets with orbital properties closely resembling those of the short-period comets. However, only comets with small inclinations and perihelia near the orbit of Jupiter are involved. This work was considered a promising approach to account for short-period comets.

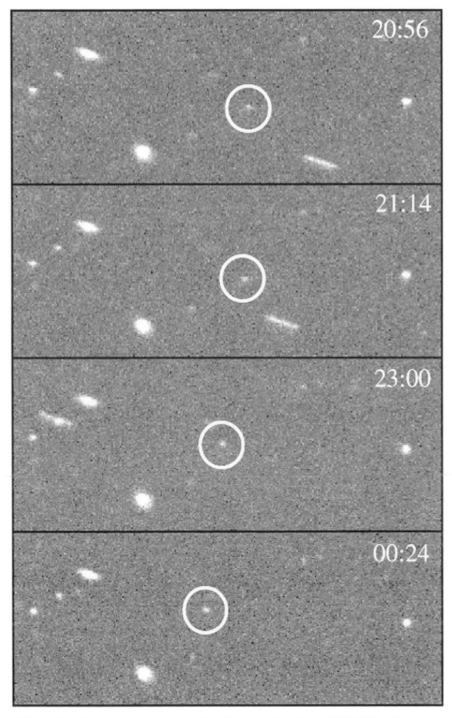

Fig. 1.16. Discovery image of the first Kuiper belt object, 1992 QB1. (Courtesy of D. Jewitt.)

Starting in the 1940s, researchers began to speculate on the existence of bodies beyond the orbits of Neptune and Pluto. Fernandez (1980) suggested a belt of comet-like bodies in the outer solar system as a source of short-period comets. Beginning in about 1988, Jewitt, Luu and others began an initially fruitless search for such bodies. Then in 1992, the first trans-Neptunian object, 1992 QB1, was discovered by Jewitt and Luu (1993) (Fig. 1.16). Since 1992, a growing number of objects have been discovered, and they have been dubbed Kuiper-belt objects or KBOs. (Or Edgeworth–Kuiper belt in Europe.) There is a concensus today that Pluto may be the largest known KBO, though a recent attempt to demote it from planetary status met with strong resistance in the USA. The KBOs are considered to be the most likely source of most short-period comets. We will return to this topic in much more detail in Chapter 9.

1.9 The Halley era

Comet Halley was recovered on its way to its 1986 perihelion by two astronomers using the 200-inch Hale Telescope and an ultrasensitive electronic camera on the night of October 16, 1982. The comet appeared as a faint starlike image with a magnitude of roughly 24.5. Subsequent exposures verified the rediscovery; it was right where orbital predictions said comet Halley should have been, about 11 AU from the sun, and it moved like the predictions said the comet should. Figure 1.17 shows the position of comet Halley in its orbit at the 1982 recovery and at the recovery at each of the previous three passes by the sun. This diagram is a striking illustration of the improvements in observational technology during the roughly 225 years.

We call the period of the 1980s during which comet Halley occupied the attention of researchers the Halley era because it provided so much new knowledge. The era marks the evolution of comet studies from remote sensing observations to *in situ* measurements. As early as the 1970s, scientists around the world began to make plans for space missions to comet Halley while it was near the sun in 1986. The ideal mission would have been a rendezvous, in which the spacecraft flew in formation with the comet, making measurements over a long period of time. The major problem in the case of comet Halley is the fact that its orbital inclination is 162°. That is to say, its orbit is inclined by 72° to the ecliptic, and the comet moves in that orbit in a retrograde direction. That combination makes a long-term rendezvous very difficult. In the end, it was decided to rely on fast flybys near March 10, 1986 as comet Halley passed through the descending node of its orbit.

The United States did not approve a dedicated Halley mission, so the flyby missions were executed by the European Space Agency (ESA), Japan and the former Soviet Union. The spacecraft of what we call the Halley Armada were

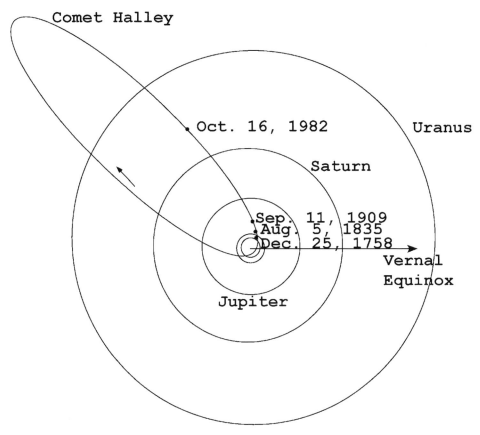

Fig. 1.17. Orbit of Halley's Comet projected on the earth's orbital plane. The distance at which the comet was recovered at each of the last four apparitions is marked on the orbit. The improvement in observing technology in the nearly 250 years is evident.

ESA's *Giotto*, the Soviet Union's *VEGA-1* and *VEGA-2*, and the Japanese *Sakigake* and *Suisei*. NASA did support the *International Cometary Explorer* (*ICE*) (described below) to comet Giacobini–Zinner (GZ) in 1985, and the International Halley Watch. A spectrograph designed to observe comet Halley near perihelion was destroyed on an early shake-down flight when the *Challenger* exploded.

The *Third International Sun–Earth Explorer* (*ISEE-3*) carried a complement of instruments to study the interplanetary medium before it interacted with the earth's magnetosphere. Robert Farquhar, then of NASA, recognized that the *ISEE-3* instruments could be useful in studying cometary phenomena. It also had a working propulsion system that had been used to move the spacecraft to its position at the L1 Lagrange point between the earth and sun. Farquhar devised an elaborate orbital maneuver that involved firing the spacecraft's engine and causing it to make five passes by the moon, the last of which sent *ISEE-3* toward its rendezvous with comet GZ. At that point, *ISEE-3* became the *ICE* mission. Fortunately, *ICE* was subsequently targeted to make a distant encounter with comet Halley as well.

The primary scientific objective of the *ICE* mission was to study the interaction between the solar wind and the atmosphere of comet GZ. The spacecraft traversed comet GZ's plasma tail on September 11, 1985, and its instruments made *in situ* measurements of particles, fields, and waves. It then passed between the sun and comet Halley in March 1986, when the spacecraft of the Halley Armada were in that comet's vicinity.

The International Halley Watch was an interdisciplinary coordination of ground-based observers worldwide. One of the disciplines was the Large-Scale Phenomena (LSP) Discipline. Through the efforts of the LSP team, a nearly complete daily record of images of the comet from September 17, 1985 through July 6, 1986 has been compiled (Brandt *et al.* 1992, 1997). The wide-field images from this effort provided an extensive database for the study of disconnection events (Chapter 6).

Plate 1.2 illustrates the geometrical characteristics of the encounters of *ICE* and the Halley Armada spacecraft with comet Halley. ESA's *Giotto* spacecraft approached closest to the comet. It and the two *VEGA* spacecraft obtained the first resolved images of the nucleus (see Figs. 7.1 through 7.4). They also explored the complex rotation of the nucleus. The images of the nucleus settled the question of its size. It is a single object with dimensions 7.5, 8.2 and 16.0 km. Its albedo is very low, between 0.02 and 0.04, making it one of the darkest bodies known. It is dangerous to generalize from one case, but the albedo measurement puts in doubt the estimates of nuclear sizes based on higher albedo estimates.

The Armada spacecraft made extensive *in situ* measurements of cometary dust, neutral molecules, plasma, and magnetic fields. ESA approved a *Giotto* extended mission (GEM), during which the spacecraft encountered comet 26P/Grigg–Skjellerup on July 10, 1992.

1.10 Summary

In this chapter, we have surveyed the history of cometary science from antiquity to the time of the *in situ* studies of comet Halley. The view of what comets are has evolved from Aristotle's idea that comets are phenomena in the earth's atmosphere to the idea that they are solar system bodies. Tycho began the evolution toward our modern ideas when he attempted to measure the parallax of the comet of 1577, and showed that the comet was well beyond the moon. Newton and Halley calculated the orbital parameters of the comet of 1682, assuming that the orbit was very close to being a parabola, when the comet was near the sun. Halley calculated the orbits of a number of comets, and found that the parameters of the orbit of the comet of 1682 was similar to the parameters of the orbits of the comet of 1531 and the comet of 1607. He proposed that these were different appearances of a periodic comet with a period of roughly 76 years, and predicted its 1758 return. That comet is now known as comet Halley.

In the 1950s, we find three major advances in cometary science. One is the icy-conglomerate model of the cometary nucleus, which has evolved to our present model. The second is the concept of the Oort cloud hypothesis for the origin of comets. This concept has been modified somewhat because of the discovery of the Kuiper belt. Today, we view the Oort cloud as the source of long-period comets, and the Kuiper belt as the source of short-period comets. The third advance is the discovery of the solar wind, which plays a major role in the theory of cometary plasma tails.

In the 1980s, the era of *in situ* comet studies began, with space missions to comet Giacobini–Zinner and comet Halley. These missions have provided a wealth of data on cometary phenomena, including the first resolved images of comet Halley's nucleus.

This is just a brief summary of the high points of cometary discovery. We also talked about a lot more concepts in the chapter. And, we will talk about even more in the remainder of the book. It is hard to decide where to cut off any discussion of history. There have been many exciting discoveries since Halley's recent return. In fact, among the many comets discovered in the 1990s, there have been three particularly notable ones.

Comet Shoemaker–Levy 9 (D/1993 F2) was discovered after a close approach to Jupiter. It was broken into approximately 21 fragments by the tidal effects of the giant planet. It was probably captured into a Jovian orbit in the 1970s. The fragments impacted Jupiter in July 1994. The naked-eye comet Hale–Bopp (C/1995 O1) appeared in 1997. Specific comet Hale–Bopp discoveries included the sodium tail and the detection of many new molecules in the radio wavelengths. The bright comet Hyakutake (C/1996 B2) passed near the earth in 1996. Among the many important discoveries was the detection of x rays by *ROSAT*. Subsequent study has shown that many comets emit x rays. On the night of March 24–25, 1996, Hyakutake approached close to the Earth ($\sim$0.1 AU), and its motion was discernable to the unaided eye.

With this teaser, let's get on with the story. As we will see later, these three objects have added significant information to our understanding of comets. The discovery process is ongoing, and there is no foreseeable end to the excitement.

2

Discovery and dynamics of comets

2.1 Discovery of comets

In the 1990s, more than 20 comets were discovered or recovered each year. By July 1999 over 1000 individual comets were known, 400 of which were discovered before the invention of the telescope. The *Catalogue of Cometary Orbits*, published periodically by the International Astronomical Union's (IAU) Central Bureau for Astronomical Telegrams and Minor Planet Center, is now available both electronically and in paper editions. The 14th edition, Marsden (2001), gives orbital parameters for over 1700 cometary apparitions, which result from 1036 individual comets. About 890 of these are called *long-period comets*, a group of comets that have periods greater than 200 years. The catalog also recognizes 140 periodic comets that have periods of less than 200 years. These periodic comets account for the remaining catalog entries.

Over the years, comets have been discovered by both amateur and professional astronomers. Many amateur astronomers search the sky systematically using wide-field, low-magnification telescopes or binoculars. Some of the brightest, recent comets have been discovered by amateurs. An example is comet Hyakutake, which was a naked-eye comet in 1996; it was discovered by a Japanese amateur. Comets discovered by professionals are often the fainter ones, which are frequently recorded by accident on wide-field plates taken for other purposes. Lubos Kohoutek discovered two comets in 1973 while he was taking plates to study asteroids. Of these two comets, one remained very faint and the other reached naked-eye brightness.

The recovery of comet Halley in 1982 was the result of a search by professional astronomers using the 5-m Hale Telescope and a modern detector system. They had the advantage of knowing where to look. The discovery of a comet is communicated to the Central Bureau for Astronomical Telegrams, Smithsonian Astrophysical

Circular No. 7812
Central Bureau for Astronomical Telegrams
INTERNATIONAL ASTRONOMICAL UNION
Mailstop 18, Smithsonian Astrophysical Observatory, Cambridge, MA 02138,
U.S.A.
IAUSUBS@CFA.HARVARD.EDU or FAX 617-495-7231 (subscriptions)
CBAT@CFA.HARVARD.EDU (science)
URL http://cfa-www.harvard.edu/iau/cbat.html   ISSN 0081-0304
Phone 617-495-7440/7244/7444 (for emergency use only)

COMET 2002 C1
    Word has been received of the independent visual discovery of
a comet by Kaoru Ikeya (Mori, Shuchi, Shizuoka, Japan; 0.25-m
reflector, 39x; communicated by S. Nakano, Sumoto, Japan; coma
diameter 2' with weak condensation; motion about 5' northeastward
in 30 min) and by Daqing Zhang (near Kaifeng, Henan province,
China; 0.2-m reflector; communicated by J. Zhu, Peking University;
coma diameter 3').

    2002 UT   R.A. (2000) Decl.   m1   Observer
    Feb. 1.408   0 08.9   -17 42   9.0   Ikeya
          1.47   0 09   -17 30   8.5   Zhang

V838 MONOCEROTIS = PECULIAR VARIABLE IN MONOCEROS
    N. N. Samus, Institute of Astronomy, Moscow, informs us that
the designation V838 Mon has been given to this unusual variable
star (cf. IAUC 7786, 7789, 7791, 7796).
    T. Zwitter, Department of Physics, University of Ljubljana;
and U. Munari, Padova and Asiago Astronomical Observatories, report
on CCD echelle spectroscopic observations (range 460-930 nm,
resolution 18 000) of V838 Mon obtained on Jan. 26.0 UT with the
Asiago 1.82-m telescope. The spectrum is that of a heavily
reddened cool giant K-type star at heliocentric velocity +53 km/s.
Deep P-Cyg profiles mark the Ca II triplet at 849-866 nm, S I
(869.3 and 675.7 nm), Si I (614.2 nm), Na I (589.0 and 589.6 nm),
Ca I (650.0, 585.7, and 527.0 nm), Ni I (677.2 nm), Y II (566.3
nm), Zr II (611.5 nm), and C I (661.1 and 667.2 nm), as well as
many Cr I, Gd II, Fe I, and Ti I (but not Ba II) lines. The
average terminal heliocentric absorption velocity in the P-Cyg
profile is -458 km/s, and the minimum is at -252 km/s. H-alpha
appears as a narrow absorption at +54 km/s (FWHM 41 km/s,
equivalent width 0.064 nm), superimposed on a marginal emission at
+72 km/s. Interstellar Na I and K I lines have two components at
+47 and +73 km/s with component intensities corresponding to E(B-V)
= 0.56 and 0.24, respectively, using the calibration of Munari and
Zwitter (1997, A.Ap. 318, 269). The total extinction E(B-V) = 0.80
is confirmed by the 0.029-nm equivalent width of the diffuse
interstellar band at 862.1 nm. The reddening suggests a distance >
3 kpc, judging from extinction maps of Neckel and Klare (1980,
A.Ap. Suppl. 42, 251).

Fig. 2.1. An International Astronomical Union announcement telegram for comet Ikeya–Zhang. (Courtesy Central Bureau for Astronomical Telegrams.)

Observatory (Cambridge, Massachusetts) and then announced by an International Astronomical Union Circular, an example of which is shown in Fig. 2.1.

2.1.1 Comet observers

The history of comets is replete with highly successful comet hunters and observers. Among the most celebrated discoverers were Charles Messier (1730–1817), who discovered 21 comets, Caroline Herschel (1750–1848) with 8, Lewis Swift (1820–1913) with 11, Giovanni Donati (1826–73) with 6, William Brooks (1844–1921) with 20, and E. E. Barnard (1857–1923) with 19. Caroline Shoemaker, who is still active, has discovered or co-discovered at least 32 comets. There is considerable confusion concerning the number of comets discovered by some of the individuals listed; hence, the numbers given are only approximate.

The grand champion is Jean Pons (1761–1831), who is generally credited with the independent discovery of 37 comets; because communications at the time were slow and often difficult, some of the comets were independently discovered by others. Pons began his career as the concierge of the Marseilles Observatory, and his reputation earned him the nickname of the Comet's Magnet. He searched for comets with telescopes figured and constructed entirely by himself; one of these had a field of view of about 3° and was called the Grand Chercheur. Unfortunately, Pons's descriptions of comets' positions were not accurate. His name lives through comet 7P/1819 L1, which is now called comet Pons–Winnecke. But Pons's most famous discovery does not bear his name; he discovered Encke's Comet in 1805 and again in 1818. Encke himself always referred to it as the Comet of Pons; fortunately, Pons had so many comet discoveries to his credit that one could be spared.

Charles Messier was a keen comet searcher, but we remember his name for his catalog of nebulous objects. He compiled the catalog to avoid being misled by celestial objects that could be mistaken for comets, and so he introduced his famous M numbers into the astronomical language. His enthusiasm for comet discoveries earned him the nickname of the Ferret of Comets by King Louis XV of France.

In the nineteenth century, the discovery of comets was encouraged by medals and cash prizes. For instance, the King of Denmark offered a gold medal and an award for such a discovery. Among the historically interesting recipients of this medal was Maria Mitchell of Nantucket Island for her discovery of comet C/1847 T1. Later in the century, a prize of $200 was offered by H. H. Warner for comet discoveries. E. E. Barnard won this prize several times and used the money to pay for his home in Nashville, Tennessee, which came to be known as "Comet

Fig. 2.2. Reproduction of one of Janssen's photographs of the Great Comet of 1881 (C/1881 K1). (From *Annuaire du Bureau des Longitudes*, 1882; Courtesy of the Observatoire de Paris, Meudon.)

House." For many years, beginning in 1890, the Astronomical Society of the Pacific awarded the Donahue Medal for the discovery of a comet. In 1999, the Smithsonian Astrophysical Observatory (SAO) began administering the Edgar Wilson Award for comet discoveries by amateur astronomers. The first annual award was divided among seven recipients.

Barnard's observations of comets were important, particularly his work on comet Morehouse in 1908. He is also credited with the first discovery of a comet by means of a photographic plate on October 12, 1892. His photograph of comet D/1892 T1 showed traces of a tail. This credit excludes the "eclipse comet" seen only on photographs of the solar corona taken at the total solar eclipse on May 17, 1882.

Photography of comets began with Donati's Comet in 1858. An English commercial photographer (Mr. Usherwood) obtained a small overall view of the comet showing the tail, using a 7-s exposure made with an $f/2.4$ camera. Two days after Usherwood's success, astronomer G. P. Bond (1825–65) obtained a photograph that showed only the nucleus with a 360-s exposure through an $f/15$ lens.

Completely successful photographs were obtained for Tebbutt's Comet (C/1881 K1), which is now listed as the Great Comet of 1881 by Marsden. The success was partly due to the comet's exceptional brightness and the development of dry (silver bromide) plates. On June 30, 1881, P. J. C. Janssen (1824–1907) obtained a beautiful photograph (half-hour exposure) showing approximately 2.5° of tail (see Fig. 2.2).

Fig. 2.3. The Great Comet of 1882 (C/1882 R1) as photographed by Gill in November 1882. (Courtesy of the Royal Astronomical Society.)

Henry Draper also obtained a photograph of this comet (exposure time: 2 hr 42 min) showing approximately 10° of tail. This comet was also the first to have its spectrum recorded photographically.

The bright comet of 1881 was followed by another in 1882. Among its observers was David Gill (1843–1914) at the Cape of Good Hope Observatory. He used an $f/4.5$ camera borrowed from a local photographer and mounted it on the observatory's equatorial telescope. The photographs obtained had exposures in the range 20 min to 2 hr. These photographs (see Fig. 2.3) were important for two reasons. First, they confirmed that comets could be photographed easily with cameras or lenses of large angular aperture (small focal ratio); the linear aperture did not matter. From this time on, comets would have their pictures taken at several observatories around the world whenever they appeared. Second, the photographs showed a large number of faint stars. This fact convinced Gill that large-scale stellar photography was entirely practical. He decided to use photography to obtain the Map of the Southern Heavens. Under Gill's direction the plates were taken by R. Wood. The measurements (for precise position and approximate magnitude) were carried out under the direction of J. C. Kapteyn. This effort produced the first great modern catalog, the *Cape Photographic Durchmusterung* (1896–1900).

Eugene Shoemaker (1928–1997) had a lifelong interest in impact craters on earth and other bodies of the solar system, notably the moon. He was one of the early supporters of the idea that Meteor Crater in Arizona is an impact crater. This interest in impact craters led him to study the objects that form them. In collaboration with a colleague, Shoemaker began to search for near-earth asteroids with the 0.46-m Schmidt telescope at Palomar. They found their first such asteroid in 1973. Gene's wife, Carolyn, became involved with the project in 1980, scanning the films, and measuring images. From our point of view, their work is most interesting because of a by-product; they discovered or co-discovered 32 comets, some of them with David Levy. Comet Shoemaker–Levy 9 must have been a particularly satisfying discovery. When discovered, it had at least 20 separate components to its nucleus. We will talk more about this event later. For now, we will just say that the comet's components all crashed into the planet Jupiter.

Since 1979, not all comets have been named after human discoverers. In the 13th edition of The *Catalogue of Cometary Orbits* there are 6 comets named after the *SOLWIND* mission, 5 named after the *Infrared Astronomical Satellite* (*IRAS*), 10 named after the *Solar Maximum Mission* (*SMM*), and 70 named after the *Solar and Heliospheric Observatory* (*SOHO*), all of which were orbiting spacecraft. Another 29 comets in the catalog are named LINEAR, after the Lincoln Laboratory Near-Earth Asteroid Research project. That listing is complete up through July 1999. Just to give you an idea of how rapidly *SOHO* is discovering comets, the list of comets it discovered in 2001 contains 80 new comets.

Among the instruments on board the *SOLWIND*, *SMM* and *SOHO* spacecraft were coronagraphs. These devices were fitted with an occulting disk that blocked the sun's photospheric light, in order to allow observations of the faint emissions from the corona. Since there was little or no atmosphere to scatter light at the altitudes of the spacecraft, the corona could be seen out to several solar radii. *SOHO* was launched into an orbit around the earth–sun L1 Lagrangian point which allowed it to observe the sun without interruption. *SMM* and *SOLWIND* were in low earth orbits. All the comets discovered using these three satellites were seen in the coronagraphs as the comets passed perihelion. What is interesting, is that all but three of these comets had very similar orbital elements, making them members of the Kreutz group of comets. We will talk more about this later. At the end of 2000, *SOHO* was still operating, and scientists had discovered about 12 more comets beyond those published in the catalog. SMM has reentered the atmosphere, and *SOLWIND* was used as a test target for the Department of Defense (DOD)'s anti-satellite technology.

The *IRAS* was a joint mission of NASA, the Netherlands, and the United Kingdom. The mission was designed to obtain an all-sky survey in four broadband infrared channels. *IRAS* contained a 0.6-meter helium-cooled telescope. An array of detectors was used to measure the flux in the four infrared bands centered at 12, 25, 60, and 100 micrometers. Since comets are dusty objects, they are bright at infrared wavelengths. As a result, scientists using *IRAS* were able to discover a number of them.

The LINEAR project is a continuing effort. It has found quite a few more comets since the publication of the 13th edition of the *Catalogue of Cometary Orbits*. The project now uses two 1-meter telescopes at the White Sands Missile Range near Socorro, NM, the first of which was originally designed to monitor earth-orbiting satellites. Its objective now is to find new Near-Earth Objects (NEOs), comets, and main belt asteroids. It is also looking for what are called "Unusual Objects." At the time of this writing, it had found over 59 000 new asteriods and nearly 400 NEOs. And, of interest here, are the nearly 40 new comets.

2.1.2 Naming conventions

2.1.2.1 Historical designation conventions

Ordinarily, comets are named after their discoverer – or codiscoverers if several nearly simultaneous reports are received. However, a few comets have been named after the persons who computed their orbits. Well-known examples are Halley's Comet, Encke's Comet, and, more recently, Crommelin's Comet.

The names of some multi-discoverer comets can be cumbersome. Two shorter designation systems were devised, over the years. When a comet was first discovered, it was designated by the year of discovery and the order of discovery in that year, indicated by a lower-case letter. Thus, the first comet discovered in 1978 was called 1978a, the second 1978b, and so on. If more than 26 comets are discovered in one year, then later ones are indicated with a subscripted letter. Thus in 1991, we had comets 1991z, $1991a_1$, $1991b_1$, ..., $1991h_1$. After the cometary orbits were calculated and their times of perihelion passage were known, they were designated according to the year and order of perihelion passage. The first comet to pass perihelion in 1978 was 1978 I, the second 1978 II, and so on.

Unfortunately, confusion crept into these designations, especially in the case where improved orbit calculations changed the predictions of order of perihelion passage or even the year. As examples, three comets observed in 1618 passed perihelion in the order 1618 I, 1618 III, 1618 II, and comet 1798 II passed perihelion on January 1, 1799. In addition, periodic comets were given a new designation at each perihelion passage. As a result, the International Astronomical Union in 1995

adopted a resolution for a new naming system, similar to the naming convention
for minor planets.

2.1.2.2 The revised 1995 designation convention

The IAU resolved that the year/letter and year/Roman numeral systems be replaced by one in which each cometary discovery is given a designation consisting
of the year of observation, an upper-case code letter identifying the half-month
of observation during that year, and a consecutive numeral to indicate the order of discovery announcement during that half-month (Marsden 1995). Thus a
comet discovered between January 1 and January 15 would have the code letter A,
between January 16 and January 31 would have the code letter B, and so on. The
letters I and Z are omitted. The IAU Central Bureau for Astronomical Telegrams
will affix the designation when the discovery is announced in one of its *Circulars*.

The nature of the named comet is indicated by an initial prefix. The acceptable
prefixes are P/ for a periodic comet (defined to have a revolution period of less than
50–60 years or confirmed observations at more than one perihelion passage) and
C/ for a comet that does not meet that criterion. The prefix X/ is used for a comet
for which a meaningful orbit cannot be computed and D/ for a periodic comet that
no longer exists or is deemed to have disappeared. Just to cover all eventualities,
the IAU proposed a prefix A/ for an object given a cometary designation, which
turned out to be a minor planet. So far, this designation has not been used.

A comet that has been established to be periodic, will have the P/ (or D/) prefix
preceded by an official sequential number. So, for instance, Halley's Comet is 1P,
Encke's Comet is 2P, Biela's Comet is 3D, and so on. Recall that Biela's Comet
broke into two pieces during its 1846 return, and has never been seen since the
1852 return. A list of periodic, or defunct, comets is maintained by the Minor
Planet Center and published in the *Minor Planet Circulars*. The IAU charged the
Minor Planet Center with providing designations for all historical comets on this
new system.

In the case of a comet that has separated into discrete components, those components are distinguished by appending -A, -B, etc., to the designation (or to the
P/ or D/ periodic comet number). Thus, comet Shoemaker–Levy 9 carries the designations D/1993 F2-A through D/1993 F2-W. The letters I and Z are not used
for components, though Shoemaker–Levy 9 is the only comet for which that has
been an issue, so far. As we will see, the designation for comet Shoemaker–Levy 9
carries the D/ prefix because the fragments were destroyed when they collided with
Jupiter, and the comet is defunct.

Table 2.1 *Asteroid surveys*

Survey	Year	Identifier
Palomar-Leiden	1960	P-L
First Trojan survey	1971	T-1
Second Trojan survey	1973	T-2
Third Trojan survey	1977	T-3

2.1.2.3 Asteroid designation conventions

The IAU adopted a designation system in 1925 and the Minor Planet Center assigns provisional designations to an asteroid after reports of two nights of observations of an object that cannot be identified immediately with some already designated object. The designation is built up as follows:

- A four-digit number, followed by a space, indicates the year of the discovery.
- A letter indicates the half month of the discovery, for instance if the discovery was made during January 1 to 15, the letter would be "A", and so on. The letter "I" is omitted and "Z" is unused.
- A second letter indicates the order within that half-month, with the letter "I" omitted. If, as is sometimes the case, more than 25 discoveries are made in a given half-month, the cycle of second letters is repeated, with an optional number to indicate the number of times the second letter has been repeated in that half-month period. An asteroid that is famous for the fact that it passed very close to the earth was designated 1997 XF11. So the asteroid was discovered between December 1 and December 15 in 1997, and the second letter was in its 11th recycle. In other words it was the 281st asteroid discovered during that half-month.

This scheme has been extended to pre-1925 discoveries. The designation is designed as above, but the initial digit of the year is replaced by the letter "A". So, for instance, A904 OA is the first object designated that was discovered in the second half of July 1904. Just to give you some idea of the evolution of asteroid hunting technologies, 106 provisional designations were assigned in 1925, while in the year 2000, a total of 71 514 were assigned.

Four special asteroid searches were undertaken between 1960 and 1977. Asteroids discovered in those surveys have designations that consist of a number identifying the order within that survey, a space and a survey identifier. The survey identifiers are as indicated in Table 2.1. Example designations are 2040 P-L, 3138 T-1, 1010 T-2 and 4101 T-3. Other designations have been used in the past, as well.

Once an asteroid has had its orbit well determined, it may be assigned an official number and a name. There is a wide variety of names based on organizations, real people, fictitious people, and so on. For instance, asteroid number 2309 is named Mr. Spock. This step is the responsibility of the IAU.

2.2 Cometary motion

2.2.1 Basic orbit theory

In the remainder of this chapter, we will present a brief sketch of the dynamics of comets only in sufficient detail to foster an understanding of their motion. There are many treatises on celestial mechanics that the reader can consult for a complete and rigorous treatment of the subject. Examples are Brouwer and Clemence (1961), Moulton (1914) and Szebehely and Mark (1998).

The theory of celestial mechanics and orbit determination was developed starting with Newton, and progress has continued unabated until now. The motivation for progress in the eighteenth century was the need to develop streamlined techniques for numerical calculations. Today, the advent of super computers has changed the need for some of those techniques. The calculations of the motions of the planets, comets and interplanetary spacecraft being done at NASA's Jet Propulsion Laboratory, for instance, use numerical integrations of the basic differential equations with carefully chosen time step sizes. Many of these calculations can be carried out by a modern mini computer.

2.2.1.1 Cometary orbits

In the absence of nongravitational effects and planetary perturbations, a comet will orbit the sun on a path that is a conic section with the sun at one focus. We define:

$$\mu = GM_0, \tag{2.1}$$
$$l = r^2\dot\theta, \tag{2.2}$$

where r and θ are the polar coordinates, and,

$$E = \frac{v^2}{2} - \frac{\mu}{r}, \tag{2.3}$$

where E, the sum of kinetic and gravitational potential energies, is the total energy per unit mass of the comet. Here v is the comet's velocity, r is its heliocentric distance, M_0 is the mass of the sun, G is the gravitational constant, and l and E are the total angular momentum and total energy per unit mass, which are constants of motion.

The general equation for a conic section orbit in polar coordinates is

$$r = \frac{q(1+e)}{1 + e\cos\theta}, \tag{2.4}$$

Table 2.2 *Cometary orbit shapes*

Shape	Eccentricity	Energy
Ellipse	$0 \le e < 1$	Negative
Parabola	$e = 1$	Zero
Hyperbola	$e > 1$	Positive

where r is the distance from the focus that contains the sun, q is the perihelion distance, e is the eccentricity, and θ is measured from perihelion. In terms of the parameters defined above, the eccentricity of the conic is:

$$e = \left(1 + \frac{2El^2}{\mu^2} \right)^{1/2}. \tag{2.5}$$

If the total energy of the comet is negative, it is bound to the sun and will travel in an ellipse. If the total energy is positive, the comet is unbound and will move in a hyperbola. The case of exactly zero total energy corresponds to a parabolic orbit. A comet that falls toward the sun with initial zero velocity at infinity describes a parabola. Comets perturbed out of the Oort cloud by passing stars meet this criterion to a high approximation and should be observed to travel in nearly parabolic orbits. The correspondence between total energy and orbital shape is listed in Table 2.2. The different types of orbits are illustrated in Fig. 2.4.

In the case of an elliptical orbit (2.4) can be recast in the form

$$r = \frac{a(1 - e^2)}{1 + e \cos \theta}, \tag{2.6}$$

where a is the semi-major axis. The perihelion distance q and the aphelion distance Q can be found from the expressions

$$q = a(1 - e), \quad \text{and} \quad Q = a(1 + e). \tag{2.7}$$

The theory of the two-body problem shows that, for elliptical orbits, the total energy per unit mass and the semi-major axis are related by the expression

$$E = \frac{\mu}{2} \cdot \frac{1}{a}. \tag{2.8}$$

Equation (2.8) is also valid for parabolic orbits ($1/a = 0$) and hyperbolic orbits ($1/a < 0$). Thus, the velocity of the comet can be found by combining (2.3) and (2.8) to be

$$v^2 = \mu \left(\frac{2}{r} - \frac{1}{a} \right). \tag{2.9}$$

Discovery and dynamics of comets

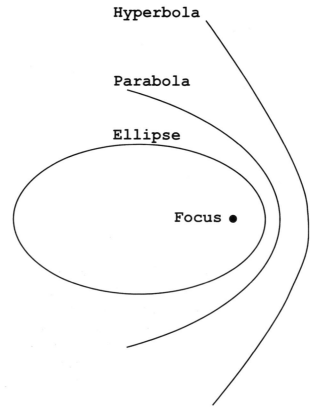

Fig. 2.4. Comparison of elliptical, parabolic and hyperbolic orbits.

At perihelion

$$v_p^2 = \frac{\mu}{a} \cdot \frac{1+e}{1-e}.$$
(2.10)

The analogous expression for the velocity for aphelion is easily found and can be combined with (2.10) to yield the simple relationship

$$v_p = v_a \frac{1+e}{1-e}.$$
(2.11)

For a parabolic orbit $E = 0$, and therefore (2.3) gives the velocity at any point in the solar system

$$v^2 = \frac{2\mu}{r}.$$
(2.12)

This velocity can be interpreted as the velocity a body needs to escape from the solar system when it is at distance r from the sun. Conversely, it is the velocity at

r of a body that has fallen from infinity. At the earth's orbit this escape speed or parabolic speed is 42 km s^{-1}.

2.2.1.2 Orbital elements

The general solution of the two-body problem can be characterized by six constants of integration that fully specify the size and shape of the orbit, its orientation in space, and the position of the body in the orbit. From the discussion above, it is clear that the total energy and angular momentum are possible parameters. For astronomical applications, a useful set of orbital parameters are those used to predict future positions (or ephemerides) of the comet. These parameters are:

1. q, the perihelion distance in astronomical units,
2. e, the eccentricity,
3. T, the time of perihelion passage,
4. i, the inclination of the orbital plane to the plane of the ecliptic,
5. Ω, the longitude of the ascending node (measured east from the vernal equinox),
6. ω, the angular distance of perihelion from the ascending node, also called the argument of perihelion.

These elements are illustrated in Fig. 2.5, which shows the orbit of comet Halley as an example.

The quantities q and e measure the size and shape of the comet's orbit in its orbital plane: T fixes the comet in time along its orbit. The orientation of the orbit in its plane is specified by ω. Thus, given the four parameters q, e, T and ω one can ascertain the position of a comet in its orbital plane. In order to find the comet's position in space, we must specify the orbital plane's orientation in space. This is specified by the two angles, i and Ω. The quantities i, Ω and ω are referred to the position of the vernal equinox at a specific date, say 2000.0. The comet's motion in its orbit is direct (counterclockwise when seen from above the earth's North Pole) if $0 < i < 90°$ and retrograde if $90° < i < 180°$. For an elliptical orbit, the semi-major axis a and the period P are easily derivable from (2.6) and Kepler's third law. The energy of an orbit and changes in energy are often described in terms of the reciprocal of the semi-major axis $1/a$ which is positive for an ellipse, zero for a parabola, and negative for a hyperbola (2.8). Sample orbital elements are given in Table 2.3, along with a variety of computed quantities.

The simple geometrical shapes are strictly applicable only for the ideal case of a comet in orbit around a sun with no perturbing bodies such as planets. Accurate orbital calculations include planetary perturbations, and the results are expressed in terms of the so-called osculating elements that accurately reproduce the comet's position and velocity at a given instant in time, called the *epoch of osculation*. The osculating orbit touches – or kisses – the actual orbit at the epoch of osculation. This time would be chosen, for example, to begin an accurate integration of an orbit

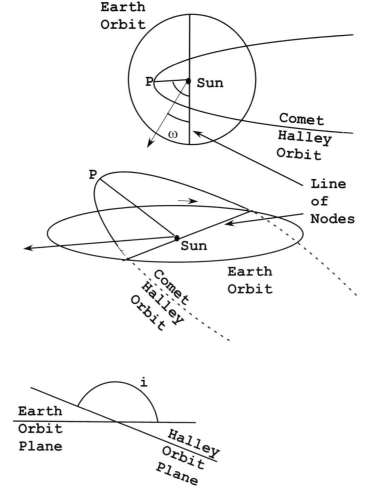

Fig. 2.5. Schematic of the orbit of Halley's Comet as seen (top to bottom) from the North Ecliptic Pole, from an oblique angle, and from an extension of the line of nodes.

with perturbations included. Osculating elements are a useful approximation when the perturbations cause only slow changes in the orbit. Obviously, they are of little value during a close planetary encounter.

2.2.2 Three-body problem

The general three-body problem has not been solved analytically. However, solutions in special cases have been found. One such case is the so-called *restricted three-body problem*. Consider two finite mass bodies revolving around one another

Table 2.3 *Sample orbital elements*

Quantity	1P/Halley	2P/Encke	21P/ Giacobini–Zinner	Hyakutake C/1996 B2
q	0.587 AU	0.331 AU	1.034 AU	0.220 AU
e	0.967	0.850	0.706	0.999 76
T	Feb. 9.4, 1986	Feb. 9.5, 1995	Nov. 21, 1998	May 1.4, 1996
i	162.2°	11.9°	31.8°	124.9°
Ω	58.9°	334.7°	195.4°	188.0°
ω	111.9°	186.3°	172.5°	130.2°
P	76 yr	3.28 yr	6.61 yr	29 650 yr
a	17.94 AU	2.21 AU	3.52 AU	951 AU
v_p	54 km s^{-1}	70 km s^{-1}	38 km s^{-1}	87 km s^{-1}
v_a	0.9 km s^{-1}	6 km s^{-1}	6.5 km s^{-1}	0.01 km s^{-1}
v_p/v_a	60	12	6	8700
Epoch	Feb. 19, 1986	Feb. 17, 1994	Nov. 3, 1998	Apr. 27, 1996

in circular orbits. A third body with infinitesimal mass moves under the influence of the two finite bodies. Since Jupiter moves in a nearly circular orbit ($e = 0.048$), a sun–Jupiter–comet system is approximately described by the restricted three-body problem.

2.2.2.1 Tisserand criterion

During a cometary encounter with a planet, for example Jupiter, an approximation called the *Tisserand criterion* holds true. If the elements of a cometary orbit are determined before and after encounter, the Tisserand criterion is

$$\frac{a_j}{a_1} + 2\left[\frac{a_1}{a_j}\left(1 - e_1^2\right)\right]^{1/2}\cos i_1' = \frac{a_j}{a_2} + 2\left[\frac{a_2}{a_j}\left(1 - e_2^2\right)\right]^{1/2}\cos i_2'. \qquad (2.13)$$

Equation (2.13) is a special case of Jacobi's integral of energy, which can be derived using the assumptions of the restricted three-body problem. In this equation a_j is the semi-major axis of Jupiter's orbit, a is the comet's semi-major axis, e is the eccentricity of the comet's orbit, i' is the inclination to Jupiter's orbit, and the subscripts 1 and 2 refer to before and after encounter. In the past, (2.13) was used to test the identity of two comets appearing at two different epochs before laborious calculations were begun to follow the orbit in detail. At present, it is used to help clarify the physical processes involved in the origin of short-period comets. The value of the expression on either side of the equal sign is sometimes called the Tisserand parameter, T.

2.3 Orbital determinations and ephemerides

In order to determine the six orbital parameters for a comet- or for any other body orbiting the sun – a minimum of three observations is required. Often, for a newly discovered comet, one initially assumes that the orbit is a parabola. The calculation of the elements then proceeds in a relatively straightforward fashion. After the comet has been observed many times over a period of a few months, a more accurate orbit at any epoch can be found: such an orbit is called a *definitive* orbit. The standard reference works on the techniques of orbital determination were written by Herget (1948), Dubyago (1961) and Escohal (1965). The fact that these three references are relatively old is an indication of the maturity of the topic. They are still quite up-to-date. If you are interested in setting up the orbital determination process on your personal computer, a good reference is Boulet (1991).

Given the six orbital elements of a comet, one can calculate its position in space. To find the position of the comet on the plane of the sky, one must know the earth's position in space as well. A table of geocentric positions of a comet is called an *ephemeris*. For bright or otherwise interesting comets, updated orbits and ephemerides are published regularly. NASA's Jet Propulsion Laboratory has an excellent website with links to all kinds of space science information, including information on comets, comet missions, asteroids, and asteroid missions, orbital elements and ephemerides. Its URL is: www.jpl.nasa.gov.

2.3.1 Astrometry of comets

In order to calculate a preliminary orbit for a newly discovered comet or a definitive orbit for a comet that has been around for a while, one must have high-quality positions. If the positions are to be used for a preliminary orbit, it is important that the measurements of the plate (or CCD image) and the reduction of the measurements be made in a timely fashion. This means a matter of hours rather than days. After all, one wants to begin following the comet as soon as possible in order to watch its phenomena evolve.

An observer must make a number of decisions before carrying out the actual observations. If one is using a photographic camera such as a wide-angle Schmidt, one must carefully select the emulsion to provide optimum images for measurement. In the case of a CCD detector, use of an appropriate filter may be necessary. If the camera is tracked at the sidereal rate, the image of the nucleus will be trailed. If the camera can track the nucleus, then the stars will be trailed. In either case, the exposure may be paused briefly to provide a fiducial mark on the trailed image(s). Of course, it is very important to record the time of the fiducial. The process of measurement and reduction of the observations is fairly straightforward, and we will not present any details here. The reader is referred to a standard textbook on the subject, such as Smart and Green (1977) for more information. Like the orbit determination process, the measurement and reduction processes in astrometry are well developed.

Recently, most astrometry has been designed to provide positions on the International Celestial Reference System (ICRS). The ICRS was adopted by the IAU as the official reference system in 1995. The International Celestial Reference Frame (ICRF) is tied to the coordinates of a set of extragalactic objects. The *Hipparcos Catalog* contains right ascensions, declinations, proper motions, parallaxes, magnitudes and other information for over 118 000 stars spread over the entire sky. The positional data has a median astrometric precision of around 1 milliarcsec. This makes the *Hipparcos Catalog* useful for defining the ICRS. However, the density of stars in the catalog is low, so the Tycho catalog, with over a million entries, is more useful for astrometric reference stars.

2.3.2 Orbit determination processes

2.3.2.1 Methods to determine initial elements

As we discussed in the previous chapter, Newton devised a method to determine the characteristics of the orbit of a comet assuming that it is parabolic in shape. Halley used the method on the comet that now bears Halley's name. Since that time, others have tackled the problem of finding a preliminary orbit using at least three observations, but not necessarily assuming the orbit is a parabola. Today, the approaches in use are based on one of the traditional methods derived from the work of Laplace or Gauss. The process is not merely the inverse of the method of finding an ephemeris.

Herget (1948) describes Laplace's method in detail. As he says, the method uses a solution of the fundamental differential equation of the two-body problem in terms of a Taylor's series. Herget derives the basic equations of the method, and describes how the calculations can be carried out on a manual, mechanical calculator. This is one of the only areas where the reference is out of date; programs now exist to carry out the calculations.

There are a number of variants of Gauss' method. Marsden (1985) discusses two of them – the standard Gauss Method and what he calls the Moulton–Väisälä–Cunningham method. Marsden points out that the methods have drawbacks which have been overcome with the advent of microcomputers. The reader is referred to the bibliography in Marsden's 1985 article and to Danby (1988) and Boulet (1991).

2.3.2.2 Detailed orbit determination

The methods mentioned briefly in the preceding section allow one to derive the elements of a cometary orbit from as few as three observations. After the comet has been observed for some time, one may have up to several hundred high-quality observations from which it is desired to derive a set of definitive elements. These observations may be from the single pass of a nearly parabolic comet, or from several revolutions of a periodic comet. The process is to begin with the preliminary elements and use the totality of observations to derive differential corrections to the elements, using processes described in the references mentioned above.

2.3.2.3 Ephemerides

Given a set of six orbital elements of a comet, one can calculate positions, or an ephemeris, at least for a short time into the future. If the elements are preliminary ones, the ephemeris may only be accurate enough to keep track of the comet for a short time while more positional observations are made. A better ephemeris can be calculated from definitive elements. Even then, the calculated positions cannot be pushed too far into the future unless planetary perturbations and nongravitationl forces are taken into account.

In (2.4), we used the more or less standard notation for the polar angular co-ordinate, the Greek letter, θ. Celestial mechanicians call that coordinate the *true anomaly*, and frequently symbolize it with the Greek letter v.

For elliptical orbits, a second angular variable, the *eccentric anomaly*, E, is extremely useful. Figure 2.6 shows the relationship between v and E. In that figure we first circumscribed around the ellipse a concentric circle, with radius equal to the ellipse's semi-major axis. We next drew a perpendicular to the semi-major axis, through the comet, P, and extending to the circumscribed circle at C. The eccentric anomaly is measured at the center of the circle.

The comet's mean motion, n, is defined by

$$n = \frac{2\pi}{P},\qquad(2.14)$$

and the *mean anomaly* at time t is

$$M = n(t - T),\qquad(2.15)$$

where T is the time of perihelion passage. One can show that

$$M = E - e \sin E.\qquad(2.16)$$

Equation (2.16) is known as Kepler's equation. It is important because the rectangular coordinates of the comet on the orbital plane, in a coordinate system centered

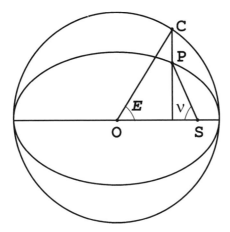

Fig. 2.6. Definition of quantities and geometric relationships in Kepler's equation.

at the sun and with the positive x-axis toward perihelion are

$$x_1 = a \left(\cos E - e \right), \qquad (2.17)$$

and,

$$y_1 = a\sqrt{1 - e^2} \sin E. \qquad (2.18)$$

In the years before computers, mathematicians devised sophisticated methods to solve transcendental (2.16) for E. Today a straightforward iterative method, starting with $E = M$ is used. The method converges for $e < 0.99$.

For parabolic orbits, one can write an equation for the true anomaly as

$$\frac{1}{3} \tan^3 \frac{v}{2} + \tan \frac{v}{2} = \frac{k}{\sqrt{2q^3}} (t - T), \qquad (2.19)$$

where $k = 0.017\,202\,098\,95\ldots$ is the Gaussian constant.

Equation (2.19) has one root for any value of t. Once it is solved for the true anomaly, the rectangular coordinates of the comet can be written

$$x_1 = r \cos v, \qquad (2.20)$$

and

$$y_1 = r \sin v. \qquad (2.21)$$

Once the rectangular coordinates have been found, the remainder of the steps needed to arrive at the right ascension and declination of the comet are the same for either elliptical or parabolic orbits. These steps are:

1. Rotate the two-dimensional rectangular coordinates in the orbit plane to the three-dimensional heliocentric ecliptic system.
2. Rotate the heliocentric ecliptic coordinates of the comet to its heliocentric equatorial coordinates.
3. Translate the origin of coordinates to the earth. The result is the geocentric equatorial coordinates of the comet. Care must be taken in this step to ensure that the heliocentric equatorial coordinates and earth–sun vector are referred to the same equator and equinox.
4. Calculate the right ascension and declination of the comet, and precess them to whatever equinox is required.

The details of these steps, and Basic routines to carry them out can be found in Danby (1988).

2.3.3 Orbital statistics

The *Catalogue of Cometary Orbits* published by Marsden in 1999 provides material for statistical discussions of orbital elements. Figure 2.7 is a histogram showing the distribution of inclinations for parabolic and nearly parabolic comets.

The dotted curve on the figure indicates the expected distribution if the inclinations were random. The numbers of comets with inclination between i and $i + di$ in a random distribution is proportional to $\sin i$, as can be seen if we consider the normals to the orbit. The normals to all orbits with inclinations between i and $i + di$ intersect the sky between the small circles with angular distances i and $i + di$ from the ecliptic pole. If the distribution of normals is random, the number of intersections is proportional to the area between the small circles, which is proportional to $\sin i$. The distribution of inclinations for the long-period comets is approximately random, but there is a distinct relative excess of retrograde orbits.

The distribution of inclinations for short-period comets is illustrated in Fig. 2.8. The differences between this distribution and that for the long-period comets is striking. Almost all short-period comets are in direct orbits with inclinations less than $30°$.

The distribution of the periods for the short-period comets is shown in Fig. 2.9. A huge peak between 7 and 8 years is the main feature. The majority of these comets have aphelia near the orbit of Jupiter (5.2 AU) as shown in Fig. 2.10.

Selection effects as well as physical situations show up in the statistics of comet orbits. There is a strong correlation between q, a and e, which are related for an elliptical orbit by $q = a(1 - e)$. For the long-period comets, we observe only those with $q \sim 1$ and, hence, a large value of a requires a value of e near 1. For the

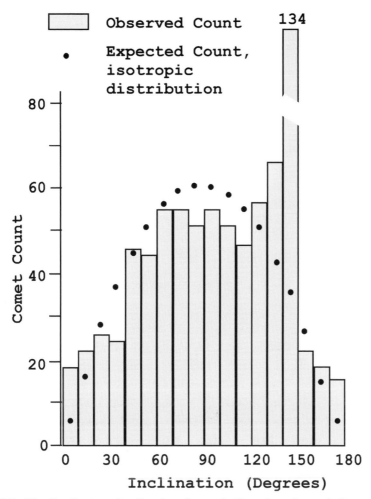

Fig. 2.7. The distribution of inclinations for parabolic and nearly parabolic comets. The peak in the 130–140 degree range is due to Kreutz group comets. (Data from B. G. Marsden, 1999, *Catalogue of Cometary Orbits*. Cambridge, MA: SAO Central Bureau for Astronomical Telegrams.)

short-period comets such as Jupiter's family a (strictly Q) is fixed, and hence there is a correlation between q and e .

The statistics of orbital parameters as illustrated in Figs. 2.7 to 2.10 were taken in the past to indicate a simple scheme for the origin of Jupiter's family of comets. The long-period comets, preferentially the direct ones, were thought to be depleted by encounters with Jupiter. This produces direct comets with low inclinations and aphelia near the orbit of Jupiter. There are specific examples of this process, such as 16P/Brooks, which encountered Jupiter and had its orbital period changed from

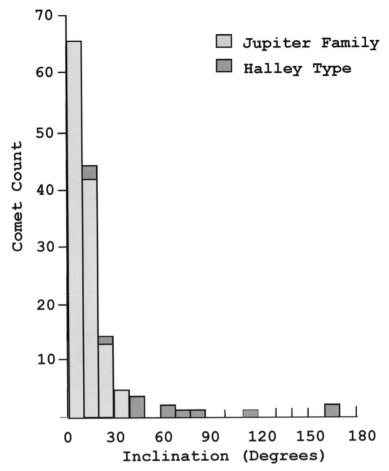

Fig. 2.8. Distributions of inclinations for short-period comets. (Data from B. G. Marsden, 1999, *Catalogue of Cometary Orbits*. Cambridge, MA: SAO Central Bureau for Astronomical Telegrams.)

approximately 29 years to approximately 7 years. Of course other results of such an encounter are possible. Comet Shoemaker–Levy 9 was captured into a Jovian orbit. The conservation of angular momentum and energy constraints could be satisfied, because the comet broke up into multiple pieces. The current view is that the short-period comets cannot arise from long-period comets interacting with Jupiter. We will return to this subject when discussing the origin of comets.

2.3.4 Comet families

Comets with nearly identical orbital elements (time of perihelion passage excepted) are members of comet groups. One example is the Kreutz group of sungrazing

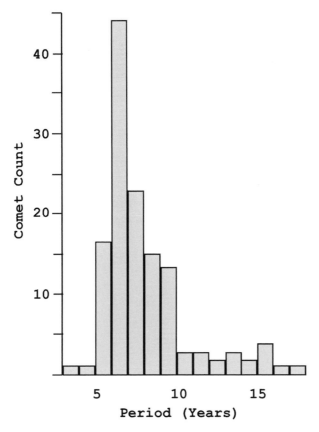

Fig. 2.9. Distribution of periods for Jupiter-family comets. (Data from B. G. Marsden, 1999, *Catalogue of Cometary Orbits*. Cambridge, MA: SAO Central Bureau for Astronomical Telegrams.)

comets which includes C/1843 D1, C/1880 C1, Ikeya–Seki (C/1965 S1) and most of the comets discovered by *SOLWIND*, *SMM* and *SOHO*. Over 200 Kreutz group comets have been discovered as of this writing. Their orbital elements are in the range $e = 1$, $q = 0.0062 \pm 0.0015$, $\omega = 77.95° \pm 9.84°$, $\Omega = 357.95° \pm 11.90°$ and $i = 143.17° \pm 2.52°$. These comets are thought to be the remains of a "super" comet that broke up some time in the past. Comet SOLWIND1 was interesting. It was seen approaching perihelion in the *SOLWIND* coronagraph, at which point the corona experienced a short-term, significant brightening (Fig. 2.11). Figure 2.12 shows the comet's orbit when it was in the vicinity of the sun. The filled circles are the observed positions of the approaching comet. The open circles are the expected positions after perihelion. However, the comet was not seen after perihelion. The circumstantial evidence suggests the nucleus was destroyed by the sun's heat as it passed perihelion. The coronal brightening was due to the cometary material.

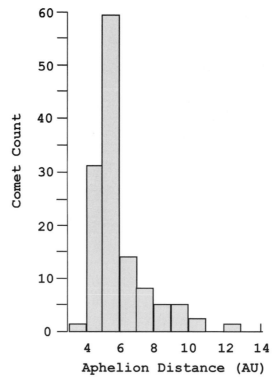

Fig. 2.10. Distribution of aphelia for Jupiter-family comets. (Data from B. G. Marsden, 1999, *Catalogue of Cometary Orbits*. Cambridge, MA: SAO Central Bureau for Astronomical Telegrams.)

Sekanina (2002) has done an analysis of the 278 sungrazing comets discovered since early 1996 by the *Solar and Heliospheric Observatory* (*SOHO*), all of which belong to the Kreutz group. The *SOHO* sungrazers he studied have all disappeared during their approach to the sun. One interesting result is the wide range in brightnesses of members of the group. The Great September Comet of 1882 (C/1882 R1) was the brightest member of the group, and, of course, was discovered from the ground. It was 20 magnitudes (10^8 times) brighter than any of the *SOHO* comets. The group members for which accurate orbits have been calculated have aphelia at 120–200 AU from the sun, which is equivalent to orbital periods of 500 to 1000 years. Sekanina (2002) gives an extensive discussion of the breakup process of the progenitor object, and describes how the pieces were distributed along their orbit. For a more detailed discussion, see Sekanina's paper. Figure 2.13 shows the comet SOHO6 near the sun.

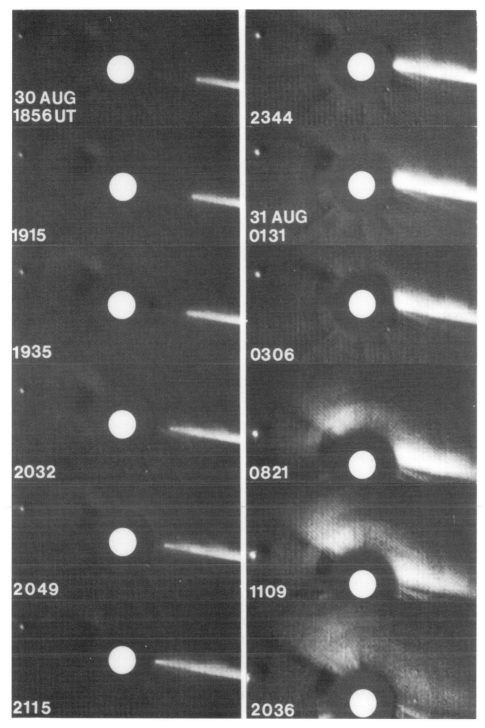

Fig. 2.11. Comet SOLWIND1 hitting the sun on August 30 and 31, 1979. Note the brightening of the corona on and after 0306 UT on the 31st. (Photographs courtesy of Naval Research Laboratory, Washington, DC.)

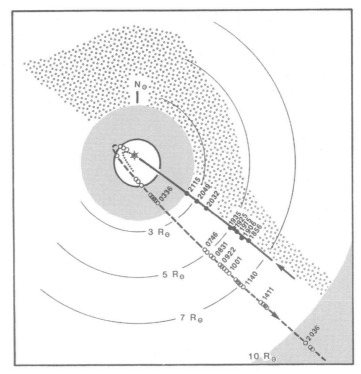

Fig. 2.12. Orbit of comet SOLWIND1 near the sun. The projected distances are measured in solar radii. The open circles are the expected positions of the comet, had it survived the passage by the sun. (Courtesy of Naval Research Laboratory, Washington, DC.)

Comet families consist of those comets with aphelia near the orbit of one of the Jovian planets. Several families were formerly recognized, but the earlier work did not fully consider that the orbital inclination would have to be close to zero for many comets to be influenced by the outer planets. Only Jupiter's family is recognized today. There are over 100 members of the Jupiter family, if one takes it to be defined by orbital periods $P < 20$ years and Tisserand parameters $T > 2$. The family includes a number of well-known comets, such as 2P/Encke, 10P/Tempel2, 19P/Borrelly, and 21P/Giacobini–Zinner.

2.4 Nongravitational forces

In our discussion of Encke's Comet (sections 1.5 and 1.6.4.1) we noted the puzzling fact that the comet's orbital period was steadily decreasing at an accelerating rate. Between 1828 when the period change was discovered and 1950 a number of comets with decreasing or increasing periods were discovered. The fact that there are both increasing and decreasing periods seemed to rule our interaction with

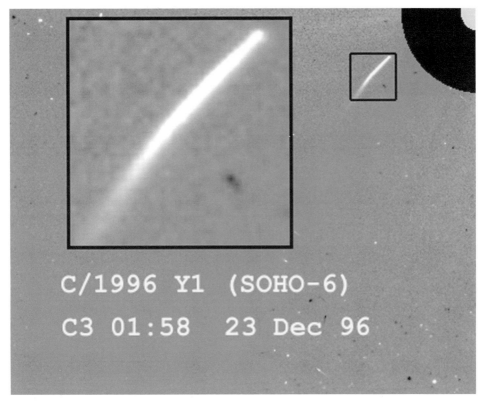

Fig. 2.13. Comet SOHO6 very near the sun. (Courtesy of *SOHO*/LASCO consortium. *SOHO* is a project of international cooperation between ESA and NASA.)

an interplanetary medium as the cause. Whipple (1950), suggested that the basic physical process is the loss of material coupled with the rotation of the cometary nucleus (Fig. 1.14).

For a nonrotating nucleus, the volatile ices sublimate primarily in the sunward direction and produce a force in the antisolar direction. In this case, the comet simply moves around the sun with a slightly reduced solar attraction. Kepler's third law, namely

$$\frac{P^2}{a^3} = \frac{4\pi^2}{\mu},\qquad(2.22)$$

tells us that the effect of the reduced solar attraction will be an increase in the ratio P^2/a^3, which will last as long as significant sublimation occurs. However, if the nucleus is rotating and there is a time lag between maximum solar flux received and maximum mass loss, nonradial forces will exist. These forces will generally have components in the transverse direction and in the direction normal to the plane of the comet's orbit. The transverse component is responsible for the long-term changes in orbits. If the rotation causes mass loss in the forward direction with respect to the comet's motion, the force generated decelerates the comet, causes the comet to spiral in toward the sun, and decreases the orbital period. If the mass loss is in the backward direction with respect to the comet's motion, the comet is accelerated and spirals away from the sun, and the orbital period is increased. Note that a relatively slow rotation rate is necessary to produce detectable nongravitational forces. Very rapid rotation would smear out any preferential direction of mass loss. The median rotation period of cometary nuclei is around 15 hours. A very rapid rotater would have a 5-hour period.

2.4.1 Early ideas

Marsden (1968, 1969) and his associates have studied the nongravitational forces on comets extensively. Originally, the nongravitational forces were included empirically and the equation of motion was written as

$$
\frac{d^2r}{dt^2} = \frac{-\mu r}{r^3} + \frac{\partial R}{\partial r} + A_1 \frac{e^{-B_1\tau}e^{-r^2/2}}{r^3}\hat{r} + A_2 \frac{e^{-B_2\tau}e^{-r^2/2}}{r^3}\hat{T} + A_3 \frac{e^{-B_3\tau}e^{-r^2/2}}{r^3}\hat{n}.
$$

$$(2.23)$$

The symbols are as follows: The As and Bs are empirically determined constants: r is the radius vector; $\mu = GM_0$; R is the planetary disturbing function; T is measured from the epoch of osculation in units of 10^4 ephemeris days: $\hat{r}$, $\hat{T}$, and $\hat{n}$ are unit vectors directed in the radial direction, in the orbital plane and 90° forward (in the sense of the comet's motion) of the radius vector, and in the direction normal to the orbital plane, respectively. The acceleration is given in astronomical units per (40 ephemeris days)^2.

Results from the earlier investigations can be summarized as follows. The radial nongravitational acceleration (in this discussion, when it can be accurately determined) is directed in the antisolar direction and is roughly an order of magnitude larger than the transverse nongravitational acceleration. This means that the lag angle is small. The transverse component is equally likely to be directed ahead of or behind the comet's motion. The nongravitational acceleration normal to the orbital plane is dynamically unimportant. The nongravitational forces seem to act continuously and may decrease slowly with time. In some cases, the nongravitational forces may have increased after a comet had a close approach to Jupiter.

2.4.2 Current ideas

Recent work has replaced the empirical nongravitational terms in (2.23) with terms based on the vaporization rate of water snow. Thus Marsden *et al.* (1973) would write

$$\frac{d^2 r}{dt^2} = \frac{-\mu r}{r^3} + \frac{\partial R}{\partial r} + A_1 g(r)\hat{r} + A_2 g(r)\hat{T}, \tag{2.24}$$

where

$$g(r) = \alpha \left(\frac{r}{r_0}\right)^{-m} \left\{1 + \left(\frac{r}{r_0}\right)^n\right\}^{-k}. \tag{2.25}$$

In (2.25), α is a normalization factor chosen so that $g(1) = 1$; r_0 is the distance beyond which the nongravitational forces (and presumably the vaporization of water snow) drop rapidly; and m, n, and k are constants. The vaporization flux can be written empirically as

$$A = Z_0 g(r). \tag{2.26}$$

Equation (2.26) is accurate to $\pm 5\%$ in representing the vaporization data. We will return to this discussion when Delsemme's model is presented, but we should mention that the total flux implied by (2.26) is compatible with values obtained by other means. The reader needs to be aware that our perception of cometary albedos has changed significantly since the 1970s, and should view (2.26) as a convenient mathematical form, which continues to be used in the discussion of nongravitational motion. Albedos are discussed in section 3.3.1.

The procedure is then to solve for the constants A_1 and A_2 and thus determine the nongravitational forces. Basically, for periodic comets, it is the transverse term or A_2 that is determined by an orbit calculation. There is a relationship between the magnitude and direction of the thermal lag angle – the angle between the nuclear subsolar point and the point on the nucleus where there is maximum outgassing. The existence of a non-zero lag angle introduces the transverse acceleration, which is either in the direction of the comet's motion or opposite to it. The sign of A_2 depends on the relative direction of the comet's motion and the direction of rotation of the nucleus.

2.4.3 Refinements

Yeomans has studied the nongravitational motion of comet 22P/Kopff quite extensively. In 1971, he (Yeomans 1971) reported that the transverse nongravitational parameter A_2 evolved from positive values prior to about 1930 to negative values in 1940 and thereafter. He argued that the simplest explanation in terms of the rotation of a water-ice nucleus is precession of the comet's spin axis. When $A_2 \approx 0$, the rotation axis would be in the plane of the comet's orbit.

Yeomans (1984) has carried out an extensive study of the historical motion of Halley's Comet, and has found that the transverse nongravitational parameter A_2 can be very well determined as far back as the beginning of the seventeenth century, and that the parameter has been essentially constant from then until now. He states that there is no evidence to suggest that A_2 has changed for the last two millennia. This suggests that Halley's nucleus has outgassed at a more or less constant rate over that time period, and that nuclear precession is not significant.

Note that (2.25) and (2.26) predict an outgassing rate and nongravitational accelerations that are symmetric with respect to perihelion passage. Sekanina (1988) has pointed out that photometric studies of comas show that cometary brightnesses tend to be asymmetric with respect to the time of perihelion, suggesting that the outgassing rate is also asymmetric. He suggested a refinement to the nongravitational acceleration model to build in that asymmetry. Equation (2.25) is replaced by $g(r')$, where $r' = r(t - DT)$. If the parameter $DT < 0$, then $g(r')$ is symmetric about a time DT days before perihelion; if $DT > 0$, then $g(r')$ is symmetric about a time DT days after perihelion. Sitarski (1994) has applied the perihelion asymmetry to the study of several comets. The study includes the parameter A_3, which describes the component normal to the plane of the orbit. Table 2.4 summarizes these results. Yeomans and Chodas (1989) propose that the observed nongravitational effect is due primarily to a radial, sun–comet acceleration that acts asymmetrically with respect to perihelion. Yeomans points out that comets like Encke that outgas preferentially before perihelion will lose orbital energy whereas comets like Halley and d'Arrest that outgas preferentially after perihelion will gain orbital energy.

When the new model is applied to comet Halley, the conclusion that A_2 has been constant over a long period of time does not change. The value of DT is found to be $+20$ days, in complete agreement with the time of the post-perihelion peak in the comet's light curve. Applying the asymmetric model to comet Kopff results in an A_2 value that remains negative throughout the time period the comet has been observed. The value of DT evolves slowly over the observation time period in a fashion that is similar to the evolution of the comet's light curve. The photometric evidence supports the model for many comets.

Another interesting result is that the nongravitational forces are important in calculating the so-called original values of orbital elements, by working backward from current elements. For the long-period comets only A_1 is well determined, and it is always in the sense of being a radial force. An empirical result is that the radial force makes the orbit more elliptical; that is, it increases $(1/a)$ as compared to the purely gravitational case. There are over 100 comets in the 1999 *Catalogue* which

Table 2.4 *Comparison of symmetric and asymmetric nongravitational acceleration models*

Comet	Interval	DT (days)	A_1	A_2	A_3
P/d'Arrest	1963–1988	0.0	+2.82	+1.14	−0.78
		+42.72	+3.40	+0.55	−0.74
P/Tempel 2	1946–1988	0.0	+0.24	−0.014	−0.017
		+29.83	+0.21	−0.005	−0.068
P/Giacobini–Zinner	1972–1987	0.0	+3.91	−0.60	−0.24
		−13.38	+3.78	−0.33	−0.99
	1959–1973	0.0	+3.74	+0.24	+0.26
		+14.04	+3.80	−0.10	+0.45
P/Kopff	1970–1990	0.0	+2.73	−1.10	−0.82
		−74.57	+3.28	−0.53	−1.28
	1970–1988	0.0	+1.88	−1.14	−0.33
		−29.82	+2.73	−0.94	−0.49
	1958–1970	0.0	+3.58	−0.81	−2.41
		−11.38	+3.68	−0.69	−2.53
	1951–1965	0.0	+6.31	−0.43	−3.40
		+7.17	+6.57	−0.56	−3.35
	1945–1958	0.0	+6.64	−0.07	−2.13
		−13.28	+6.59	−0.11	−2.19
	1932–1945	0.0	+5.34	−0.19	−1.81
		+26.38	+5.74	−0.56	−2.04
	1906–1932	0.0	+3.96	+0.26	−0.93
		+97.36	+3.93	−0.63	−2.25

A_1, A_2 and A_3 are expressed in units of 10^{-9} AU/day^2.
Source: After Sitarski (1994).

have $e > 1.0$. Of these, there are 30 or so that seem to have negative values for their original $1/a$; that is, they appear to have originally been hyperbolic. There is some discussion in the literature about whether these negative original $1/a$ values are real, or are due to other effects such as splitting, outbursts and observational inaccuracies, (see Kresák 1992, Yabushita 1996). In our opinion, the most telling argument against the interstellar origin of these comets are their small excess velocities. The maximum excess velocity of these comets over the parabolic velocity is about 0.8 km/s, while the sun's space velocity relative to the average of nearby stars is 20 km/s. One would expect excess velocities closer to that value.

An attempt was made in 1986 to estimate the mass of the nucleus of comet Halley using nongravitational forces, and Newton's second law of motion, $F = ma$. Rickman (1986) estimated the nongravitational acceleration on Halley's nucleus from the comet's motion. He then estimated the force acting on the nucleus from the mass loss rate (or observed water production rate) and the flow velocity of the material escaping from the nucleus. Knowing the force and the acceleration, a simple division yielded the mass. The result placed the mass in the range 5×10^{16} to 10^{17} g, which implied a bulk density in the range 0.1 to 0.2 g cm^{-3}. This number is below the value of 1 g cm^{-3} that we mentioned in Chapter 1; however, it is an attempt at measurement, and should be considered. See the discussion in section 7.3.5.

2.5 Comet 109P/Swift–Tuttle: an interesting case study

Comet 109P/Swift–Tuttle is an interesting case. It was observed in 1862, at what was originally thought to be its first sighting. F. Hayn did an orbit calculation in 1899 as part of his dissertation. Marsden (1973) recalculated an orbit using Hayn's data and found an orbital period of roughly 120 years, meaning it should reappear sometime between 1980 and 1983. A comet observed in 1737 was thought to be an earlier appearance of P/Swift–Tuttle, but no earlier sightings could be found. If, in fact, the comet of 1737 was P/Swift–Tuttle, then it should reappear in 1992. The comet did not appear in the 1980s, but was recovered in September 1992. With the positional data from the three apparitions, Yao *et al.* (1994) calculated good elements, and integrated the orbit backward to 703 B.C. They found two earlier apparitions – 69 B.C. and A.D. 188 in Chinese records – but none in between. As it turns out, the comet was poorly placed relative to earth for one and a half millennia, and never became bright enough to be seen.

   The nongravitational forces acting on P/Swift–Tuttle appear to be negligible, and yet the comet is as active as comet Halley. This means that the comet outgasses radially toward the sun in a fashion that is symmetric about perihelion, or has a nucleus that is more massive than Halley's, or a bit of both. Since the comet experiences no measurable nongravitational forces, we know it will return to perihelion on 12 July 2126. The comet in 2126 could pose a significant threat to earth (section 10.4).

2.6 Orbital resonances in the solar system

A variety of orbital resonances have been recognized in the solar system. Peale (1976) reviews most of the resonances known at the time of his review. These include commensurabilities of mean motions for satellites of Jupiter and Saturn, the Kirkwood gaps in the asteroid belt and other asteroid associations with Jupiter, gaps in the rings of Saturn, and Neptune–Pluto. One of the more interesting commensurabilities in the satellites of Jupiter involves the trio Io, Europa and Ganymede. If the mean motions of the three satellites are n_1, n_2 and n_3, respectively, then

$$n_1 - 3n_2 + 2n_3 = 0. \qquad (2.27)$$

Celestial mechanicians have been working, since Laplace recognized the stability of this commensurability in 1829, to explain some of the observed resonances. Of course, fascinating though they are, the resonances in satellite systems will not be discussed further here.

We will be interested in resonances in the asteroid belt and the Kuiper belt. In the asteroid belt, there are the Kirkwood gaps at $3:1$, $5:2$, $7:3$ and $2:1$ commensurabilities with Jupiter. In the first of these commensurabilities, the asteroid makes three revolutions about the sun while Jupiter makes one, and similarly for the others. The Hilda group asteroids are trapped in the $3:2$ commensurability and Thule is trapped in the $4:3$ commensurability with Jupiter. The Trojan asteroids lie in Jupiter's orbit $60°$ behind and ahead of the planet, and oscillate about the libration points. Peale (1976) also talks about the $3:2$ resonance between Neptune and Pluto. This resonance has taken on new interest with the discovery of a number of Kuiper belt objects trapped in the resonance. These objects are now called plutinos.

2.6.1 Mean motion resonances

Peale presents a quantitative description of the physical processes that take place in the resonance. A conjunction will occur between the sun, Jupiter and an asteroid at some point. If the asteroid is in a resonance, the conjunction will occur at nearly the same longitude, L, again and again. The process can have Jupiter add angular momentum for several conjunctions, then take away angular momentum for several more conjunctions, then switch back to adding angular momentum, and so on. From the asteroid's point of view, the process behaves like the oscillation of a pendulum. This behavior was recognized early on, and suggests a mathematical formulation of the problem. The equations of motion are broken up into two parts. The major part is the well-known restricted three-body equation of motion of the asteroid (see, e.g., Danby 1988). The remaining motion is represented by a series expansion in terms of a resonance variable, $\sigma = L - \varpi$, where ϖ is the longitude of perihelion of the asteroid. That part of the motion, as we said earlier, resembles the motion of a pendulum. Peale (1976) and Froeschlé and Greenberg (1989) describe the mathematical formulation of the analytical models, and the reader is referred to these authors and others for the details.

Greenberg and Scholl (1979) reviewed the state of understanding of the Kirkwood gaps as of the time of their review. They group the theories into four classes: 1. the gaps are a statistical phenomenon, 2. the gaps are formed by gravitational forces only, 3. the gaps are formed by asteroid–asteroid collisions with asteroids near the gaps, which tend to have large eccentricities, and 4. some process prevented asteroids from forming in the gaps in the first place. The hypothesis that the gaps are a statistical phenomenon goes roughly like this. A pendulum spends more time at the extremes of its swing, where its speed is slower than at the bottom of its swing. In the same sense, an asteroid would spend more time at the edges of the gap than in the center. In an ensemble of resonant asteroids, the center of the gap would be statistically more lightly populated than the edges. While this is an interesting idea, it fails to model the width and shape (density distribution) of the gaps.

One of the more successful analyses of the gaps is due to Wisdom (1982, 1983, 1985). He used a semi-analytic perturbation theory to study the motion near the 3 : 1 resonance. The key to his approach was a phase-space mapping which speeded up the computations involved by a factor of around 1000. His initial computations showed that a test asteroid in the 3 : 1 gap will continue with an eccentricity, $e <$ 0.05, for a million years, then suddenly jump to a larger eccentricity, $e > 0.3$. The asteroid then becomes a Mars crosser, and may be perturbed out of the resonance by a close encounter with Mars. In the earliest of his papers, Wisdom began with a distribution of 300 test asteroids in the vicinity of the 3 : 1 resonance. As the Mars crossers were removed, a gap appeared in the distribution. However, the model gap is too narrow to fit the observed 3 : 1 Kirkwood gap. Including inclinations and secular perturbations of Jupiter's orbit widens the gap somewhat, but not sufficiently to fit the observations. In the later papers, Wisdom also finds that the 3 : 1 resonance is associated with a zone of chaotic motion, in which trajectories cross planetary orbits. The removal of the planetary crossers fully explains the size and shape of the 3 : 1 gap. At least the 3 : 1 gap can be explained by fully gravitational phenomena. So far, the 2 : 1 and 3 : 2 gaps have eluded a similar approach. The Hilda group of asteroids is not yet understood. The questions of the capture of asteroids into the resonances, and the stability of the resonances are also of great interest. The reader is urged to consult Peale (1976) and the other references quoted here for a discussion of these issues.

Yu and Tremaine (1999) have addressed the dynamics of the plutinos. They find that the long-term stability of the plutino orbits depend on their eccentricities. Those plutinos with eccentricities near that of Pluto will be stable. Orbits with larger eccentricities will be unstable on time scales significant relative to the age of the solar system, since encounters with Pluto will drive the bodies out of the resonance. The authors suggest this instability as a possible source for Jupiter-family comets.

2.6.2 Secular resonances

Secular resonances are conceptually more complex than the mean motion reson-ances, and, as a result, are more difficult to visualize. We will summarize the mathematical background for secular resonances, then briefly discuss one of the secular resonances known to occur in the asteroid belt. Scholl *et al.* (1989) review the background of secular resonances in the asteroid belt, and discuss their effect on the distribution of asteroids. The bases of the theory are the eigenfrequencies of the main secular variations in the motions of the major planets. Brouwer and van Woerkom (1950) found numerical values for these eigenfrequencies.

Like in the case of mean motion resonances, the equations of motion of an asteroid are split into two parts. One part is the restricted three-body motion of the asteroid. The second part is the disturbing function. In the linear theory of asteroidal motion, the disturbing function is expanded in a series involving orbital inclinations and eccentricities through the parameters h, k, p and q, where

$$
\begin{aligned}
h &= e \sin \varpi, \\
k &= e \cos \varpi, \\
p &= \sin i \sin \Omega, \text{ and} \\
q &= \sin i \cos \Omega.
\end{aligned}
\tag{2.28}
$$

If we drop all short-period terms in the disturbing function, and keep second order terms, only then the differential equations of motion are integrable in closed form. The solutions are

$$
\begin{aligned}
h &= e_0 \sin (g_0 t + \varpi_0) + H(t), \\
k &= e_0 \cos (g_0 t + \varpi_0) + K(t), \\
p &= \sin i_0 \sin (-g_0 t + \Omega_0) + P(t), \\
q &= \sin i_0 \cos (-g_0 t + \Omega_0) + Q(t),
\end{aligned}
\tag{2.29}
$$

where

$$
\begin{aligned}
H(t) &= \sum_{j=1}^{10} \frac{G_j}{g_0 - \dot{v}_j} \sin(\dot{v}_j t + \beta_j), \\
K(t) &= \sum_{j=1}^{10} \frac{G_j}{g_0 - \dot{v}_j} \cos(\dot{v}_j t + \beta_j), \\
P(t) &= \sum_{j=11}^{18} \frac{G_j}{g_0 + \dot{v}_j} \sin(\dot{v}_j t + \beta_j), \\
Q(t) &= \sum_{j=11}^{18} \frac{G_j}{g_0 + \dot{v}_j} \cos(\dot{v}_j t + \beta_j).
\end{aligned}
\tag{2.30}
$$

The quantities $\dot{v}_j$ are the eigenfrequencies mentioned above. The summations are over the eight planets, excluding Pluto. The two extra terms in the solution for $H(t)$ and $K(t)$ account for the $5:2$ mean motion resonance between Jupiter and Saturn. A secular resonance occurs when a coefficient in the expansions in (2.29) becomes infinite; that is, when $g_0 \sim \dot{v}_j$ for $j \leq 10$ or when $g_0 \sim -\dot{v}_j$ for $j > 10$. Three secular resonances have been recognized in the asteroid belt, v_5, v_6, and v_{16}. An asteroid in the v_5 or v_6 resonance will experience large variations in its eccentricity. An asteroid in the v_{16} resonance will experience large variations in its inclination. Williams (1969) has carried out a more extensive solution of the equations of motion and finds a coupling between h, k, p and q. As a result, an asteroid in a particular secular resonance will experience some variations in both eccentricity and inclination.

We will look at one example of asteroidal motion in a secular resonance; we choose to look at the v_6 secular resonance, which is located at 2.05 AU. Froeschlé and Scholl (1986) have carried out numerical integrations over the time span of 1 million years to investigate the orbital evolution of an asteroid in the v_6 secular resonance. They found that the libration of the resonance argument $\varpi - \varpi_S$, where ϖ_S is Saturn's longitude of perihelion, takes place about $\varpi - \varpi_S = 180°$ and has a period of approximately 1 million years. The large increases in the eccentricity of these bodies cause them to become earth crossers, with perihelia near Venus' orbit; i.e., they become Apollo-type asteroids.

2.6.3 Shepherding resonances

High-resolution studies of planetary rings, especially those of Saturn and Uranus, show the presence of thin rings, sometimes with braids and kinks. Processes such as interparticle collisions and the Poynting–Robertson effect lead to the radial spreading of ringlets and limit their total lifetimes to times of order 10^8 years. Goldreich and Tremaine (1982) have developed the concept of shepherding to explain both the radial containment of ringlets and the kinks and braids. The Goldreich and Tremaine treatment requires two shepherding moonlets, one in an orbit interior (i.e., closer to the planet) to the ringlet and one exterior (i.e., further from the planet) to the ringlet. The shepherding moonlets induce epicyclic radial perturbations in the ring as they pass. The nature of the effects depends on whether the orbits of one or both moonlets are elliptical. Goldreich *et al.* (1995) have developed the theory of one-sided shepherding. We will not discuss the details of the theory here, but refer the reader to the literature.

The theory is of interest to us because in 1983 the *IRAS* satellite observed narrow rings in the interplanetary dust just interior to the earth's orbit. Jackson and Zook (1989) have carried out numerical integrations which show that small dust grains

(30–100 μm) can be captured into shepherding resonances with the earth, where they may stay for periods from 10^4 to 10^5 years. The interplanetary dust is discussed in section 10.2.

2.7 The Roche limit

The Roche limit is the distance from a large body at which a smaller body, with no tensile strength, will be torn apart by tidal forces. A good example, which we will encounter later, is the case of comet Shoemaker–Levy 9 (SL9). In 1992, comet SL9 passed within 0.001 AU of Jupiter and was torn apart by tidal forces (see section 10.3.2). A simplified treatment of the problem is as follows. Consider a small body of mass m and radius r, with a smaller body of mass δm at its surface. This combination is orbiting a larger body of mass M and radius R at a distance ρ. The Roche limit is the distance inside which the tidal force disrupting m and δm exceeds the gravitational force holding the two bodies together. We have

$$\frac{2GM\delta m}{\rho^3}r = \frac{Gm\delta m}{r^2}, \tag{2.31}$$

where the term on the left is the tidal force on δm by M and the term on the right is the attractive force between m and δm. Rearranging, and taking the cube root of both sides yields, for the Roche limit,

$$\rho = 1.26r\left(\frac{M}{m}\right)^{1/3}. \tag{2.32}$$

Using a reasonable dimension for a cometary nucleus of $r = 2$ km, and $m = 4 \times 10^{16}$ g, this formula yields a Roche limit for such a comet at Jupiter to be $\rho \approx 0.001$ AU, consistent with the observations for comet SL9.

2.8 Summary

The discovery of comets is an area where amateur astronomers have contributed as much as professionals. Bright comets such as comet Hyakutake were discovered by amateurs. Professionals have the advantage of large, wide-field astrographs, but many of the discoveries were made as by-products of the search for near-earth asteroids. Recently, space instrumentation has aided the discovery of comets, especially the sungrazing variety. In 1995, the IAU devised a streamlined comet-naming convention.

We discussed the dynamics of comets. Orbit theory, the determination of orbital elements and ephemerides has become a fairly cut and dried process. So, we have given a brief summary of the steps involved, and have cited references that the reader can use to find the details.

Nongravitational forces caused by nonradial outgassing from the nuclei of comets are important in the motion of the objects. We have discussed the concepts involved, and described some of the modern refinements.

Resonances are important in the study of motions in the solar system, especially for the asteroids, and as we will see later, the Kuiper belt objects. We have presented a brief summary of this complex subject, which has occupied celestial mechanicians for nearly two centuries.

3

Observing and measuring techniques

3.1 Introduction

Observing and measuring techniques fall into three broad, overlapping categories: imaging, photometry and spectroscopy. In Chapter 1, we described the fact that Donati's Comet (C/1858 L1) was the first comet to be photographed, though that photograph was of limited quality. The first fully successful photograph was of the Great Comet of 1881 (C/1881 K1). This comet was the first to have its spectrum photographed, as well. Prior to photography, cometary observations were carried out visually; even today, such observations continue. An experienced observer with a good telescope could measure accurate positions and observe structure near the nucleus (see Fig. 1.8). Drawings made by visual observers show detail that cannot be fully captured by imaging through large telescopes. This is in part because drawers can take advantage of very short-lived periods of excellent seeing. The reader might find it instructive to attempt to draw cometary structure.

However, drawings are subjective. For that reason, photography continues to be a prime source of data on comets. Photographs can vary from short, astrometric exposures designed to produce a small image suitable for positional determination to long, wide-angle exposures designed to show the extent of and structure in comet tails. For instance, when comet Halley was visible in 1985 and 1986 over 3500 wide-field photographs were made. *The International Halley Watch Atlas of Large-Scale Phenomena* (Brandt *et al.* 1992, 1997) provides a nearly daily sequence of images covering the time frame from September 17, 1985 through July 6, 1986.

Two-dimensional Charge Coupled Devices (CCDs) have replaced photography for applications that require high photometric precision. CCDs with arrays up to 2048 pixels square are readily available, and successful mosaicing techniques have been developed to build larger arrays. CCDs are popular because they:

- produce digital output,
- have higher quantum efficiencies than photographic emulsions,

- are highly linear, and
- have large dynamic ranges.

These factors make CCDs particularly useful for photometric applications. Of course, CCDs can be used in conjunction with specialized filters to isolate interesting wavelength regions. From the point of view of the cometary observer, the prime disadvantage of a CCD is its limited size. A CCD does not replace a photographic plate for large-scale imaging, which can be 5 to 10 degrees on a side, while CCDs are in the fraction of a degree range.

The signal at pixel i, j of a CCD, measured in electrons, can be converted to flux density at each pixel, by the formula (see Jewitt 1991)

$$I_{i,j}(\lambda) = \frac{S_{i,j}(\lambda) - b_{i,j} - d_{i,j}t}{s_{i,j}(\lambda)t}, \tag{3.1}$$

where $S_{i,j}(\lambda)$ is the signal in electrons, $b_{i,j}$ is a bias level, $d_{i,j}$ is the dark emission rate in electrons per second, $s_{i,j}(\lambda)$ is the sensitivity and t is the integration time. If the CCD is cooled to minimize the dark current, then it is possible to set $d_{i,j} = 0$. Observers generally begin each observing session by measuring the bias level. This is done by taking several exposures with the instrument shutter closed, and averaging the results.

3.2 Cometary observational techniques

We describe the observational techniques in this chapter, and cite a few results as examples. The more detailed discussion of results is presented in later chapters.

3.2.1 Ultraviolet and visible

Ultraviolet and visible imaging can be carried out with filter-emulsion or filter-detector combinations to isolate an interesting wavelength region or the emission from a particular molecule. An example of a filter-emulsion combination is a Kodak IIa-O plate with a UG-1 filter to produce excellent photographs of cometary plasma tails that emit strongly in the blue due to their principal constituent, CO^+.

The principal source of UV imaging is orbiting spacecraft. We provide a brief description of the contributions of the *Hubble Space Telescope* (*HST*) in section 3.4.4. It was a major disappointment when the *Challenger* accident caused the Space Shuttle program to stand down, and prevented the launch of *HST* in time to observe comet Halley.

Cometary spectroscopy can be carried out usefully at a variety of dispersions ranging from roughly 500 Å mm^{-1} with objective prisms to high dispersions of approximately 1 Å mm^{-1}. The reader must understand that these dispersions are just indications of the resolution of the spectra. The actual resolution depends, additionally, on the slit width and the detector resolution, especially in the case of photoelectric detectors. The spectra have been recorded both photographically and photoelectrically. Photoelectric spectrum scans are illustrated in Fig. 3.1. Note the large variations in the appearance of the spectra. Examples of low-dispersion spectrograms are shown in Fig. 3.2. High-dispersion spectra of the CN bands near 3880 Å are shown in Figs. 3.3 and 3.4. In addition, an example of a high-dispersion spectrogram of comet Kohoutek is shown in Fig. 3.5.

Analysis of the spectroscopic data provides several important pieces of information, including: (1) identification of atomic and molecular species, (2) abundances of these species, (3) bulk motions in comets from the doppler effect, and (4) excitation processes in the cometary atmosphere. Each of the enumerated points is useful individually. However, cometary spectroscopic analysis has proceeded to the point where attempts are made to find an internally consistent set of all the parameters. In this case, theoretical spectra are computed, with all the known relevant physical processes included. An acceptable synthetic spectrum must fit the observed spectrum in detail.

Usher (1990) has discussed the issues of wide-angle photographic photometry as applied to the images archived by the Large-Scale Phenomena Network of the International Halley Watch. He shows that, under certain conditions, standard stars in the field can be used to find a complete solution for extinction, sky brightness, and the characteristic curve. The conditions are that the standards be numerous, unsaturated and well-distributed in zenith angle and color index.

3.2.2 Infrared

Infrared observations in the range 1–20 μm with relatively coarse spectral resolution have been carried out for some time and are sufficient to establish the intensity of the continuous emission. From these data one can calculate the temperature of the grains and gain some insight into their composition, for example, by observation of the silicate signature at 10 and 18 μm in comet Bennett (C/1969 Y1), as reported by Maas *et al.* (1970). Infrared observations made of comets Bennett and Kohoutek at the same heliocentric distances are shown in Fig. 3.6. The figure shows the black body radiation temperatures and silicate features from both comets.

The near infrared (2–5 μm) is an important wavelength region for cometary studies because strong vibration transitions of many parent molecules fall there. Infrared searches for water in comets are carried out by the *Kuiper Airborne Observatory* (*KAO*), which flies above about 99% of the water vapor in earth's atmosphere. Pre-perihelion observations of comet Halley by *KAO* provided the first detection of H_2O at 2.65 μm (Weaver *et al.* 1986). Post-perihelion observations by *KAO* and

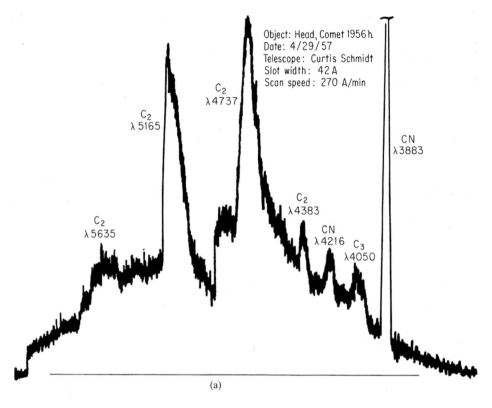

Object: Head, Comet 1956h
Date: 4/29/57
Telescope: Curtis Schmidt
Slot width: 42 Å
Scan speed: 270 Å/min

C_2 λ4737

C_2 λ5165

CN λ3883

C_2 λ5635

C_2 λ4383

CN λ4216

C_3 λ4050

(a)

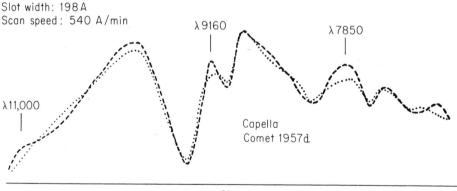

Object: Comet 1957d
Date: 8/15/57
Telescope: Curtis Schmidt
Slot width: 198 Å
Scan speed: 540 Å/min

λ9160

λ7850

λ11,000

Capella
Comet 1957d

(b)

Fig. 3.1. Photoelectric scans of two comets. (a) Comet Arend–Roland. A yellow filter was used to eliminate the effects of overlapping orders for wavelengths greater than 5000 Å. (b) Comet Mrkos. In this wavelength region, the cometary scan (dashed line) is similar to the scan of Capella (dotted line), a yellow star approximately the same color as the sun. (Courtesy of W. Liller, Instituto de Estudios de Novae, Chile.)

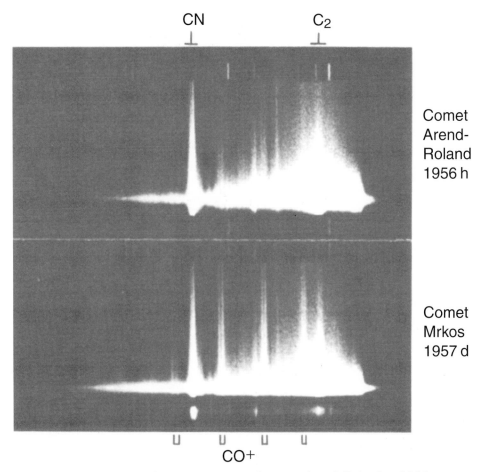

Fig. 3.2. Low-dispersion spectrograms of comets Arend–Roland and Mrkos at visual wavelengths. The CN band is at 3883 Å. (Photo © UC Regents/Lick Observatory.)

the *VEGA* IKS experiment also detected H_2O (Weaver *et al.* 1986, Moroz *et al.* 1987, Combs *et al.* 1988), definitely verifying the existence of water ice in the nucleus. The *VEGA* IKS experiment also detected CO_2, and H_2CO in the comet. See sections 7.1 and 7.3.5 for more discussion of IR observations of comets.

3.2.3 Radio

Radio wavelength studies of comets have become a major source of information since about 1970, and the field came into its own with the 1985–1986 apparition of comet Halley. Radio observations provide both unique new data and complementary

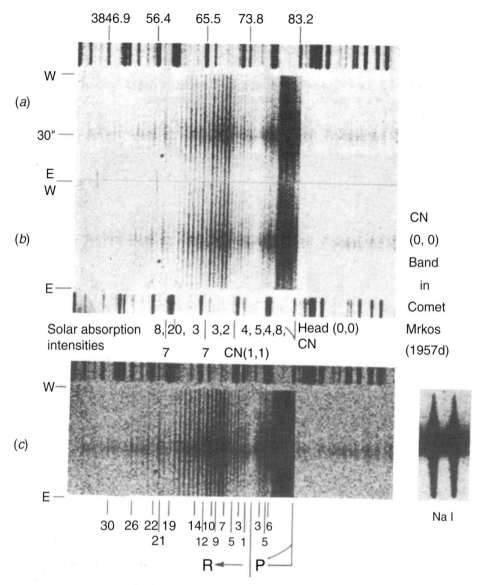

Fig. 3.3. High-dispersion spectra of the CN bands near 3880 Å in comet Mrkos. Irregularities in the CN intensities are caused by the indicated solar absorption lines. Taken by J. L. Greenstein. (California Institute of Technology.)

data to other wavelength regions. Continuum observations detect emissions from the nucleus and large dust grains. Spectroscopic observations can detect emission from parent molecules in the coma, as well as from a wide range of molecules that do not emit in other wavelength regions. A summary of the state of cometary radio studies post comet Halley can be found in Crovisier *et al.* (1991) and de

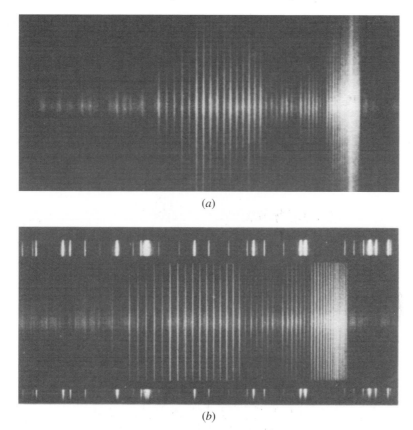

(a)

(b)

Fig. 3.4. Spectrograms of the CN bands near 3880 Å in comet Bennett, showing resolution of the band head. (a) Observatoire de Haute Provence spectrogram at original dispersion of 7 Å mm^{-1}. (b) Hale Observatory spectrogram taken by G. W. Preston at original dispersion of 4.5 Å mm^{-1}. (Courtesy of C. Arpigny, Institut d'Astrophysique, Liège.)

Pater *et al.* (1991). Figure 3.7 shows radio spectra of comet Lee (C/1999 H1) obtained by Biver *et al.* (2000). Emission features of HCN at 88.632 GHz are clearly visible.

Early continuum radio observations of comet Kohoutek (C/1973 E1) were ambiguous, with Hobbs *et al.* (1975) reporting a 4-sigma detection near 3 cm, while other groups did not detect the comet. Hobbs and his co-workers attempted to reconcile their observations with the icy grain halo model of Delsemme and Miller (1971). Detections of other comets over the intervening years have been elusive. Searches with the Very Large Array (VLA) have not detected continuum from comets at a level two orders of magnitude below the reported levels of Kohoutek. The icy grain halo model seems to be ruled out. Continuum from comets Halley, Hyakutake and Hale–Bopp has been detected at millimeter wavelengths

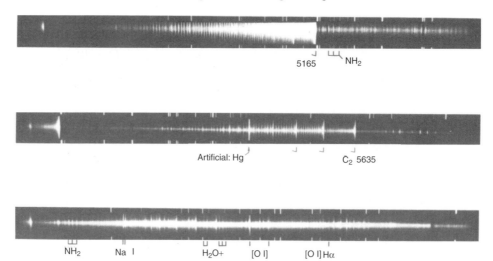

Fig. 3.5. High-dispersion spectrograms of comet Kohoutek (C/1973 E1) on January 9 and 11, 1974. (Photo © UC Regents/Lick Observatory.)

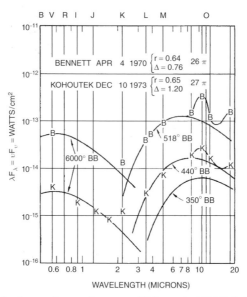

Fig. 3.6. Infrared observations of comets Kohoutek (C/1973 E1) and Bennett (C/1969 Y1) at the same heliocentric distance showing their black body temperatures and silicate features (10 to 18 μm). (Reprinted from *Icarus*, Vol. 23, by Ney, 1974. Multiband photometry of comets Kohoutek, Bennett, Bradfield, and Encke, pp. 551–60. Copyright 1974, with permission from Elsevier.)

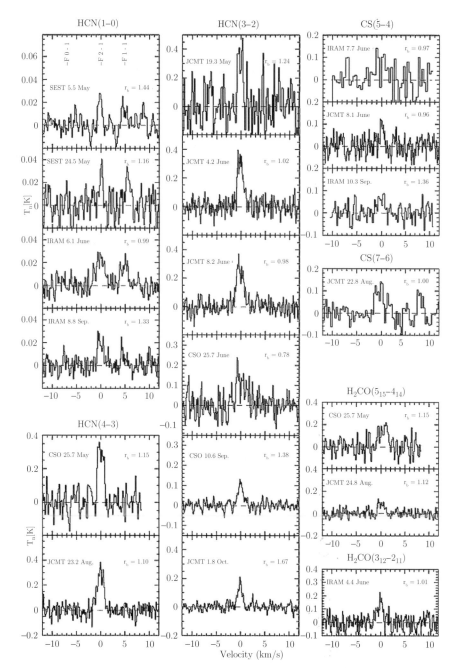

Fig. 3.7. Radio spectra of comet Lee (C/1999 H1) showing several HCN lines. Observations were made at the Swedish–ESO Submillimetre Telescope (SEST), James Clerk Maxwell Telescope (JCMT), Caltech Submillimeter Observatory (CSO), and the 30-m telescope of the Institut de Radio Astronomie Millimétrique (IRAM). (Courtesy N. Biver: from Biver, Bockelée-Morvan, Crovisier, Henry, Davies, Matthews, Colom, Gérard, Lis, Phillips, Rantakyrö, Haikala, and Weaver, 2000. Spectroscopic observations of comet C/1999 H1 (Lee) with the SEST, JCMT, CSO, IRAM and Nançay radio telescopes. *Astron. J.*, 120: 1554–70. Figure 1, spectra showing HCN (1-0), reproduced by permission of the AAS.)

(Altenhoff *et al.* 1986, 1999). Radar and radio observations of comet Halley suggest the continuum emission comes from a halo of large particles surrounding the nucleus.

One application of radio studies is the determination of production rates from observed spectral line intensities. Crovisier *et al.* (1991) presents a detailed discussion of the steps in that process which include calibration procedures, methods to ascertain the single-state molecular column densities, determination of the total molecular column density, and the total density in the coma. The authors then apply the steps to observations of the OH radical in comet Halley.

Radio interferometric imaging has also helped advance cometary science. Observations with, for instance the VLA, have included OH imaging, molecular line and continuum studies. The OH radio emission from comets is observed at a wavelength of about 18 cm. As with other molecules, the OH radical is excited from hyperfine levels of the ground state by solar UV, then decays back to the hyperfine levels of the ground state through a series of intermediate levels. Under certain circumstances, the upper levels of the transition can become over populated, and galactic background radiation will cause maser emission. More detail on VLA observations can be found in de Pater *et al.* (1991).

3.3 Photometric studies

The radiation from, for instance, the coma of a comet is a combination of two components:

- solar continuum scattered by coma dust and the nucleus, and
- emission from the atoms, radicals and ions of the gas coma.

In order for the observations to be physically meaningful, the two components must be separated. The typical method is to use a pair of filters, one of which isolates emission from a particular species and the other of which isolates a nearby continuum, selected to be as free as possible of emission lines. For example, photoelectric photometry of the CN band at 3880 Å could be carried out as follows: The cometary brightness can be measured alternatingly through an interference filter or intermediate passband filter (full width at half transmission of order 75 Å) having a maximum transmission near 3880 Å and through a filter similar to the U filter of the UBV system (full width at half transmission around 700 Å, maximum transmission around 3600 Å). The CN emission relative to the continuum can be derived from a color calculated in the usual way by

$$U - E = -2.5 \log \frac{F_u}{F_e} + \text{const},$$

where F_u and F_e are the measured fluxes in the U color and in the narrow passband near 3880 Å denoted by E, respectively. These colors can be used to define a color difference,

$$D = (U - E)_{comet} - (U - E)_s$$

between the comet and a source s (e.g., stars of spectral type G2) with a continuum distribution similar to the comet under study. When $D = 0$ no CN emission is present, and the value of D increases with increasing strength up to a maximum value depending on the details of the filter transmission curves.

The general scheme outlined above can be adapted to the study of any cometary emission line or band. Note that the projected size and location, at the comet, of the entrance diaphragm or the pixel size of the CCD used in the photometry should be specified for the measurements to be useful.

Continuum optical filter photometry of many comets has been carried out by A'Hearn and his colleagues. An example of the work is A'Hearn *et al.* (1984). A'Hearn and his co-workers relate their continuum observations to the quantity $Af\rho$, where A is the albedo, f is the filling factor of the dust grains within the field of view, and ρ is the linear radius of the field of view at the comet. If the number of grains within the field of view is $N(\rho)$ and the cross section of each grain is σ, then $f = N(\rho)\sigma/\pi\rho^2$. In a simple radial outflow model, $N(\rho)$ is proportional to ρ so that $Af\rho$ is proportional to $A\sigma$, that is, it is dependent on the reflectivity of the grains and is independent of the field of view.

Photometric studies of the nucleus itself are important in determining nuclear dimensions. Jewitt (1991) suggests some criterion to use to ascertain whether one is observing a bare nucleus: the nucleus should appear unresolved, the absolute magnitude, H_0, should appear to be constant with heliocentric distance, and, should exhibit a light curve due to rotation of the nucleus. As an example, when comet Halley was recovered in 1982, it was 11.0 AU from the sun, and appeared stellar. The image remained stellar until the comet reached 5.9 AU from the sun, and showed a coma as it approached closer to the sun. Photometric observations made before coma activity began led to estimates of the size of the nucleus as small as about 3 km. When the *Giotto* spacecraft flew by Halley, it obtained a large number of resolved images, which showed the dimensions of the nucleus to be about 15 km by 8 km. The difference between this measurement and the earlier size estimates is due to the fact that researchers were only beginning to realize the very low albedo of Halley's nucleus, which turns out to be about 0.04. Albedo (section 3.3.1) estimates of other cometary nuclei remain a major uncertainty in calculating nuclear sizes.

O'Ceallaigh *et al.* (1995) carried out CCD photometry of comet 109P/Swift–Tuttle when it was at a heliocentric distance of 5.3 AU in an attempt to measure the size of the nucleus. The authors compared the image profile of the nucleus with a nearby field star, and found that the cometary profile was similar to the stellar point-spread function except for the faint wings, where the comet was slightly brighter. They estimated that the wings were contamination from the coma that contributed about 14% to the nuclear brightness. They derived a nuclear radius of 11.8 km under the assumption that the albedo is similar to Halley's.

Frequently one can distinguish between the nuclear magnitude m_1 and the total magnitude m_2 of the cometary head when the comet is close enough to the sun for a coma to form. The nuclear magnitude refers to the brightness of the central, quasi-stellar concentration of light or photometric nucleus (sometimes called *the false nucleus*). This concentration of light may or may not be due to the comet's physical nucleus. The total magnitude, of course, refers to the entire material in the head. One must take careful note of the difficulties with visual magnitudes; the magnitude derived can depend on the size of the telescope used, contrast effects, and so on. Nuclear magnitudes are typically two to five magnitudes larger (fainter) than total magnitudes.

3.3.1 Albedos

The *albedo* of the surface of a comet or asteroid is simple to think of as the fraction of light that is reflected by the surface. But, in practice, it is not easy to determine accurately. The *Bond albedo*, A, is defined as the ratio of the total amount of light reflected from the surface to the total amount of light incident on it. This can be written as

$$A = \frac{r^2 I_0 \Delta^2}{s^2} \cdot 2 \int_0^\pi \frac{I(\phi)}{I_0} \sin\phi \, d\phi, \qquad (3.2)$$

or

$$A = p \cdot q, \qquad (3.3)$$

where

$$p = \frac{r^2 I_0 \Delta^2}{s^2}, \qquad (3.4)$$

and

$$q = 2 \int_0^\pi \frac{I(\phi)}{I_0} \sin\phi \, d\phi. \qquad (3.5)$$

Here r is the heliocentric distance, Δ is the geocentric distance, s is the radius of the body, and I_0 is the ratio of the apparent brightness of the body at distance Δ and at full phase ($\phi = 0$) to the sun's apparent brightness. The *geometrical albedo*, p, depends on geometrical factors and the brightness measured at full phase. The geometrical albedo is also the ratio of the radiation reflected from the body at full phase to the radiation that would be reflected from a flat, perfectly reflecting surface of the same cross-sectional area according to Lambert's Law.

The *phase integral*, *q*, contains all of the brightness variation with phase. The integral is carried out over the *phase angle*, ϕ, which is the sun–object–earth angle. Usually, the phase integral is the part that is observationally difficult to determine. If observations over sufficiently different phase angles are not available, a model variation must be used.

If the brightness at another phase angle needs to be used, the Bond albedo can be written as

$$A = p(\phi)\frac{I(0)}{I(\phi)} \cdot q \qquad (3.6)$$

Values of *A*, *q* and *p* usually refer to a wavelength range, e.g., the visible or infrared. If integrated over all wavelengths, the bolometric albedo is obtained. (The discussion in this section follows Russell 1916.)

3.3.2 Heliocentric brightness variations – observations

Naked-eye observations or observations through defocused binoculars are useful in determining the total magnitude of the cometary head. Such observations over a period of time lead to the cometary light curve. Experienced observers can produce a light curve that is internally consistent to within a few tenths of a magnitude, but comparison of light curves made by different observers often reveals extraordinary differences in the magnitude level of the curves. No seemingly straightforward observational quantity is (and should be) regarded with as much suspicion as cometary magnitudes.

The variation of cometary brightnesses, expressed in magnitudes, can be represented by the equation (see section 1.6.3)

$$H = H_0 + 5 \log \Delta + 2.5n \log r. \qquad (3.7)$$

The parameter *n* encapsulates the heliocentric brightness variation of the comet, which is assumed to be a power law (1.3). If data are available over a large enough range of *r*, then H_0 and *n* can be determined by a least squares fit to (3.7). If insufficient data exist for a fit, then an average value is chosen for *n*, frequently $n = 4$. In this case H_0 is referred to as H_{10} since $2.5n = 10$. The heliocentric variations are described by (H_0, n) sets. Extensive lists of (H_0, n) visually determined sets can be found in Meisel and Morris (1976, 1982). Table 3.1 lists a sample of visually determined (H_0, n) sets for selected comets. Since the introduction of the CCD detector, similar photometry has been carried out photoelectrically. Figure 3.8

Table 3.1 *Visual photometric parameters*
(H_0–n sets) for selected comets.

Comet	H_0	n
8P/Tuttle	7.97	6.01
Bradfield	7.61	2.92
C/1975 C1		
Kobayashi–Berger–Milon	7.34	3.77
C/1975 N1		
West	5.94	2.42
C/1978 A1		
Kohler	6.80	5.22
C/1977 R1		
Suzuki–Saigusa–Mori	9.72	−0.47
C/1975 T2		
38P/Stephan–Oterma	3.46	11.92
81P/Wild 2	6.51	5.58

Source: Meisel and Morris (1982).

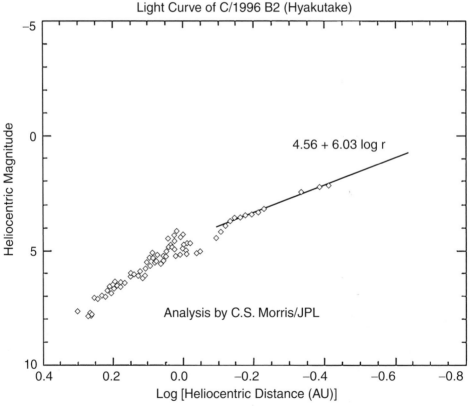

Fig. 3.8. *H_0–r* variation for comet Hyakutake (C/1995 Y1). (Courtesy C. Morris,
NASA–Jet Propulsion Laboratory.)

illustrates the H_0–r variations of comet Hyakutake. One must be wary when comparing (H_0, n) sets determined visually and photoelectrically. So far, researchers have not successfully simulated the wavelength dependence of the human eye with a CCD-filter combination. We will return to the theoretical basis of heliocentric variations in Chapter 5.

3.4 Space-based studies

3.4.1 High flying aircraft

Since 1974, extraterrestrial particles thought to be cometary in origin have been collected in the atmosphere at altitudes of 20 km by U-2 aircraft. When pure aluminum oxide (Al_2O_3) spherical particles generated by solid fuel rocket engines are excluded, approximately half the collected particles have elemental abundances that closely match the abundances of carbonaceous chondrite meteorites. Scanning electron microscope pictures of one of these particles are shown in Fig. 3.9. Notice the porous nature of the structures. This class of particles, according to Brownlee *et al.* (1977:134), "seems to come from the gentle fragmentation of a single type of parent-body material: a black aggregate of grains mainly 1000 Å in size." The particles do not appear altered or strongly heated by entry into the atmosphere.

The extraterrestrial origin of these particles seems definite because helium atoms implanted by the solar wind have been detected. Although these particles probably come from comets, this conclusion is still tentative.

3.4.2 Ultraviolet observations from above the atmosphere

On January 14, 1970, the first ultraviolet observations of a comet were made from the second *Orbiting Astronomical Observatory* (*OAO-2*) when the spectrum of comet Tago–Sato–Kosaka (C/1969 T1) was recorded. The observations clearly showed the huge cloud of Lyman-α emission. Observations of the comet were continued throughout January 1970. Comet Bennett (C/1969 Y1) was observable from space in February 1970, and it was observed from *OAO-2* and by two experiments on board the fifth *Orbiting Geophysical Observatory* (*OGO-5*). The *OGO* experiments were with wide-field photometers that were more sensitive than the *OAO-2* instrument. In addition, *OGO-5* was at a high altitude, above most of the earth's hydrogen cloud (geocorona). As a result, it was possible to observe much fainter emission with *OGO-5* than with *OAO-2*. Comet Encke was detected by *OGO-5* in January 1971, but comet Toba (C/1971 E1) was not. Extensive observations of the hydrogen cloud surrounding comet Kohoutek were obtained in 1973–74. A discussion of the significance of these observations is given in section 5.2.

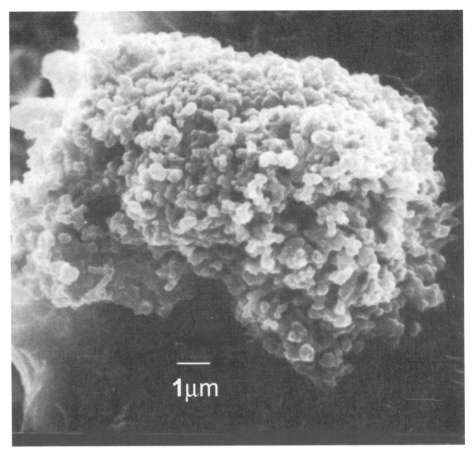

Fig. 3.9. Scanning electron microscope picture of possible cometary dust particle collected in the earth's atmosphere. (Courtesy D. Brownlee, University of Washington.)

3.4.3 Kohoutek and Skylab

Comet Kohoutek was discovered roughly 10 months before it passed perihelion, which gave researchers adequate time to plan a coordinated research program. The lead time allowed NASA to reprogram some of the time of the astronauts aboard the manned *Skylab* spacecraft, and the comet was observed extensively by the astronauts during December 1973 and January 1974. Instrumentation on board *Skylab* was used to obtain photometry of the Lyman-α halo. Sketches showing Gibson's view from *Skylab* are shown in Fig. 3.10.

Scientist–astronaut E. G. Gibson made visual observations of the comet near perihelion. These observations showed the development of a prominent antitail (Fig. 3.10). The antitail was observed for comet Kohoutek both from *Skylab* and from the ground (Fig. 3.11). According to Sekanina (1977), the only requirements

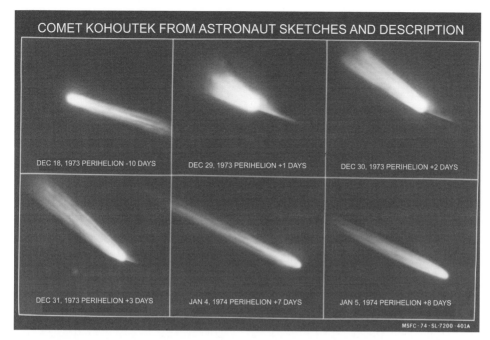

Fig. 3.10. Sketches of comet Kohoutek (C/1973 E1) made by astronaut E. Gibson from *Skylab*. (Courtesy Johnson Space Center, Houston, Texas.)

for an antitail or sunward-appearing dust tail are (1) a favorable geometry (i.e., observer situated near the orbit plane) and (2) the presence of millimeter-sized particles. The antitail was observed at post-perihelion times. The favored explanation is that millimeter-sized particles were released by the nucleus and dragged away by near-nucleus gas flows. The particles experienced relatively small repulsive forces because of their masses, and were left behind in the orbit plane as the comet moved away. When earth was near the orbit plane the dust gave the appearance of an antitail. This dust becomes part of a meteoroid stream along the comet's orbit.

3.4.4 International Ultraviolet Explorer

The *International Ultraviolet Explorer* (*IUE*) was launched into a geosynchronous orbit in January 1978, with a lifetime goal of five years. It was turned off in September 1996 after operating for nearly 19 years, when the instrumentation had degraded to a point where it was difficult to obtain good scientific data. The spectrophotometers on *IUE* collected both high (0.1Å to 0.3Å) and low (7 Å) resolution spectra between 1150Å and about 3200Å. Because it was in a geosynchronous orbit, *IUE* could be operated in real time from the USA and Europe.

Fig. 3.11. Photograph of comet Kohoutek (C/1973 E1) on January 14, 1974. A faint antitail is visible at the arrow. (Joint Observatory for Cometary Research, NASA–Goddard Space Flight Center and New Mexico Institute of Mining and Technology.)

One of the strengths of *IUE* was the ability of the whole system to respond rapidly to targets of opportunity such as novae, supernovae, and, of course, comets. Over 26 comets were studied using the observatory. One of the most significant discoveries was made from spectra of comet IRAS–Araki–Alcock taken soon after it was first seen in May 1983. The observations led to the first discovery of diatomic sulfur, S_2, in a comet. *IUE* also found that the spectra of all comets observed were very similar in the UV, and were dominated by the hydrogen Lyman α at 1216 Å and the hydroxyl (OH) bands at 3090 Å. Figure 3.12 shows an example of two *IUE* spectra of comet Halley; one taken with the short-wavelength camera and one with the long-wavelength camera. The spectra show emissions of O I, C I, S I, CO and OH.

3.4.5 Hubble Space Telescope

NASA's *Hubble Space Telescope* was launched into low earth orbit by the Space Shuttle in April 1990. Since then, the observatory has studied a large number of comets. Because of its small field of view, *HST* has produced much data on

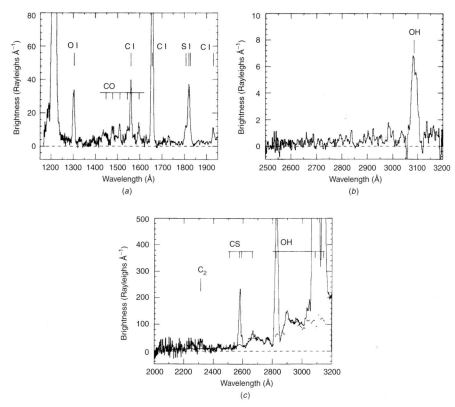

Fig. 3.12. *International Ultraviolet Explorer* (*IUE*) spectra of comet Halley in 1985 and 1986. (a) Spectrum on March 9, 1986; the very strong line at 1200 Å is the Lyman-α line of neutral hydrogen. (b) Spectrum on September 12, 1985. (c) Spectrum on March 11, 1986. (Courtesy of P. D. Feldman, Johns Hopkins University.)

the nuclei and inner comas of the comets. The list of observations includes the impact of the fragments of comet Shoemaker–Levy 9 (D/1993 F2) with Jupiter, and the subsequent evolution of impact sites on the planet. Among other comets studied are 4P/Faye (Lamy and Toth 1995), 19P/Borrelly (Lamy *et al.* 1998), Hale–Bopp (C/1995 O1), 45P/Honda–Mrkos–Pajdusakova, and Hyakutake (C/1996 B2) (e.g., Weaver *et al.* 1999, Richter *et al.* 2000). Plate 3.1 is a mosaic of images of comets observed by *HST* after the initial correction of its optics in 1993. Figure 3.13 is a Goddard High-Resolution Spectrograph (GHRS) spectrum of comet Hyaku-take made in 1996 when the comet was at a heliocentric distance of 0.88 AU. The spectrum shows the CO Fourth Positive Group, the CO Cameron system and monatomic S.

In the case of 4P/Faye and 19P/Borrelly, the observers were able to separate the signals from the nucleus and the coma, when the comets were about 0.6 AU

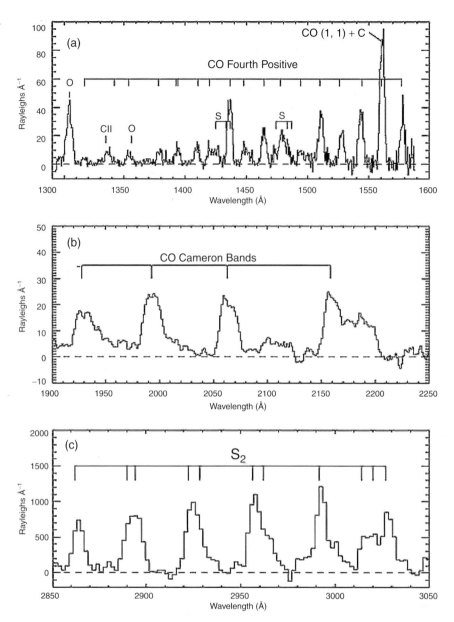

Fig. 3.13. *Hubble Space Telescope* spectra of comet Hyakutake, taken on April 1, 1996. The comet was at a heliocentric distance of 0.88 AU and a geocentric distance of 0.26 AU. (a) Goddard High-Resolution Spectrograph (GHRS) data taken through a 1.74 arcsec square aperture, showing the CO fourth positive group. (b) Faint Object Spectrograph (FOS) data taken through the 4.3 arcsec aperture, showing the CO Cameron bands. (c) FOS data taken through the 0.86 arcsec aperture, showing lines of S_2. (Courtesy H. Weaver, Johns Hopkins University) (From Weaver. 1998. Comets. In Brandt, J. C., Ake, T. B. III, and Petersen, C. C. (eds). *The Scientific Impact of the Goddard High Resolution Spectrograph.* Astronomical Society of the Pacific Conference Series Volume 143, page 213, Figure 6. HST/GHRS/FOS spectra of comet Hyakutake. Used with the kind permission of the Astronomical Society of the Pacific.)

from earth, using, respectively, the Planetary Camera with a 702 nm filter, and the newer WFPC2 with a 670 nm filter. See section 7.3.1 for a further discussion of this approach. The results show that comet 4P/Faye has a roughly spherical nucleus with a radius of 2.68 km and comet 19P/Borrelly has a prolate spheroidal nucleus with semi-axes of 4.4 and 1.8 km. In September 2001, NASA's *Deep Space 1* spacecraft flew by comet Borrelly and took close up images of the nucleus (see Chapter 7). The preliminary announcement placed the axes of the nucleus at 8×4 km, quite consistent with the semi-axis numbers above.

3.4.6 X-rays

Observations of comet Hyakutake (C/1996 B2) by the *Röntgen Satellite* (*ROSAT*) and the *Extreme Ultraviolet Explorer* (*EUVE*) led to the discovery of variable x-ray emission (Lisse *et al.* 1996, Mumma *et al.* 1997). See section 6.4.4. Since then, x rays have been detected from several additional periodic comets (2P/Encke, 45P/Honda–Mrkos–Padjusakova, 55P/Tempel–Tuttle) and long-period comets (C/Levy (1990 K1), C/Tsuchiya-Kiuch (1990 N1) and others).

The *ROSAT* High Resolution Imager (HRI) is a grazing incidence telescope with a microchannel plate detector. The instrument has an effective area of about 20 cm^2 at 0.09–0.75 keV, a 38 arcminute circular field-of-view, and a 6 arcsecond angular resolution. Additional details of the HRI instrument can be found in Lisse *et al.* (1996).

The *EUVE* payload consists of four telescopes with seven instruments (Bowyer and Malina 1991). The scanner telescopes, which made the comet Hyakutake observations, are mounted at a right angle to the spin axis of the spacecraft. Two of the three scanning telescopes are of the Wolter–Schwarzschild type I design with 5 degree diameter FOV. The third scanning telescope is a Wolter Type II design, with a 4 degree diameter circular FOV. All the telescopes have microchannel plate detectors and various filters to provide specific bandpasses. More details of the instrument characteristics can be found in the above references. The physical mechanisms of x-ray emission are discussed in section 6.4.4.

3.5 *In situ* observations

3.5.1 Comet Giacobini–Zinner

In 1985, the study of comets took a giant step forward from *remote sensing* observations to *in situ* measurements, when the *International Cometary Explorer* (*ICE*) spacecraft flew through the plasma tail of comet Giacobini–Zinner (GZ) and made the first *in situ* measurements of the cometary environment. This step made it possible to measure, among other things, the properties of cometary dust, gas and plasma, magnetic fields, plasma waves, high energy particles, and other parameters

(Mendis 1988). We will give a brief description of the measuring techniques, with references for further reading.

ICE began on 12 August 1978 as part of a trio of spacecraft designed to study the solar wind interaction with the earth's magnetosphere. At launch it was called the *Third International Sun–Earth Explorer* (*ISEE-3*), and it was placed in a halo orbit around the Lagrangian point, L_1. Instruments on board the spacecraft were designed to monitor the solar-wind input to the magnetosphere from a location toward the sun. *ISEE-1* and *ISEE-2* were located inside the magnetosphere where they measured responses to changes in the solar-wind input. The *ISEE* mission spacecraft completed four years of highly successful operation in this role.

A great deal of study went into a dedicated mission to comet Halley. When NASA decided not to develop a Halley mission, a study was undertaken of the possibility of diverting the *ISEE-3* spacecraft to study a comet. Robert Farquhar, then of the Goddard Space Flight Center, devised an elaborate orbital maneuver that involved firing the spacecraft's engines and sending it into an orbit around the earth and moon, in which it made five close encounters with the moon. On fifth encounter the spacecraft skimmed only 120 km above the lunar surface, and travelled off to encounter comet Giacobini–Zinner (GZ). Because the mission of *ISEE-3* had been so drastically changed, the name was changed to the *International Cometary Explorer. ICE*'s trajectory to the cometary intercept is shown in Fig. 3.14.

Closest approach to comet GZ took place on 11 September 1985 at 11:02 UT, when the spacecraft passed 7800 km from the nucleus toward the comet's tail, at a relative speed of 21 km s^{-1}. The tailward intercept of the comet was chosen because the instrumentation was nearly ideal for probing the environment of the plasma tail and because other nations' missions to comet Halley were being planned to pass on the sunward side of its nucleus. Researchers were concerned that relatively high-speed impacts with cometary dust particles might degrade the solar cells and cause a dangerous reduction in spacecraft electrical power. However, *ICE* came through the encounter unscathed. Table 3.2 summarizes the *ISEE-3/ICE* payload (Reinhard 1990). The results from the *ICE* encounter are discussed in Chapter 6.

3.5.2 The Halley Armada

Comet Halley is interesting to scientists for a number of reasons, not the least of which is being the first comet recognized to be periodic. It is also highly active. In our earlier edition we talked about some of the plans for a NASA mission to comet Halley, and speculated on some of the cutting-edge technologies, such as

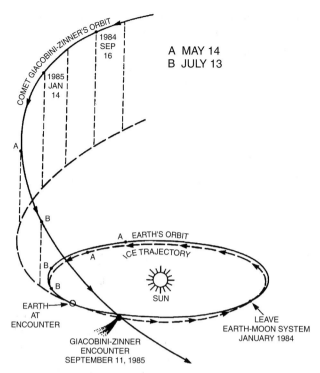

COMET GIACOBINI-ZINNER'S ORBIT

1984 SEP 16

1985 JAN 14

A MAY 14
B JULY 13

A

B

B

B

A EARTH'S ORBIT

ICE TRAJECTORY

A

SUN

EARTH AT ENCOUNTER

LEAVE EARTH-MOON SYSTEM JANUARY 1984

GIACOBINI-ZINNER ENCOUNTER SEPTEMBER 11, 1985

Fig. 3.14. *International Cometary Explorer* trajectory to comet Giacobini–Zinner, showing the encounter circumstances. (NASA–Goddard Space Flight Center.)

solar electric propulsion, that could enable an exciting rendezvous mission. The spacecraft, as planned, could have maneuvered in the vicinity of the comet for extended periods of time, collecting data on the evolution of cometary phenomena. Those plans were soon scaled back, and were replaced with a flyby mission. The disadvantage of a flyby mission to comet Halley is the comet's 163° inclination, which means the comet moves through the solar system in a retrograde sense.

An interplanetary spacecraft is launched in a direct orbit, because the launch uses the earth's orbital velocity to provide the necessary velocity to the spacecraft. The relative flyby velocity of the comet and a spacecraft is high; nearly 70 km s^{-1}. The European Space Agency (ESA), Japan, and the former Soviet Union also planned flyby missions. NASA and ESA formed a collaboration for an *International Comet Mission* (*ICM*) to comets Halley and Tempel 2. The ESA contribution was to be a direct entry probe to make *in situ* observations of Halley. In the end, the United States failed to approve any dedicated Halley mission (see Chapter 12), and the ESA portion of the *ICM* became the *Giotto* mission. NASA's role in cometary studies in the Halley era consisted of the *ICE* mission to comets Giacobini–Zinner

Table 3.2 ICE *scientific payloads*

Acronym	Name	Description	Principal investigator
	Solar Wind Plasma	Measure ion velocity distributions in both two and three dimensions; measure electron velocity distributions also in two and three dimensions. Ions were measured in 32 channels from 237 eV/charge to 10.7 keV/charge.	Dr. Robert T. Gosling, Los Alamos National Laboratory
	Vector Helium Magnetometer	A boom-mounted triaxial vector helium magnetometer to measure the steady magnetic field and its low-frequency variations.	Dr. Edward J. Smith, Jet Propulsion Laboratory
	Radio Mapping of Solar Wind Disturbances	Measure the trajectories of type III solar radio bursts using two perpendicular dipole antennas.	Dr. Jean-Louis Steinberg, Observatoire de Paris, Meudon
	Plasma Waves Spectrum Analyzer	Electric dipoles and a boom-mounted magnetic search coil were used to measure magnetic and electric field wave levels from 17 Hz to 1 kHz and electric field levels from 17 Hz to 100 kHz. A third spectrum analyzer with three bands between 0.316 and 8.8 Hz was included for measurement of the magnetic field.	Mr. Eugene W. Greenstadt, TRW Systems Group
	Solar Wind Ion Composition	Measure the charge state and isotopic constitution of the solar wind.	Dr. Keith W. Ogilvie, Goddard Space Flight Center
HOH	Low-energy cosmic rays	Measure energetic ions in numerous bands within the energy range 2 keV/charge to 80 MeV/nucleon, and electrons in four contiguous bands from 75 to 1300 keV.	Dr. Dieter K. Hovestadt, Max-Planck-Institut für Extraterrestriche Physik
EPAS	Energetic Particle Anisotropic Spectrometer	Measure the energy spectrum of low-energy protons in 8 channels, and the 3-dimensional angular distribution of protons in the energy range 0.035 to 1.6 MeV with a basic time resolution of 16 s.	Dr. Robert J. Hynds, Imperial College, London

Table 3.3 *Dates/times of*
spacecraft encounters with Halley

Spacecraft	1986 date
VEGA 1	March 6
Suisei	March 8
VEGA 2	March 9
Sakigake	March 11
Giotto	March 14
ICE	March 25

and Halley, described above, and NASA's leadership in the International Halley Watch.

Then, in early 1986, NASA's role in the study of comet Halley received another blow; namely, the *Challenger* disaster. A spectrograph designed to observe comet Halley while it was near the sun was being tested on *Challenger*, and was destroyed. And, the stand down of the Shuttle fleet made impossible extensive observations of the comet planned for early March 1986.

The international space community sent five spacecraft to intercept comet Halley, in what we call the Halley Armada. The experts planning the encounters realized that a spacecraft would carry the largest possible payload if it intercepted the comet as it crossed the earth's orbital plane near a node of the comet's orbit. On March 10, 1986, comet Halley moved across the earth's orbital plane from north to south; that is, at the descending node. As we can see in Table 3.3, the spacecraft encounters all clustered around this date. Figure 3.15 shows the orbital geometry of *Giotto*'s trajectory and the descending node concept. After the bad luck encountered in the planning stages, scientists enjoyed exceptionally good fortune with the Halley Armada; all the launches were successful and most of the experiments on the spacecraft functioned as designed. The path of the Armada spacecraft in the vicinity of comet Halley are illustrated in Plate 1.2.

The two Soviet *VEGA* spacecraft had been planned originally to just fly by the planet Venus. The name *VEGA*, in fact, consists of the first two letters of the Russian name for Venus (*Venera*) and the Cyrillic spelling of Halley (*Galley*). The orbit of the spacecraft took them first past Venus, where they dropped off instrument packages and where the spacecraft picked up additional energy to propel them to the comet. The dates of the comet Halley flybys for *VEGA-1* and *VEGA-2* were 6 March and 9 March 1986, and their minimum distances were 8890 and 8030 km, respectively. Each of the spacecraft carried camera systems to collect images of Halley's nucleus and spectrometers to record emissions, including thermal emissions, from the nucleus and emissions from the coma. The spacecraft also

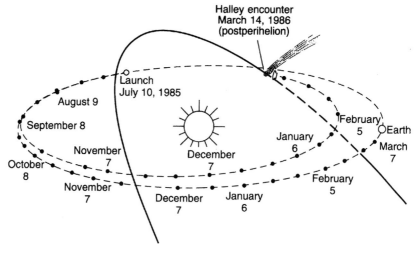

Fig. 3.15. *Giotto*'s trajectory to comet Halley. The spacecraft's position is shown from launch on July 10, 1985 to its encounter on March 14, 1986. (Courtesy R. Reinhard, European Space Agency.)

carried several experiments designed to measure the chemical composition, quantity, and the distribution of particle sizes of the dust. And, of course, instruments measured the composition of neutral and ionized gases and the energy spectrum of electrons. Finally, additional instruments measured the magnetic field and plasma waves around the comet. Table 3.4 summarizes the payloads of the two *VEGA* spacecraft (Reinhard 1990).

The Japanese also sent two spacecraft to Halley; *Sakigake* ("pioneer" or "forerunner") and *Suisei* ("comet"). The two spacecraft are nearly identical but they carried different scientific experiments (Hirao 1987). They were Japan's first mission into interplanetary space. The instruments on *Sakigake* measured basic properties of the solar wind near the comet, radio emission from the coma, and plasma waves. *Suisei* observed the hydrogen cloud around the comet, and solar wind and cometary ions. *Suisei* flew by Halley on 8 March 1986 and *Sakigake* flew by on 11 March 1986. Their minimum distances were 1.51×10^5 km and 6.99×10^6 km. Table 3.5 summarizes the *Sakigake* and *Suisei* payloads (Reinhard 1990).

ESA's *Giotto* spacecraft was named after the Italian painter Giotto di Bondone, who depicted the "star of Bethlehem" as a comet in one of the Arena Chapel frescoes in Padua (Plate 3.2). *Giotto* was also ESA's first interplanetary mission. The *Giotto* scientific instrument's imaging of the nucleus and inner coma achieved a resolution of 100 meters at a distance of 500 km. *Giotto*'s instruments also measured the composition of neutral atoms, ions, and the amount of dust and the distribution of dust particles. Photometric measurements determined the brightness of the coma. Plasma instruments measured the properties of ions and electrons in a variety of

Table 3.4 VEGA *scientific payloads*

Acronym	Name	Description
TVS	Television System	Imaging of nucleus and inner coma.
IKS	Infrared Spectrometer	Measure IR radiation from the coma in the range $2.5 < \lambda < 12\mu m$.
TKS	Three-Channel Spectrometer	Measure coma emissions in the range $0.12 < \lambda < 19\mu m$.
PHOTON	Shield Penetration Detector	Detect large dust particles.
DUCMA	Dust Particle Detector	Measure dust particle flux and mass spectrum for $m > 1.5 \times 10^{-15}$ g.
SP-1	Dust Particle Counter	Measure dust particle flux and mass spectrum for $m > 10^{-16}$ g.
SP-2	Dust Particle Counter	Measure dust particle flux and mass spectrum for $m > 10^{-16}$ g.
PUMA	Dust Mass Spectrometer	Measure dust composition.
ING	Neutral Gas Mass Spectrometer	Measure composition of neutral gas in the comet.
PLASMAG	Plasma Energy Analyzer	Measure ion flux composition and the energy spectrum of ions and electrons.
TÜNDE	Energetic Particle Analyzer	Measure energy and flux of accelerated cometary ions.
MISCHA	Magnetometer	Measure cometary magnetic field.
APV-N	Wave and Plasma Analyzer	Measure plasma waves in the range 0.01 to 1000 Hz, and plasma ion flux fluctuations.
APN-V	Wave and Plasma Analyzer	Measure plasma waves in the range 0 to 300 kHz, and plasma density and temperature.

Source: Grard, R., Gombosi, T. I., and Sagdeev, R. Z., 1986, in *Space Missions to Halley's Comet* ESA SP-1066.

energy ranges, and a magnetometer measured the comet's magnetic field (Reinhard 1990). The broad goal of these plasma experiments and those on the other spacecraft was to study the interaction of the solar wind with comets. *Giotto* flew by Halley on 14 March 1986 at a minimum distance of 596 km. Table 3.6 summarizes the *Giotto* payload (Reinhard 1990).

Designers solved the potential problem of dust damage to the spacecraft in a variety of ways. *Suisei* and *Sakigake* were far enough from the nucleus to avoid the dust. Models predicted a very low hazard at distances of roughly 200 000 km. The more vulnerable parts of the *VEGA* spacecraft were protected, but the science

Table 3.5 Suisei *and* Sakigake *scientific payloads*

Acronym	Spacecraft	Name	Description
UVI	*Suisei*	Ultraviolet Imager	UV optical system for imaging and photometry.
ESP	*Suisei*	Energy Analyzer of Charged Particles	Measure physical parameters of solar-wind plasma and cometary ions in the range 30 eV/q to 16 keV/q, with $\Delta E / E = 0.06$.
PWP	*Sakigake*	Plasma Wave Probe	Measure electric and magnetic field oscillations in the frequency ranges 5 kHz to 200 kHz and 70 Hz to 2.8 kHz.
SOW	*Sakigake*	Solar Wind Ion Detector	Measure flow direction, bulk velocity, density and temperature of the solar wind. Dynamic range of ion current is 10^{-11} to 10^{-8} amp.
IMF	*Sakigake*	Interplanetary Magnetic Field	Measure interplanetary magnetic field oscillations in the frequency range 0 to 3 Hz with a resolution of 0.032 nT.

Source: Hirio, K., 1986, in *Space Missions to Halley's Comet* ESA SP-1066.

team also relied on the redundancy of having two spacecraft. The close-encounter distance of *Giotto* demanded special precautions. Encounter speeds were roughly the same for all the spacecraft: *Giotto*'s, for example, was 68 km per second. At this speed, a dust particle with mass of 0.1 g could penetrate an aluminum sheet 8 centimeters thick.

The *Giotto* team's solution to the dust penetration problem was a double bumper shield consisting of a thin front sheet, a large gap, and a thick rear sheet. Incidentally, that shield was the brain-child of Fred Whipple. Figure 3.16 shows the shield and the rest of the spacecraft. A dust particle striking the front sheet would vaporize and expand into the gap. When it reached the rear sheet, its energy would be dissipated over a large area. Estimates indicated that the shield could withstand an impact from a dust particle with mass of roughly 1 gram. However, the impact of a much smaller particle on the rim of the shield could cause the spacecraft to wobble and potentially lose contact with earth. Such an event, which was expected, occurred 14 seconds before closest approach, and data were received only intermittently for the next 32 minutes, as the *Giotto* rocket jets worked to de-wobble the spacecraft.

At the high flyby speeds, dust impacts were clearly hazardous to the spacecraft. On *Giotto*, the capabilities of several experiments were degraded or the instruments

Table 3.6 Giotto *scientific payload*

Acronym	Name	Description	References
HMC	Camera	Image the nucleus and inner coma.	Schmidt, *et al.* (1986)
NMS	Neutral Mass Spectrometer	Measure the mass and energy of neutral gas atoms and molecules.	Krankowsky, *et al.* (1986)
IMS	Ion Mass Spectrometer	Measure the ion composition.	Balsiger, *et al.* (1986)
PIA	Dust Mass Spectrometer	Measure dust particle flux and composition.	Kissel (1986)
DIDYS	Dust Impact Detection System	Measure dust particle flux and mass distribution.	McDonnell, *et al.* (1986)
JPA	Plasma Analysis 1: Fast Ion Sensor (FIS) Implanted Ion Sensor (IIS)	Measure 3-D ion velocity distribution for 10 eV to 20 keV. Measure ion flux, mass and velocity distribution for 90 eV, to 90 keV, 1–45 AMU/q.	Johnstone, *et al.* (1986)
RPA	Plasma Analysis 2: Electron Electrostatic Analyzer (EESA) Positive Ion Cluster Composition Analyzer (PICCA)	Measure 3-D electron velocity distribution for 10 eV to 30 keV. Measure composition of cold ions 10–203 AMU.	Rème, *et al.* (1986)
PEA	Energetic Particle Analyzer	Measure energy and flux of electrons and accelerated ions >20 keV.	McKenna-Lawlor, *et al.* (1986)
MAG	Magnetometer	Measure magnetics fields in the range 0.004 to 65 536 nT.	Neubauer, *et al.* (1986)
OPE	Optical Probe Experiment	Measure coma brightness in four continuum bands and four discrete emissions (OH, CN, CO^+, C_2).	Levasseur-Regourd, *et al.* (1986)
GRE	Radio-Science Experiment	Measure dust and gas column densities in the coma.	Edenhofer, *et al.* (1986)

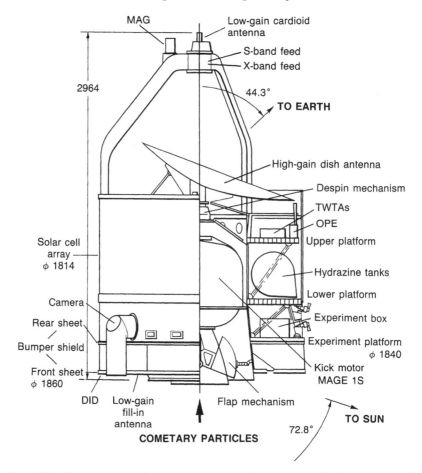

Fig. 3.16. *Giotto* spacecraft and meteoroid shield (bumper shield at lower left). (Courtesy R. Reinhard, European Space Agency.)

were knocked out completely. The dust impacts caused some instruments to fail on the *VEGA*s, and the power available from the solar cells was reduced by about 50 percent. Even *Suisei* was hit near closest approach by two dust particles with masses of several milligrams, despite its distance of 1.51×10^5 km from the nucleus. *Sakigake* and *ICE* made their passes unscathed.

Targeting *Giotto* to pass within 600 km of Halley's nucleus required exceptional (± 50 km) targeting accuracy. Traditional ground-based measuring techniques were sufficiently inaccurate that their use might have caused the spacecraft to pass on the dark side of the nucleus or even collide with it. Thus was born the *Pathfinder* Project, a fine example of international cooperation. NASA determined the positions of the *VEGA* spacecraft by carrying out Very Long Baseline Interferometry (VLBI) using its Deep Space Network at Goldstone, CA, and antennas at Madrid, Spain, and

Canberra, Australia. In turn, the *VEGA*s determined the position of the comet at their times of closest approach on 6 and 9 March. The positional data was processed and communicated rapidly to ESA for the *Giotto* encounter on 14 March. The result was a predicted flyby distance of 605 ± 40 km. The actual flyby distance was 596 km. We think this is a shining testament to international cooperation. The only negative of the *Giotto* encounter was the fact that the camera system was damaged by the dust in the vicinity of the nucleus.

In addition to the *in situ* measurements, valuable data has come from other sources. Spacecraft in earth orbit, such as the *International Ultraviolet Explorer* (*IUE*), the *Solar Maximum Mission* (*SMM*), and the *Dynamics Explorer 1*, also contributed. The *Pioneer Venus Orbiter* in Venus orbit provided important measurements, as have numerous rocket flights and the *Kuiper Airborne Observatory*. Last, but emphatically not least, are the ground-based networks of the International Halley Watch. We will have much more to say about the interpretation of the measurements in later chapters.

After *Giotto* completed its observations of comet Halley, controllers executed a series of small orbital corrections so that it would return to earth almost exactly five years after launch. In April 1986, the spacecraft was put to sleep. Then, in February 1990, controllers sent *Giotto* a wake up call, and despite their worries, the spacecraft returned an answer. An assessment of its health showed that three scientific instruments were fully operational, and four others were useable. The *Giotto* team decided to send the spacecraft to comet P/Grigg–Skjellerup in what became the *Giotto Extended Mission* (*GEM*). *Giotto*'s orbit was altered as it passed earth in July 1990. The spacecraft was then put back to sleep until July 1992. The Grigg–Skjellerup flyby occurred about 215 million km from earth when *Giotto* crossed the bow shock and entered the dust coma. *Giotto* passed Grigg–Skjellerup's nucleus within 100 to 200 km. It was the closest ever cometary flyby. Unfortunately, the camera was not one of the working instruments.

3.5.3 Measuring techniques

3.5.3.1 Plasma detectors

Mass spectrometers of several different designs flew on the spacecraft of the Halley Armada to measure the properties of the plasma, neutral gas, and dust. As we shall see, a mass spectrometer operates on ions; therefore, neutral gas and dust must be converted to plasma in the measurement process. Figure 3.17 illustrates the functioning of an idealized mass spectrometer. A beam of ions passes through a slit into the first stage, which is a velocity filter. In some cases, the velocity filter is preceded by a pair of slits in metal plates, separated by a potential difference, to accelerate the ions. The velocity filter consists of crossed electric and magnetic

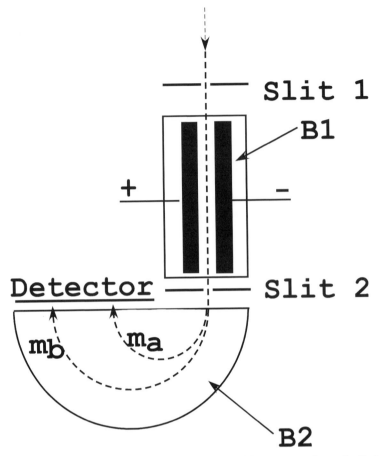

Fig. 3.17. Idealized mass spectrometer. A beam of ions passes through slit 1 and then through a velocity filter consisting of a perpendicular electric field (positive and negative plates) and magnetic field, B1. Ions with speeds equal to E/B pass through slit 2 into a region of magnetic field B2. The ions spiral to a point on the detector depending on their mass. The magnetic fields B1 and B2 are perpendicular to the plane of the diagram.

fields, $\mathbf{E}$ and $\mathbf{B}_1$. Ions will be deflected by the fields unless the forces acting on them by the electric and magnetic fields are equal, that is, unless $e\mathbf{E} = e\mathbf{vB}_1$. Ions for which the equality holds have velocity $\mathbf{v} = \mathbf{E}/\mathbf{B}_1$. They pass through an exit slit from the velocity filter into the second stage of the mass spectrograph. In this second stage ions move in a magnetic field $\mathbf{B}_2$, and travel in a semicircular path to a detector. Since all the ions now have the same velocity, the radius of curvature is proportional to m/q. The detector counts the number of ions in various channels. The basic design dates back to the early twentieth century, when the detector was a photographic emulsion. The *Giotto* instrument used digital detectors (e.g., Channeltrons).

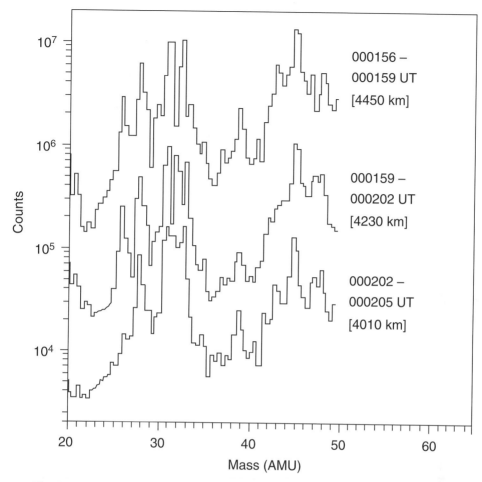

Fig. 3.18. Three consecutive mass spectra of comet Halley from *Giotto*'s Positive Ion Cluster Composition Analyzer (PICCA) instrument. Each trace is labeled with the spacecraft event time and cometocentric distance. (Used by permission © *Nature*.)

The ion mass spectrometers flown on the Halley Armada spacecraft provided a wealth of data. However, the data can be difficult to interpret, because what is measured is the distribution of m/q. As Mendis (1988) points out, the identification of some of the peaks in the m/q distribution are fairly clear cut. For instance, a peak at $m/q = 12$ is undoubtedly due to C^+. However, a peak at $m/q = 18$ is likely to be due to both H_2O^+ and NH_4^+. A sample mass spectrum from the *Giotto* PICCA instrument is shown in Fig. 3.18. This figure, from Korth *et al.* (1989) suggests the presence of $C_3H_3^+$ at 39 AMU. We will talk about the significance of the results in later chapters.

A neutral mass spectrometer is similar to the ion mass spectrometer. A beam of cometary gas first passes through a device which rejects incoming ions, then ionizes the neutral gas with an electron beam. The resulting ion beam can be studied in a modified mass spectrometer.

Dust composition in comet Halley was measured in much the same way, because of the high relative velocity of the flyby. Dust particles impacted on a target and flashed into plasma, which was deflected into a mass-spectrometer-like device known as a time-of-flight detector. The device has very high mass resolution and can measure isotope ratios such as $^{13}C/^{12}C$.

3.5.3.2 Magnetometers

Over the years, several spacecraft have had magnetometers as part of their payloads. Typical magnetometers use sets of three sensors to measure the three orthogonal components of a magnetic field. Generally a magnetometer is mounted on a boom of some sort to place it a distance from the spacecraft to minimize magnetic effects of the spacecraft itself. The rotation of the spacecraft is often used to separate the field to be measured from any spacecraft fields.

The magnetometer on the *Giotto* was a triaxial fluxgate magnetometer with dynamic ranges of $\pm 16, \pm 64, \ldots, \pm 65\,536$ nT. The proper range was selected automatically. When it was in the vicinity of comet Halley, it sampled the magnetic field at a rate of 28 vectors s^{-1}. More details of the instrument and the operation of the sensors can be found in Mason (1990a). Similar devices have been flown to Jupiter on the *Galileo* satellite. The interested reader is referred to the *Galileo* page of the JPL website for additional details of the operation of that instrument.

3.5.3.3 Dust detectors

In addition to the composition of the dust, researchers are interested in the physical parameters of the particles: mass, density, size distribution, and the like. Thus the Halley Armada spacecraft payloads contained instruments to measure these parameters. One example is the Dust Impact Detector System (DIDSY) system on *Giotto*.

The DIDSY system had three components: a Meteoroid Shield Momentum (MSM) sensor, an Impact Plasma and Momentum (IPM) sensor, and a Capacitor Impact Sensor (CIS). The MSM detected large dust particles impacting on the spacecraft shield using piezoelectric microphones. The CIS measured the flux of small particles penetrating an aluminum coated mylar sheet which acted as a large capacitor. Particles with sufficient energy would penetrate the capacitor and flash into plasma which discharged it. The CIS had a limited counting rate, since it had to recharge between impacts. The IPM could count at a very high rate. It collected the plasma generated electrons from impacts and measured the charge density. It also had a piezoelectric microphone to measure the impact momentum.

3.6 Summary

Charge Coupled Devices (CCDs) have become a major tool for cometary imaging because of their significant advantages over the photographic plate. The CCD has high photometric precision, is linear and outputs a digital signal. The main disadvantage of the CCD is the lower number of picture elements. But CCD arrays with 2048 pixels square are now commonly available.

Comets have been studied remotely with ground based and space based instrumentation at wavelengths from the x ray to the radio. Comet Hyakutake (C/1996 B2) was the first to be observed at x-ray wavelengths. The 1986 return of Halley's Comet stimulated scientists to undertake *in situ* studies of that comet and two others – comets Giacobini–Zinner (GZ) and P/Grigg–Skjellerup. The *Third International Sun–Earth Explorer* (*ISEE-3*) was diverted from its original position at the sun–earth L_1 point to a flyby of comet GZ. The spacecraft, now named the *International Comet Explorer* (*ICE*) enabled studies of that comet. The complement of scientific instruments, designed to study the solar wind plasma and fields, were also very capable of studying the cometary plasma and fields.

Earth orbiting spacecraft contributed significantly to the studies of comets. The *International Ultraviolet Explorer* (*IUE*) led to the first detection of S_2 from comet IRAS–Araki–Alcock. Images made with the *Hubble Space Telescope* led to the resolution of the nuclei of several comets including 19P/Borrelly. The latter result was confirmed by the *in situ* imaging of *Deep Space 1*.

The Halley Armada consisted of European, Japanese, and Soviet spacecraft. The spacecraft flew to the vicinity of Halley's comet and collected a wealth of *in situ* data on dust, composition of neutral and ionized gas, plasma phenomena, energetic particles, magnetic fields. Of special interest, data on the nucleus of comet Halley were obtained, showing it to be a monolithic object.

For more on observing small bodies and the zodiacal light, see Edberg and Levy (1994).

4

Tails

A bright comet tail stretching across the sky is a spectacular sight. While the nucleus is the source of all cometary phenomena, the tails visible to the naked eye are historically responsible for the interest in comets. In broad outline, there are two kinds of tails. Plasma tails are called type I and dust tails are called type II (or type III).

Figure 4.1 and Plate 4.1 show a comet with both a plasma and a dust tail. The type I or plasma tails are straight, have considerable fine structure, are dynamic and appear blue on color photographs; these tails are also referred to as ion tails. The type II or dust tails are usually curved, broad, relatively featureless and slowly changing; they appear yellow (or whitish yellow) on color photographs. Type III tails are dust tails that are strongly curved; the type III designation is not commonly used now.

There are also less obvious types of tails. A narrow sodium tail was discovered in comet Hale–Bopp. It was long and oriented close to the prolonged radius vector. Comets far from the sun may display tails composed of ice grains. In this chapter, our discussion will concentrate on the appearance of comet tails as displayed in remote-sensing observations, particularly wide-field imaging.

4.1 Dust tails and features

4.1.1 Morphology of dust tails

Dust tails have been observed to reach lengths of roughly 10^7 km. They are basically flat structures, based on observations made when the earth is near the orbital plane of the comet. Changes in shape or structure of a dust tail are usually slow. While the general appearance of a dust tail is a relatively featureless broad arc, some structure or fine structure does exist. The boundary or edge of the dust tail is sharper on the convex side (near the plasma tail) than on the concave side (away from the plasma tail). Two kinds of bands in dust tails are known. (1) *Synchronic*

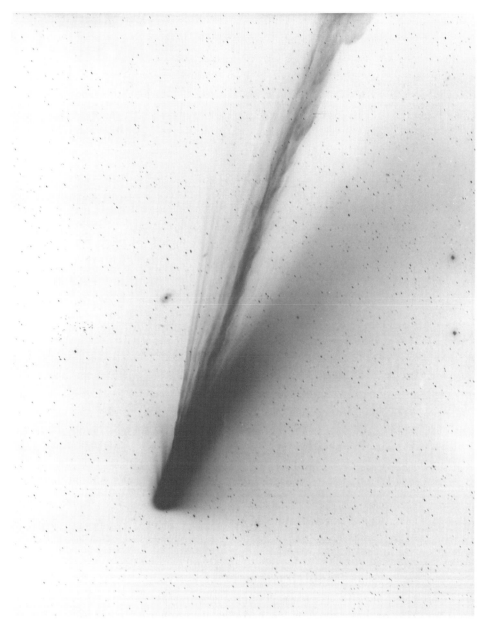

Fig. 4.1. Image of comet Mrkos on August 24, 1957. The structured tail (above) is the plasma tail, and the smooth tail (right) is the dust tail. (California Institute of Technology.)

bands point toward the nucleus and have orientations departing from the radius vector. (2) *Striae* are a series of narrow, parallel bands that do not point toward the nucleus. They are usually seen far from the nucleus.

Other features are the *dust trails*, narrow features extending over large arcs as detected in the infrared. *Sunward spikes* or *fans* are dust features that appear to extend toward the sun. Finally, we have the *neck-line structure*, a long, narrow feature extending in the antisolar direction.

4.1.2 Theory of dust tails

All of the dust features described appear to be consistent with dust grains released from the nucleus, primarily on the sunward side, as sublimation of the ices by solar heating occurs, followed by subsequent evolution of the spatial dust distribution as described below. Details of the calculations can be complex, but the basic approach seems sound.

Consider a dust particle liberated from the nucleus. It has an initial speed which depends on the details of the gas flow near the surface (section 5.3.1.5). The dust particles are accelerated by the drag of the outflowing coma gases from essentially zero to a few tenths of a kilometer per second. The nucleus itself is in orbit around the sun, but its mass is small and its gravitational field does not influence the motion of the dust tail particles. Because the dust tails indicate particle trajectories streaming away from the sun, forces other than solar gravity must be acting on these particles. This idea is introduced into the equations of motion by the parameter μ, the ratio of the net force on the tail particle to the gravitational force. The non-gravitational force is invariably repulsive. Its magnitude is given by $(1 - \mu)$, in units of the gravitational force. The value of $(1 - \mu)$ determines the form of the dust particle's orbit. For the usual case of the repulsive force being large enough to make the total energy positive, the particle's orbits are hyperbolas. Thus, if $(1 - \mu) < 1$, the solar gravitational force is diminished and the particle moves off along a hyperbola concave to the sun. For the special case of $(1 - \mu) = 1$, the effective gravitational force is zero and the particle moves along a straight line. If $(1 - \mu) > 1$, the effective force is repulsive and the particle moves off along a hyperbola convex to the sun. For dust particles, the repulsive force is solar radiation pressure (section 1.6.1).

If the dust particles all experience the same force and are emitted continuously from the nucleus, the shape of the tail is determined by $(1 - \mu)$. The dust tails on this simple picture are tangent to the radius vector at the head (nucleus) and the curvature of the tail increases for decreasing values of $(1 - \mu)$. Note that the observed shape of the tail is not the shape of the particle orbits. At a given point

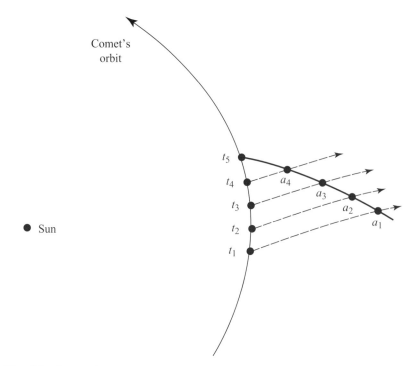

Fig. 4.2. Comparison between particle orbits and tail shape. Particles emitted at times t_1, t_2, t_3, and t_4 are at positions a_1, a_2, a_3, and a_4 at time t_5 to form the tail.

in time, the observed tail shape is traced out by the locations of dust particles previously emitted from the head. See Fig. 4.2.

The case just described is one limiting case of the Bessel–Bredichin theory started by Bessel at the time of the apparition of Halley's Comet in 1835 and extended by Bredichin. The syndyname or *syndyne* (same force) just described would be a good description of a dust tail composed of a small size range of particles emitted continuously. Sample syndynes for comet Arend–Roland are shown in Fig. 4.3.

The other limiting case marks the simultaneous emission of particles with a broad range of sizes or $(1 - \mu)$ values. This is a *synchrone* (same time). It has a rectilinear form and makes an angle with the radius vector that increases with time. Sample synchrones for comet Arend–Roland are shown in Fig. 4.3.

Until the 1950s, syndynes were regarded as the explanation for the shapes of dust tails, but there were problems. Osterbrock (1958) studied comets Baade (C/1954 O2) and Haro–Chavira (C/1954 Y1). These comets, which were observed at heliocentric distances between 4 and 5 AU, had essentially no plasma tail. The orientation of the dust tail lagged the radius vector by about 45°. However, brighter, less distant comets such as comet Mrkos had dust tails tangent to the radius vector near the

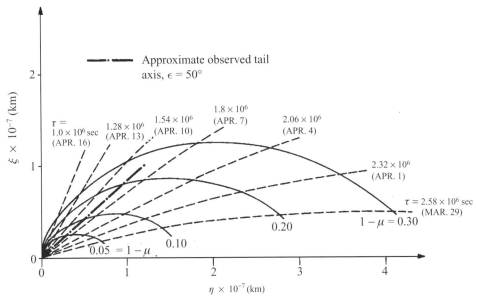

Fig. 4.3. Syndynes (solid curves) and synchrones (dashed curves) for comet Arend–Roland on April 27.8, 1957. The coordinate ξ is in the radial direction, and η is perpendicular to ξ directed opposite to the comet's motion and in the orbital plane. Dates given refer to the time of emission for the synchrones. (Courtesy of M. L. Finson and R. Probstein, Massachusetts Institute of Technology.)

head. This turned out to be caused by a drag force exerted on the dust tail by the streaming plasma in the plasma tail close to the head. In comet Mrkos, the dust and the plasma are decoupled within about 10^6 km of the nucleus. Belton (1965) found that the approximately 45° orientation held for all pure dust tails regardless of heliocentric distance. Thus, syndynes could not explain dust tails and synchrones could only match the observations some of the time (see section 4.2.4.1).

The solution to the dilemma was, in principle, fairly simple. Finson and Probstein (1968a,b) dropped the limiting cases of the Bessel–Bredichin theory and allowed particles of all sizes to be emitted continuously with the rate allowed to vary with time. Finson and Probstein applied their approach to comet Arend–Roland. They assumed that the initial speed was given by conditions in the expanding coma, namely, dust-molecular collisions close to the nucleus. Observations of expanding dust shells in the inner coma indicate a speed of ≈ 0.1 to 1 km s^{-1}, interpreted as the "initial speed" for the dust particles. Superposition of synchrones broadened by the initial speed and with a time-varying dust-ejection rate produced an excellent fit to the observations (Fig. 4.4).

The detailed calculation can be carried out two different ways: (1) calculate the surface density of dust in the tail for one particle size and then integrate over all relevant particle sizes (i.e., a differential syndyne approach); or (2) calculate the

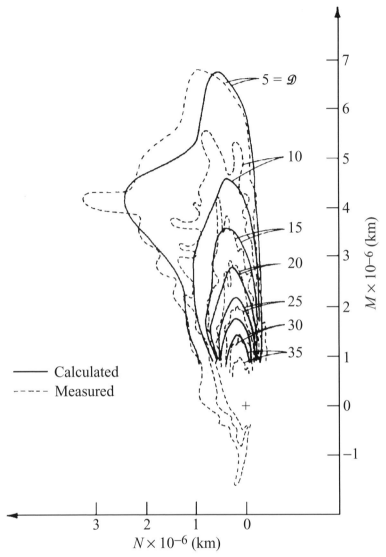

Fig. 4.4. Calculated and measured isophotes for comet Arend–Roland on April 27.8, 1957, showing the agreement. The calculations were not extended to reproduce the sunward spike, but, in principle, could have. (Courtesy of M. L. Finson and R. Probstein, Massachusetts Institute of Technology.)

surface density of dust in the tail for all particles emitted at one time and then integrate over all times of emission that contribute to the tail (i.e., a differential synchrone approach). Here we outline the differential syndyne approach.

To compare models with images of the dust tail, the model calculations need to yield the surface density of dust particles in the tail times the efficiency for light scattering, or in other words the intensity of the light as a function of position in the dusty tail. We can write

$$1 - \mu = \frac{F(\text{radiation})}{F(\text{gravity})}. \tag{4.1}$$

The radiation pressure has the form $F_{\text{rad}} \sim Q_{\text{pr}} a^2$, where Q_{pr} is the scattering efficiency for radiation pressure and a is the radius of the (assumed) spherical particles. Note that the radiation pressure (proportional to the solar radiation flux) and gravity vary as r^{-2}. Thus, $1 - \mu$ is independent of distance. Similarly, $F_{\text{grav}} \sim \rho_d a^3$, where ρ_d is the density of the dust particles. Hence, we have

$$1 - \mu = \frac{D}{\rho_d a}, \tag{4.2}$$

where

$$D = 0.6 \times 10^{-4} \, Q_{\text{pr}} (\text{g cm}^{-2}). \tag{4.3}$$

Here D contains various constants in the expression for the force of radiation pressure and gravity except for Q_{pr}.

The amount of scattered light is proportional to $(\rho_d a)^2 \, g(\rho_d a)$, where $g(\rho_d a)$ is the distribution function for dust grain size as a function of $(\rho_d a)$. Changing the independent variable to the more convenient $(1 - \mu)$ gives

$$(\rho_d a)^2 \, g(\rho_d a) \, d(\rho_d a) \propto (1 - u)^{-4} f(1 - \mu) d(1 - \mu), \tag{4.4}$$

where $f(1 - \mu)$ is the distribution function for particle sizes as a function of $(1 - \mu)$. A straightforward derivation shows that the surface density of dust in a differential syndyne tail is proportional to

$$\dot{N}_d f(1 - \mu) d(1 - \mu) \left[2 v_i \tau \frac{dx}{d\tau} \right]^{-1}. \tag{4.5}$$

The quantities are: $\dot{N}_d$ the rate of dust emission; v_i, the initial injection speed; τ, the time since the emission (varying along the syndyne); and x, the distance along the axis of the syndyne. Physically, $\dot{N}_d f(1 - \mu) d(1 - \mu)$ gives the effective surface density of scattering particles at a point. The terms $2 v_i \tau$ and $dx/d\tau$ give the reduction in surface density due to the dispersion in the lateral and longitudinal directions, respectively, with respect to the tail axis.

To carry out the integration of (4.5) over $(1 - \mu)$, three functions must be specified, namely, $\dot{N}_d(t)$, $f(1 - \mu)$, and $v_i(1 - \mu, t)$. Finson and Probstein initially assumed these functions and adjusted them to produce excellent agreement with the observations of comet Arend–Roland (Fig. 4.4). Numerical experiments showed that the solutions were essentially unique. The rate of dust emission peaked before perihelion. This result is consistent with the observation that comets approaching the sun for the first time are generally dusty and tend to be dustier before perihelion. The size distribution peaks at particle diameters $\sim 1\mu m$ for $\rho_d = 3$ g cm^{-3}. This is consistent with photometric and polarimetric data on dust tails. The total light emission from the dust tail yields dust emission rates of $\sim 10^{18}$ particles per second or 10^8 g s^{-1}. The gas emission rate can be estimated from the values needed to produce the empirically determined values of v_i. This gives an emission rate $\dot{N}_g$ of approximately 1.5×10^{30} molecules s^{-1}.

Finson and Probstein also applied their approach to comet van Gent (C/1941 K1), and Sekanina and Miller (1973) applied it to comet Bennett (C/1969 Y1). Jambor (1973) applied it to comet Seki–Lines (C1962 C1), an interesting case of a split tail; this case of the split-tail phenomenon was the result of a sudden burst of dust release. These investigations were successful. Not only could this approach explain the observations with sensible physics, it also provided information on the nature of the dust particles.

Nevertheless, the Finson–Probstein approach contained assumptions and simplifications that could limit its application. First, the efficiencies for radiation pressure Q_{pr} and scattering Q_{sc} are assumed to be constant, but actually vary with the size and nature of the particles. Second, the assumption that the particles are spherical is highly suspect. Third, the density ρ_d is not likely to be a constant. Fourth, there are details of the calculations that can lead to ambiguities. Fifth, we know that the dust emission is probably concentrated on the sunlit side of the nucleus (see the discussion of the images of the nucleus of comet Halley) and is unlikely to be isotropic.

Hence, some modifications were needed to this valuable theory and this has been done by Fulle (1989). In fact, Fulle reversed the approach by using sample grains and computation of their orbital motion to model the cometary dust tails as well as the dust loss rates and the time-dependent size distribution. The approach was applied to comet Seki–Lines (C/1962 III) and comet Kohoutek (1973 XII) with good results.

4.2 Other dust features

Once the dust has been released from the nucleus and moves beyond the inner coma, other observable tail features can be formed. These are briefly described in this section.

4.2.1 Sunward spikes and fans

The most famous sunward spike (or antitail) of the twentieth century was observed on comet Arend–Roland in April of 1957 (Larsson–Leander 1957).

Fig. 4.5. Comet Arend–Roland on April 22, 1957 showing the sunward fan. (Courtesy of R. Fogelquist, Bifrost Observatory.)

Figure 4.5 shows the fan on April 22, 1957 and Fig. 4.6 shows the spike on April 25, 1957. These "sunward" appendages do not really point sunward. The geometrical circumstances are shown in Fig. 4.7. The apparently normal tail points away from the sun and consists of smaller particles that were released from the nucleus recently. Larger or heavier particles would experience a small or nearly negligible repulsive force and, instead of being sent rapidly in the antisolar direction, are dispersed along or near the orbit of the comet. Of course, the historical association between comets and meteoroid streams supported this idea. Sekanina (1976) has shown that this picture is entirely plausible and has estimated diameters in the range 0.1 to 1.0 mm for the particles in the spike. When seen at small angles to the orbital plane, these particles form a fan. When seen nearly edge-on, they appear as a sunward spike or antitail. The thickness of the spike may not give a reliable estimate of the initial injection velocities because some of the material could be particles ejected months earlier and now approaching the orbital plane of the comet again.

Fig. 4.6. Comet Arend–Roland on April 25, 1957, showing the sunward spike. (Photo © UC Regents/Lick Observatory.)

Sunward spikes have been observed in many comets including comet Kohoutek (Fig. 4.8), and comet Halley in 1986 (Fig. 4.9). We note again that the apparently sunward appendages generally lie outside the orbits of the comets because the radiation pressure is a repulsive force. A large initial injection velocity could overturn this statement and the feature might not lie in the plane of the comet's orbit.

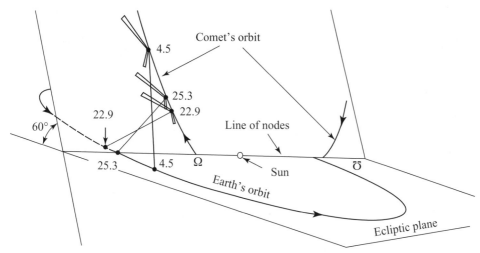

Fig. 4.7. Diagram showing the relative positions of earth and comet Arend–Roland in April and May 1957. On April 22.9, the earth is somewhat away from the comet's orbital plane, and the particles producing the anomalous tail appear as a fan-shaped appendage as shown in Fig. 4.5. The earth crossed the orbital plane of the comet on April 25.7. Thus, on April 25.3, the earth was quite close to the plane and the particles producing the anomalous tail appeared as a spike as shown in Fig. 4.6. By May 4.5, the earth had moved well away from the orbital plane. The anomalous particles were not seen as an apparently sunward appendage. (Adapted from a figure by N. B. Richter.)

A faint, sunward, spike-like feature was observed in comet Halley from late April to early June 1986 (Sekanina *et al.* 1987). This feature has nothing in common with the "normal" sunward spike or antitail observed just after perihelion (Fig. 4.9). It extends to a projected distance of about 700 000 km from the nucleus and appears to be produced by high particle ejection speeds. Because we see this feature only in projection, it could extend inside the comet's orbit. The grains are thought to be dielectric or slightly absorbing and much less than 0.1 μm in size.

Because the orientation of this spike-like feature does not line up with the neck-line structure (next section) thought to be in the comet's orbital plane, the orientation (which remained nearly constant for about 40 days) could be attributed to a property of the nucleus. A possible candidate is the plane perpendicular to the angular momentum vector of the nucleus. In essence, the centripetal acceleration in this plane from the rotation of the nucleus could assist the ejection of dust grains. Possibly, jets on the surface of the nucleus can eject a fraction of the dust at high speeds to form a relatively stable feature. But, this view is not universally accepted. Cremonese and Fulle (1989) found that the antitail and the neck-line structure (below) lined up to within 1° and consider the sunward extension as part of the neck-line structure.

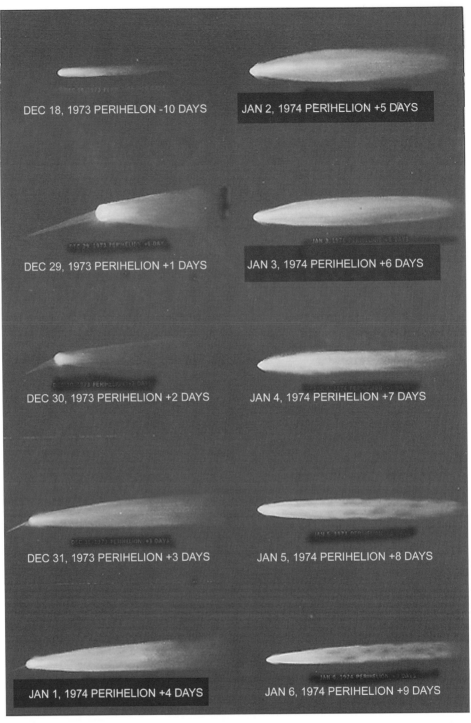

Fig. 4.8. Astronaut sketches of comet Kohoutek near perihelion as seen from *Skylab*. The sunward spike is clearly shown. (Courtesy of E. Gibson, NASA–Johnson Space Center.)

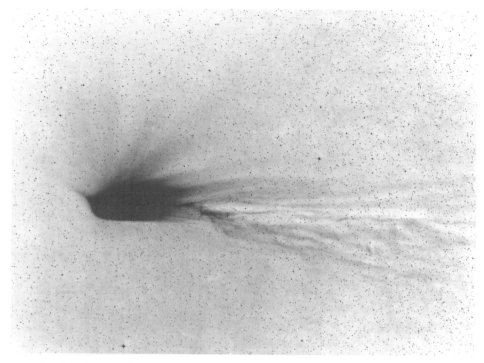

Fig. 4.9. Image of comet Halley on February 22, 1986, showing a sunward spike (left), synchronic bands (above), and a disconnected plasma tail (right). (Original plates taken with the UK Schmidt Telescope in Australia. Copyright © 1986, Royal Observatory, Edinburgh.)

4.2.2 The neck-line structures

The neck-line structures (NLS) are two-dimensional dust features that can be detected when the comet is post-perihelion and the earth is close to the comet's orbital plane. Basically, these structures result from spherical dust shells with low initial speeds that collapse onto the comet's orbital plane. This happens because the particles in the shells around the nucleus are in orbit around the sun and their orbits return to the orbital plane.

While papers on the neck-line structures had been in the literature for years, the concept outlined above became widely known with the Cremonese and Fulle (1989; also see Fulle 1987) analysis of the NLS in comet Halley. The NLS was a stable feature for comet Halley from early May through June of 1986 and one was recorded in comet Hale–Bopp (Fig. 4.10).

Another NLS is known for comet Bennett (C/1969 Y1). Understanding the NLS for comets contributes to knowledge of the dust distribution function. The results generally confirm the strong excess of large dust particles suggested by data from

Fig. 4.10. The neck-line structure in comet Hale–Bopp on June 6, 1997. (Image taken by G. Pizarro, European Southern Observatory.)

Giotto (McDonnell *et al.* 1987). Bear in mind, however, that there are significant differences in the details of the dust size–distribution function.

4.2.3 Dust trails

Infrared observations provide another approach for tracing the fate of large dust particles ejected at low velocities (perhaps meters per second) from the nucleus. Sykes *et al.* (1986) examined sky maps taken by the *Infrared Astronomical Satellite* (*IRAS*) in wavelength ranges from 12 to 100 μm and found many long, narrow trails of dust extending on the order of 50°. Several were found to coincide with the orbits of comets, but some could not be associated with the orbit of any known periodic comet. Bright dust trails were found for comets Tempel 2, Encke, and Gunn, and these comets exhibit trail material in front of and behind the comet's position. An *IRAS* image of comet Tempel 2 is shown in Fig. 4.11 and Plate 4.2. Estimates indicate that for comet Tempel 2 the particles ahead of the comet are at least one centimeter in radius and the particles behind the comet are at least one millimeter in radius. The time necessary to disperse the dust particles along the comet's orbit is estimated at several hundreds of years. Thus, these trails can be considered as the dust buildup from emission over many orbits.

Specific links between observed trails and known meteor streams have not been established. Still, the possibility that these dust trails are an intermediate step between comets and meteor streams – a meteor stream in the making – is suggestive. We discuss meteor streams in section 10.1.

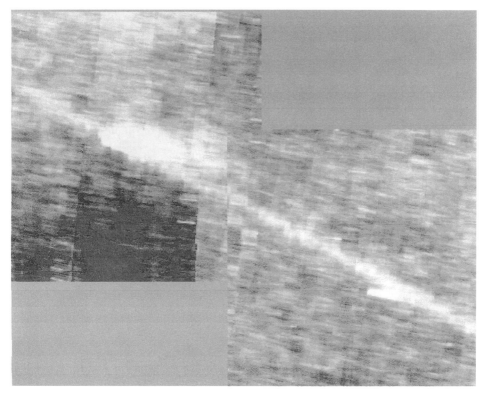

Fig. 4.11. *Infrared Astronomical Satellite* (*IRAS*) observations of comet Tempel 2 and its dust trail on September 6, 1983. The comet (large, fishlike object, left) consists of particles tens of microns in radius. The trail (the streamlike object passing through the comet and sloping downward, left to right) is composed of particles at least a centimeter in radius ahead of the comet (left) and at least a millimeter in radius behind the comet (right). (Courtesy of Mark Sykes, University of Arizona.)

4.2.4 Fine structure

Dust tails exhibit two kinds of reasonably common structure and we discuss these here.

4.2.4.1 Synchronic bands

Recall the discussion of synchrones in section 4.1.2. These are produced by the instantaneous emission or burst of dust particles of all sizes covering the full range of repulsive force magnitude. Synchrones have a rectilinear form, point toward the nucleus, and their angle with the radius vector increases with time. If there are multiple outbursts, multiple synchronic bands are expected.

This behavior was seen in comet West (C/1975 V1) and a splendid example was observed in comet Halley on February 22, 1986 (Fig. 4.9). The synchronic bands appear to be understood in terms of the relatively simple Bessel–Bredichin theory.

4.2.4.2 Striae

The striae or striations (Koutchmy *et al.* 1979) are a system of narrow, linear, and parallel bands that are found at large distances from the comet's head. Extensions of the striae intersect the radius vector on the sunward side of the nucleus. The striae appear at heliocentric distances greater than 1 AU and always after perihelion.

Striae were seen in several comets including comet West in 1976 (Sekanina and Farrell 1980) and comet Hale–Bopp in 1997. The striae in comet Hale–Bopp were seen for a long period of time, the prime time period being in March of 1997. The observations collected and analyzed by Pittichová *et al.* (1997) should advance our physical understanding.

Figure 4.12 shows the striae in comet Hale–Bopp. The formation mechanism is still an unsolved problem. Large particles ejected from the nucleus could fragment as suggested by Sekanina and Farrell (1980). If so, the striae could be synchrones starting at the location of the fragmentation rather than at the nucleus. The difficulty lies in finding a plausible mechanism that produces the fragmentation essentially simultaneously over substantial dimensions. Sekanina and Pittichová (1997) have investigated this problem and found that the fragmentation times cannot be distributed over a period more than about 2 to 3 days if they are to be consistent with the straie's width of about 150 000 km. An alternative mechanism has been proposed by Lamy and Koutchmy (1979) involving charged dust particles being organized by the solar wind's magnetic field. At present, the basic mechanism responsible for the formation of striae is simply not understood.

4.2.5 Icy tails

Some properties of the type II tails of distant comets such as Baade (C/1954 O2) and Haro–Chavira (C/1954 Y1) are consistent with being composed of icy water grains as proposed by Sekanina (1973, 1975). These particles could also be ice covered dust or dust with embedded ice, etc. Such particles if sufficiently large (at least 0.01 cm in size) would have very low values of $(1 - \mu)$ and could fit the observed tail orientations as discussed by Osterbrock (1958); see section 4.1.2. This idea gains strength from the general belief that the principal volatiles in a comet nucleus are water ices. As sublimation occurs, these grains would certainly be liberated from the surface of the nucleus. If this occurs between 2 to 3 AU, the sublimation could occur by water itself. If it occurs at larger distances, a substance more volatile than water, such as carbon monoxide or methane, is required to lift the grains off the surface. Sekanina (1973) has shown that the grains themselves

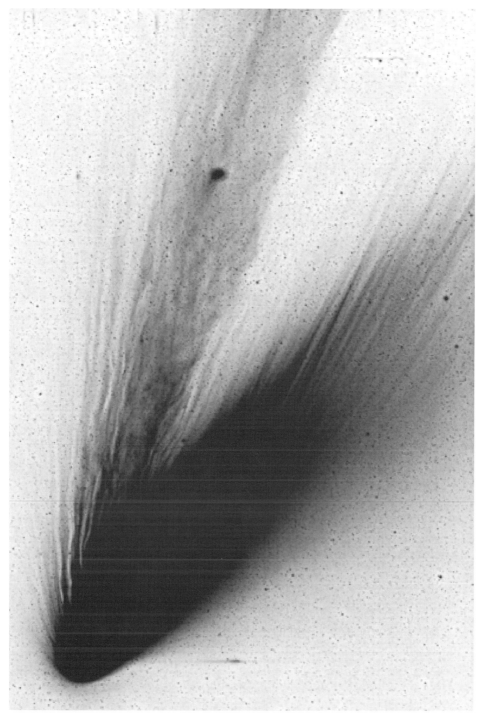

Fig. 4.12. Image of comet Hale–Bopp on March 17, 1997 showing well-defined striae in the dust tail (right). (Courtesy of Kurt Birkle, Max Planck Institute for Astronomy, Heidelberg, Germany.)

would survive sublimation for comets with perihelia near 4 AU or beyond. The number of comets to which this idea may apply is small and the total data set is not extensive. Still, this approach is reasonable and may apply to distant comets.

4.2.6 Unusual dust tails

We conclude the discussion of dust tails and their features by mentioning two examples of unusual tails that serve as a reminder that our understanding is far from complete.

The first is comet Ikeya–Seki (C/1965 S1), which had a perihelion distance on October 21.2, 1965, of only 0.0077 AU from the center of the sun. A sample image is shown in Plate 4.3. The spectra showed bright sodium emission and the color of the tail was yellowish near the head and bluish-green down the tail. The comet was still quite near the sun circa November 1, 1965, and the complex spiral structure remains a challenge to interpretation. The sodium emission (and yellow color) probably results from the intense heating of the nucleus and the dust grains by the solar radiation.

The second example consists of the small comets observed very close to the sun by *SOLWIND*, the *Solar Maximum Mission* (*SMM*), and the *Solar and Heliospheric Observer* (*SOHO*). At least 300 of these sungrazing comets are known and none are known to have survived perihelion passage intact. A sample sequence of images is shown in Fig. 4.13. The full understanding of a comet's approach to graze the sun, formation of a dust tail, and complete disintegration should provide valuable information on the dust particles and structure of the nucleus. This should lead to insights on the disruption history of these comets or on the formation of comets in general, and specifically these presumably small comets. Their orbits are mostly similar and many, but not all, are believed to be members of the Kreutz family of comets. For example, three comets discovered by *SOHO* (C/1997 H2, C/1997 L2, and C/1998 J1) are not members of the Kreutz group.

4.3 Sodium gas tail

An unexpected discovery in comet Hale–Bopp was a gas tail dominated by the element sodium (Na). Cremonese *et al.* (1997; see also Cremonese 1999) found a long, narrow sodium tail; it was almost 7° long and about 10 arcmin wide. This tail is shown in Fig. 4.14. Other observers (e.g., Wilson *et al.* 1998) also found a sodium tail, but not the same one reported by Cremonese *et al.* In fact, there are two sodium tails. The first is the narrow sodium tail illustrated in Fig. 4.14. The second is the diffuse sodium tail superimposed on the dust.

The large size of comet Hale–Bopp's nucleus (see section 7.3.1) undoubtedly contributed to the visibility of the sodium tails. Also contributing is the nature of the

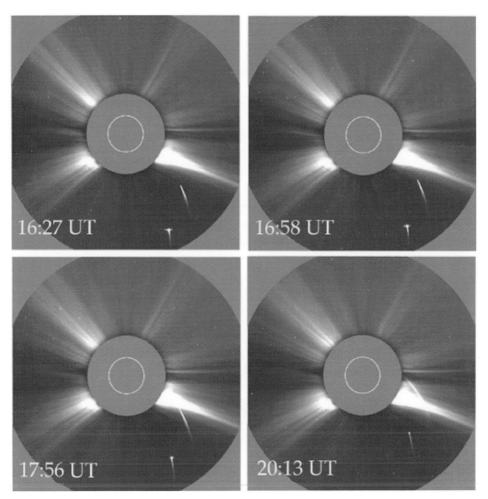

16:27 UT

16:58 UT

17:56 UT

20:13 UT

Fig. 4.13. A series of *SOHO* LASCO images taken on June 1, 1998, showing twin comets approaching the sun. The comets were not seen after passing the sun. (*Solar and Heliospheric Observatory* (*SOHO*). *SOHO* is a project of international cooperation between ESA and NASA.)

sodium atom. Its resonance scattering efficiency is one of the strongest in nature, meaning that it can be detected with a smaller column density than many other species. This strong interaction via the absorption and reemission of solar radiation in the sodium D lines produces a high radiation pressure and the long, straight tail. In the terminology of dust tails, the value of $(1 - \mu)$ was large.

Spectra taken along the narrow tail confirm that the composition is sodium. The source for the narrow sodium tail is thought to be the dissociation of sodium-bearing molecules in the inner coma. The exact molecule is yet to be determined. The source for the diffuse sodium tail is probably the dust, perhaps by photon sputtering. Comet Hyakutake also had a sodium tail as reported by Mendillo *et al.* (1998). Sodium tails may well be a common feature of comets.

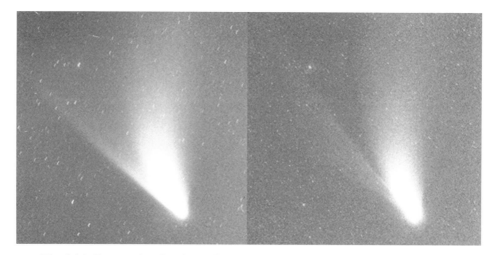

Fig. 4.14. Images showing the sodium tail in comet Hale–Bopp (April 1999). The left-hand image records the fluorescence emission (D-line) from the sodium atoms and clearly shows the thin, straight sodium tail. The traditional plasma and dust tails are shown in the right-hand image. (Courtesy of Gabriele Cremonese, Padova Astronomical Observatory, and the Isaac Newton Team.)

4.4 Plasma tails and features

4.4.1 Morphology of plasma tails

Plasma tails, also known as ion or type I tails, are straight and show considerable, ever changing, fine structure. These tails receive their name from the fact that spectroscopic studies show that they consist of ionized molecules, principally H_2O^+, OH^+, CO^+, CO_2^+, CH^+, and N_2^+. In images taken with photographic emulsions, the bands of CO^+ are dominant and responsible for the blue color.

Widths of plasma tails are in the range 10^5 to 10^6 km. The visible lengths routinely reach tenths of AU and occasionally exceed 1 AU. Of course, the lengths recorded depend on the sensitivity of the imaging system. Spacecraft measurements made by *Ulysses* have detected the signature of comet Hyakutake's plasma tail (in the magnetic field and ions) at distances of 550 million km (Jones *et al.* 2000; Gloeckler *et al.* 2000). While there may be special circumstances for these measurements(see section 6.4.2), they suggest that the structure of the plasma tail may extend far beyond the visible tail. The orientation of plasma tails near the head is as follows. The tails lag behind the comet's motion and make an angle of a few degrees with the prolonged radius vector.

Plasma tails are replete with fine structure. Perhaps the most basic form is the *tail ray* or *streamer*. These thin bundles of material form the plasma tail. They have radii in the range 2000 to 4000 km. Fine structure is characteristic of magnetized plasmas in nature, e.g., the solar corona. See Fig. 4.15 for cometary examples.

Fig. 4.15. Fine structure in cometary plasma tails. (a) The near-nucleus region of comet Morehouse in 1908, showing the tail streamers. (From the *Atlas of Cometary Forms*, by J. Rahe, B. Donn, and K. Wurm 1969. (b) The head of Halley's Comet on June 4, 1910, showing the tail streamers. (From the *Atlas of Comet Halley 1910 II*, by B. Donn, J. Rahe, and J. C. Brandt, NASA-488, 1986.)

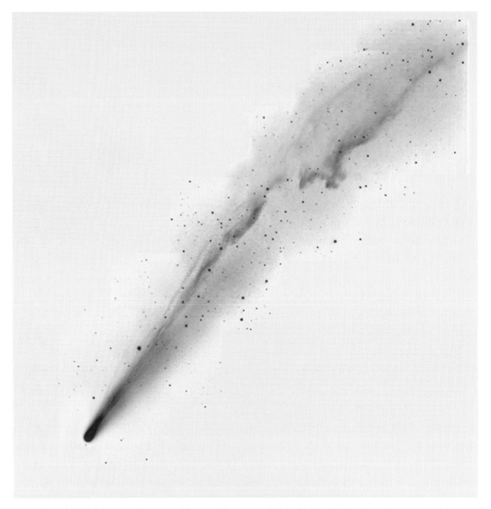

Fig. 4.16. Image of comet Ikeya–Zhang on March 11, 2002 showing extensive knots and kinks. (Courtesy of Gilbert Jones, Three Buttes Observatory, Arizona.)

Plasma tails also have an extensive variety of other fine structure that generally shows motion and acceleration away from the head. These are knots and kinks (Fig. 4.16). Typical velocities range from 10 km s^{-1} near the head to 250 km s^{-1} far from the head. Accelerations in terms of the quantity $(1 - \mu)$ are about 100, but with wide variations. A long-standing problem concerns the interpretation of these motions. Are they real physical motion or wave motion? Doppler measurements (e.g., Spinrad *et al.* 1994) show speeds of 30 km s^{-1} or more in the near-nucleus parts of plasma tails out to 3×10^5 km. The accelerations derived from the velocities indicate a systematic increase out to distances of about 4×10^5 km. Almost surely both bulk motions and waves exist in plasma tails.

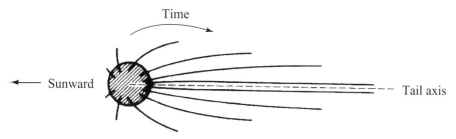

Fig. 4.17. Schematic diagram showing the lengthening and turning of tail streamers.

The origin of the CO^+ plasma and the evolution of tail rays have been studied observationally. Some tail streamers near the head are shown in Fig. 4.15. The CO^+ plasma (and presumably the other ionized molecular species) appears to originate very close to the nucleus on the sunward side. Jets of material seem to point initially sunward, but then bend back, lengthen, and merge into the main body of the tail. This process is illustrated schematically in Fig. 4.17 and with images of comet Kobayashi–Berger–Milon (Moore 1991) in Fig. 4.18. The process has been likened to a picture of a folding umbrella.

Plasma tails are normally found only in comets with heliocentric distance less than 1.5 to 2.0 AU. There are exceptions to this rule. In addition, comets that pass too close to the sun such as comet Ikeya–Seki (Plate 4.3) do not show plasma tails. The reason is not understood.

We conclude this section on morphology with a brief discussion of two topics that require an understanding of the solar wind and its interaction with comets. The first concerns the variation with solar latitude of the numbers of disturbances such as knots or kinks seen in plasma tails. Plasma tails in the equatorial or near-ecliptic region show a much higher level of disturbed appearance than plasma tails in the polar or far-from-the-ecliptic region. This cometary behavior reflects major differences between the equatorial and polar regions of the solar wind; see section 6.4.2.

The second concerns the most spectacular phenomenon in plasma tails, the disconnection event or DE. These events involve the complete severing of the plasma tail from the head and were described in the literature long ago by E. E. Barnard (1920). Figure 4.19 gives sample images over part of the twentieth century and Fig. 6.19 shows a time sequence of the great DE in Comet Hyakutake.

The disconnection event follows a general, systematic sequence illustrated in Fig. 4.20. Strong rays are often associated with the beginning of a DE sequence. There is general agreement that the DE is produced by the interaction of solar-wind features with the comet. There is less agreement on the specific feature or whether more than one kind of solar-wind feature is involved. We return to DEs in Chapter 6 where we treat the comet/solar-wind interaction.

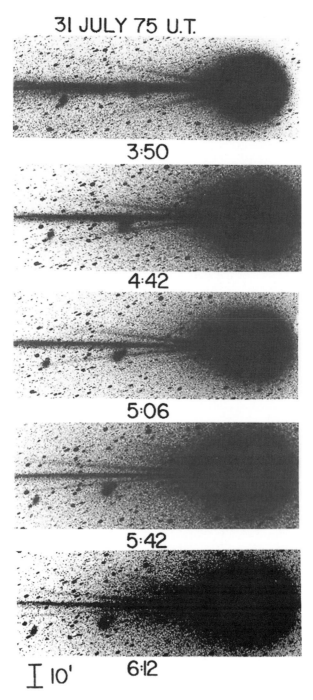

Fig. 4.18. Sequence of images of comet Kobayashi–Berger–Milon showing the turning of tail rays or streamers to the tail axis. (Photographs from the Joint Observatory for Cometary Research taken by E. P. Moore and K. Jockers. See Moore 1991.)

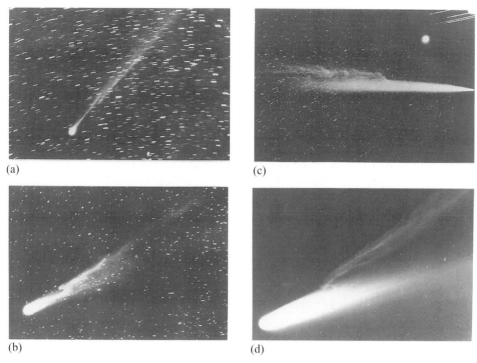

(a)

(c)

(b)

(d)

Fig. 4.19. Examples of disconnection events (DEs). (a) Comet Borrelly, July 24, 1903 (Yerkes Observatory photograph); (b) Halley's Comet, June 6, 1910 (Yerkes Observatory photograph); (c) Halley's Comet, May 13, 1910 (Lowell Observatory photograph); (d) Comet Bennett, April 4, 1970 (Photograph by K. Lübeck, Hamburg Observatory.) (Courtesy of M. B. Niedner, Jr., NASA–Goddard Space Flight Center, and J. C. Brandt, University of New Mexico.)

4.4.2 Introduction to the theory of plasma tails

The images of plasma tails show fine structure that strongly implies the existence of magnetic fields in comets. The radii of streamers in the range 2000 to 4000 km are a major constraint. For any reasonable value of the kinetic temperature in the tail, a CO^+ ion would move a distance comparable to the streamer diameter in at most a few hours. Thus, absent some confining mechanism, the thermal motions should wash out any fine structure. One possible way out is to consider a very low temperature for the tail plasma. The required temperature is so low, less than 1 K, that this possible explanation is clearly unreasonable.

The second possibility is that magnetic fields thread the plasma and constrain the CO^+ ions to move along a helical path. The radius would be given by the Larmor radius

$$r_L = \frac{v_\perp m}{qB}. \tag{4.6}$$

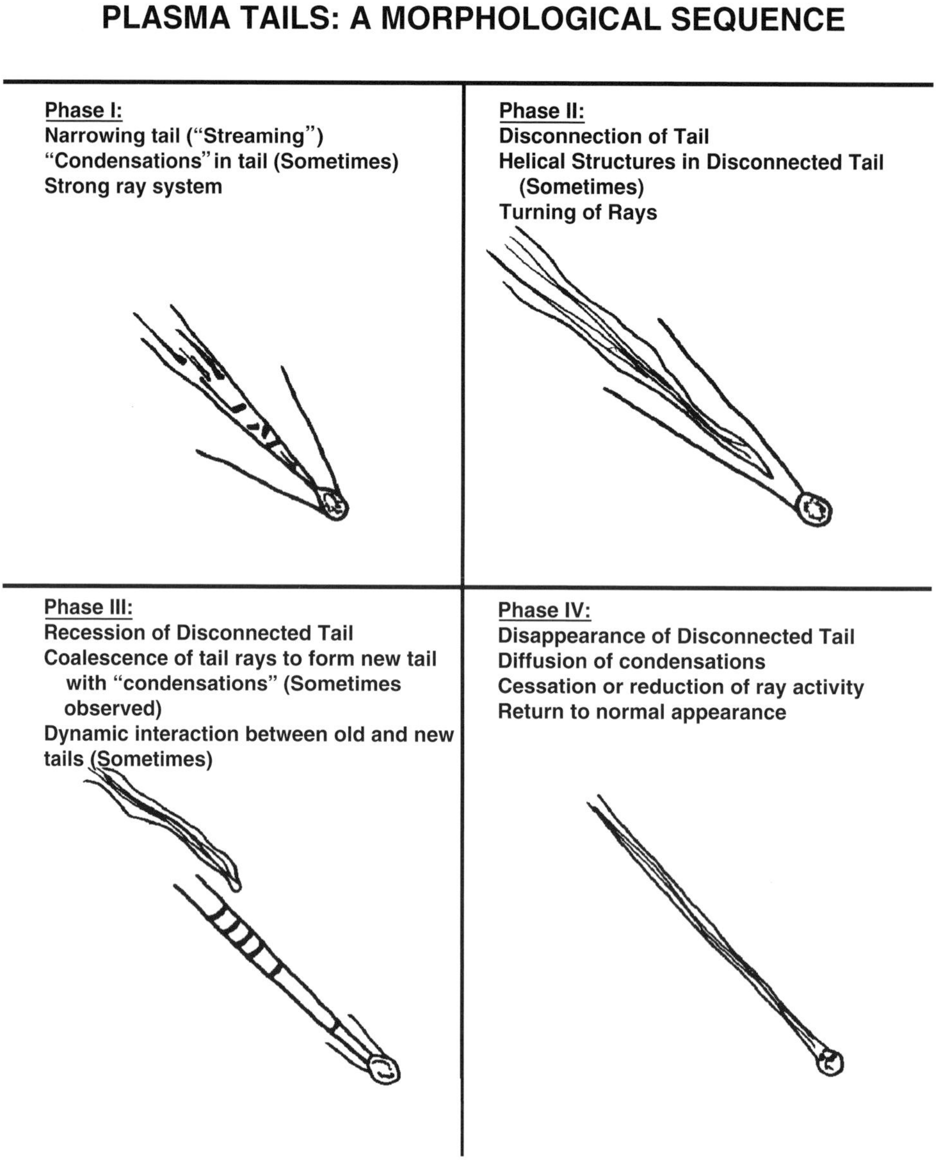

PLASMA TAILS: A MORPHOLOGICAL SEQUENCE

Phase I:
Narrowing tail ("Streaming")
"Condensations" in tail (Sometimes)
Strong ray system

Phase II:
Disconnection of Tail
Helical Structures in Disconnected Tail
   (Sometimes)
Turning of Rays

Phase III:
Recession of Disconnected Tail
Coalescence of tail rays to form new tail
   with "condensations" (Sometimes
   observed)
Dynamic interaction between old and new
tails (Sometimes)

Phase IV:
Disappearance of Disconnected Tail
Diffusion of condensations
Cessation or reduction of ray activity
Return to normal appearance

Fig. 4.20. Schematic of phases in a disconnection event (DE). (Courtesy of M. B. Niedner, Jr., NASA–Goddard Space Flight Center, and J. C. Brandt, University of New Mexico.)

In (4.6), $v_\perp$ is the component of the ion velocity perpendicular to the magnetic field lines, m is the mass of the ion, q is the charge in electromagnetic units, and B is the magnetic field in gauss. If we assume a $v_\perp$ corresponding to the thermal speed for 10^4 K and $B = 5 \times 10^{-5}$ gauss (or 5 γ), a value comparable to the quiet solar-wind field, we find $r_L \sim 10^2$ km. Almost any set of plausible parameters yields a value

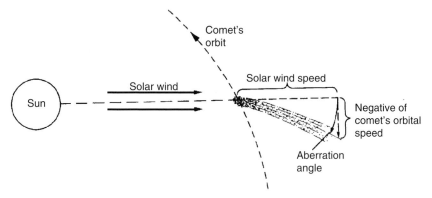

Fig. 4.21. Schematic showing the aberration effect and the orientation of plasma tails. The comet's motion through the solar wind causes the tail to lag behind the prolonged line from the sun to the comet.

of r_L much less than the observed streamer diameters. Thus, the plasma in comet tails is a magnetized plasma, an important concept for development of theory.

Evidence that the solar wind was important came from Biermann's (1951) interpretation of the orientations of plasma tails as measured by Hoffmeister (1943). The orientations lagged the prolonged radius vector in the sense opposite the comet's motion as shown schematically in Fig. 4.21. This orientation is simply explained if the plasma tail acts as a wind sock in the solar wind. The lag or aberration angle is given by

$$\tan \varepsilon = \frac{V_\perp}{w},\tag{4.7}$$

where $V_\perp$ is the component of the comet's orbital velocity perpendicular to the radius vector and w is the radial solar-wind speed. The aberration angle ε is about $5°$ and corresponds to a w of about 450 km s^{-1}. The lag is produced by the comet's motion through a medium that produces a component of the wind seen by a hypothetical observer riding on the comet.

The aberration effect gave very strong early evidence for the existence of the solar wind. Biermann's (1951) investigation is widely credited as being the discovery of the solar wind. This work emphasized that understanding the solar wind was necessary for a sound physical picture of the plasmas in comets.

Alfvén (1957) showed how the capture of the solar-wind magnetic field by comets could naturally explain many observed features. Figure 1.16 shows the basic picture. If the solar wind with its frozen-in magnetic field (see section 6.1) encounters ionized molecules from a comet, these ionized molecules cannot cross the field lines; they spiral around the field and, in essence, are attached to the field lines. Thus, the field lines near the comet are "loaded down" with the heavy ions, but the

Fig. 4.22. Comet Humason on August 6, 1962. Photographed by R. Rudnick and C. Kearns who also noted the spiral structure. The comet was more than 2.5 AU from the sun. (California Institute of Technology.)

field lines well away from the comet are not. This causes the field lines to wrap around the comet and produce the ray folding discussed just above. Of course, this introduction of the magnetic field into the cometary plasmas naturally explained the narrow, straight streamers that make up the plasma tails.

The basic picture proposed by Alfvén (1957) has been verified by spacecraft sent to comets and forms the framework for our understanding of plasma tails. The full comet/solar-wind interaction is complex and we return to this subject in Chapter 6.

4.4.3 Unusual plasma tails

We conclude the discussion of plasma tails with several examples of unusual behavior or circumstances. These remind us of facts to always bear in mind or exceptions to general rules that serve to illustrate the bounds of our knowledge.

Comet Humason (C/1960 M1) showed strong CO^+ at distances out to 5 AU, well beyond the normal distance for plasma tails of 1.5 to 2.0 AU or smaller heliocentric distances. Comet Humason (Fig. 4.22) also showed an extremely turbulent or disrupted appearance. The observations of comet Humason remain a challenge to our understanding of plasma tails generally, and, specifically, to the mechanism responsible for producing the ionized molecules.

The observations of comet Bradfield (C/1979 Y1) reinforce the importance of the role of the solar wind in determining plasma tail behavior and show just how rapid changes can be. Figure 4.23 shows a time sequence of images of comet Bradfield

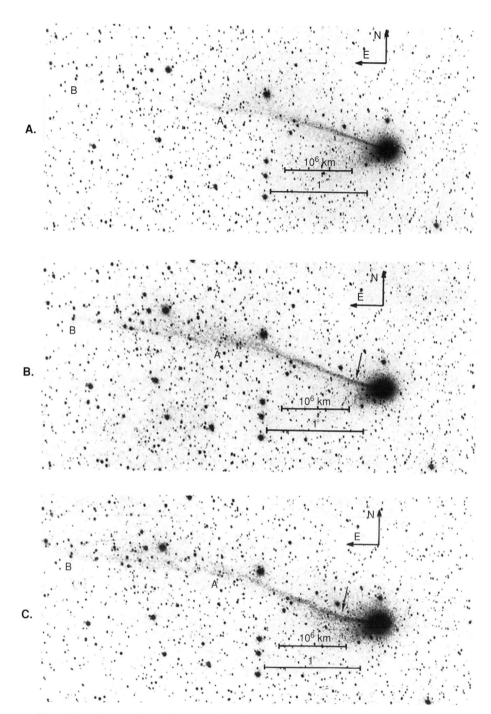

Fig. 4.23. Observations of comet Bradfield on February 6, 1980 showing a very rapid turning of the plasma tail. The mid-exposure times are: (a) $2^h32^m30^s$; (b) $2^h48^m00^s$; (c) $3^h00^m00^s$. The position angle of the inner tail (marked by arrow) turns by $10°$ in 27.5^m while the outer tail (segment marked A–B) remains nearly constant. (Joint Observatory for Cometary Research photographs.)

on February 6, 1980 (Brandt *et al.* 1980). The images cover only 27.5 minutes in time and the plasma tail has changed orientation by 10°. This corresponds to a turning rate of about 22° per hour. The event was probably due to a solar-flare generated change in the direction of the solar wind passing over the comet. This event shows the need for closely spaced observations for understanding cometary plasma tails.

Finally, we mention comet Hale–Bopp (Plate 4.1). This well-observed comet has the distinction of being the largest and most active comet seen in modern times. The observations involve the latest techniques applied over a wide range of heliocentric distances. Studies of comet Hale–Bopp are extraordinarily important and are rivaled only by the worldwide campaign and the armada of spacecraft sent to study comet Halley in 1985–1986. Of particular interest to plasma tails is the large inclination of comet Hale–Bopp's orbit; it was close to 90° and the comet sampled the environments of the equatorial and polar solar-wind regions.

4.5 Summary

Tails are the striking features of comets and come in three types. (1) Dust tails consist of particles liberated from the nucleus and pushed anti-sunward by solar radiation pressure. They usually change slowly, but exhibit fine structure. Sometimes, sunward spikes and fans and tailward neck-line structures are seen. (2) Recently, tails of sodium have been discovered. (3) Plasma tails consist of molecular ions and electrons pushed anti-sunward by the solar-wind interaction. They exhibit fine structure and can change rapidly. The most spectacular change is the disconnection event (DE), where the entire plasma tail detaches from the head and a new tail forms.

5

Coma and related phenomena

5.1 Introduction

The coma is composed of neutral radicals and molecules and dust particles in a more or less spherical volume centered on the nucleus. Over a century of remote sensing observations, augmented by the spectacular *in situ* data collected by the spacecraft sent to comets Giacobini–Zinner and Halley, have provided a wealth of information about the composition of the coma. Ultimately, all the material in the coma arises from the nucleus. But, by the time we observe the gas, it has been transformed by the solar UV and a variety of gas-phase reactions into many daughter products. There is also a source of neutral species within the coma itself. This source is undoubtedly due, at least in part, to dust grains that produce neutral species by sublimation or outgassing. These species then may be transformed further by gas-phase reactions. One of the major objectives of cometary science is to work backward through the maze of reactions to ascertain both the parent molecules that make up the volatile fraction of the nucleus as well as the nature of the dust.

In Chapter 1, we described the Delsemme and Swings clathrate–hydrate model of the nucleus, where the parent molecules are entrapped in a crystalline cage made up of six H_2O molecules. It is quite clear, now, that this model is not correct. In fact, the nucleus is made up of amorphous water ice (see also section 9.3.1). Impurities and small dust particles are embedded in the nuclear amorphous ice. When solar visible and IR radiation heats the nucleus, the ice sublimates releasing H_2O molecules, and other molecules frozen in the ice. In the clathrate–hydrate case, the total production rate of all other molecules should be one-sixth of the production rate of water molecules, if all the ice is a clathrate, and all the cavities are full. When *Giotto* flew through the coma of comet Halley, it found that about 1 in 20 of the gas molecules was not water. The volatiles are not affected by the small gravitational field of the nucleus, and flow away, dragging the dust along.

5.2 Hydrogen–hydroxyl cloud

The huge hydrogen clouds now known to surround comets were predicted in 1968 by Biermann (1968). He assumed a total production rate of 10^{30} to 10^{31} molecules s^{-1} and, based on the ionization time scale for neutral hydrogen, estimated the cloud radius at approximately 10^7 km. In the region where both destruction of hydrogen by ionization and production of hydrogen by dissociation of parent molecules can be neglected, the density can be approximated by

$$n_H \approx \frac{Q_H}{4\pi R^2 v_H} \text{(atoms cm}^{-3}), \qquad (5.1)$$

where Q_H is the hydrogen production rate in atoms per second, R is the cometo-centric distance, and v_H is the average outflow speed of the hydrogen. The outflow speed was assumed to correspond to the speed of mean thermal energy for a gas with $T = 2000$ K, a value chosen by analogy with observations of the earth's daylight-side hydrogen inner coma. (Note that the temperature of the hydrogen can be significantly different from that of the parent water, which sublimates at 230 K.) This yields $v_H \approx 6$–7 km s^{-1}. If we also take $Q_H = 4 \times 10^{30}$ s^{-1}, (5.1) becomes

$$n_H \approx 5 \times 10^5 \left(\frac{10^{10} \text{cm}}{R} \right)^2 \text{(cm}^{-3}). \qquad (5.2)$$

In the region of validity, the column density along a ray that passes the nucleus at a distance R varies as

$$N_H \approx \frac{16 \times 10^{23}}{R(\text{cm})} \text{(atoms cm}^{-2}). \qquad (5.3)$$

The value for the absorption cross section at Lyman α of the cometary hydrogen gas should be $\approx 10^{-13}$ cm^2. Hence, in this case, the gas has an optical depth unity where $N_H \approx 10^{13}$ atoms cm^{-2}, or at $R \approx 2 \times 10^6$ km. Biermann also found a large number of hydrogen atoms outside of this quoted distance. Thus, he concluded that "comets should, therefore, be very conspicuous objects in Lyman α." The total extent of the hydrogen cloud follows from the lifetime of hydrogen for ionization τ_H (due to photoionization and charge transfer) of 10^6 s (at 1 AU) and the speed v_H of ~ 10 km s^{-1} and is $\sim 10^7$ km. The radiation pressure due to the solar Lyman α was found to be large enough to expect considerable asymmetry in the outer shape of the hydrogen cloud. Thus, Q_H, v_H, τ_H and the effects of radiation pressure need to be specified in order to calculate a model of cometary hydrogen distribution.

5.2.1 Discovery

In Chapter 3, we described briefly the first ultraviolet observations of several comets made from above the earth's atmosphere – with *OAO-2*, *OGO-5* – in the 1970s.

The observations clearly showed the huge Lyman-α emission cloud surrounding the comets (Fig. 5.1). In addition, spectral scans of comet Bennett (Fig. 5.1b), revealed hydroxyl (OH) and oxygen (O I) emission as well as Lyman α. It is clear today that H_2O is the parent molecule for H, OH, and O; the O and some of the H would result from the dissociation of OH.

The terrestrial neutral hydrogen above the *OAO-2* spacecraft had an optical depth in Lyman α of approximately 0.6 and a width (due to the thermal doppler effect) corresponding to a temperature of approximately 1000 K. The relative velocity between the earth and comet Tago–Sato–Kosaka varied in such a way that the cometary Lyman-α emission line moved in wavelength across the telluric absorption as the comet passed perigee. Analysis of the decrease in observed Lyman-α emission gives a doppler width corresponding to a temperature of 1600 K for the cometary hydrogen.

5.2.2 Physics

The central part of the cometary hydrogen clouds is optically thick and detailed radiative transfer treatments are necessary for the interpretation of the multiple scattered radiation. Monte-Carlo investigations have been carried out and the results applied to comet Bennett. Also, note that the region of maximum Lyman-α intensity may not coincide with the cometary nucleus. Treatment of the outer part of the hydrogen cloud is relatively simple because only single scattering is involved. Much of our quantitative data on the hydrogen cloud refers to the outer regions and follows the general outline given below.

The model of the hydrogen cloud we shall describe is the fountain model as generalized by Haser (1957) to include the effects of solar radiation pressure, lifetimes of molecules, and different velocity distributions for the outflowing molecules. We stress again that this model strictly applies only to the optically thin outer region, typically beyond cometocentric distances of 10^6 km. To proceed, we assume an isotropic point source that produces neutral hydrogen atoms at the rate Q_H. These atoms leave the source with speeds corresponding to a Maxwellian distribution, namely,

$$f(v)dv = \frac{4v^2}{\pi^{1/2}v_0^3}\exp[-(v/v_0)^2]dv. \tag{5.4}$$

Here, v_0 is the most probable speed that is related to the mean speed by

$$\langle v \rangle = \frac{2}{\pi}v_0. \tag{5.5}$$

The quantity $\langle v \rangle$ is usually identified with the outflow speed of the hydrogen atoms v_H. Neutral hydrogen atoms are destroyed (ionized) by photoionization or charge

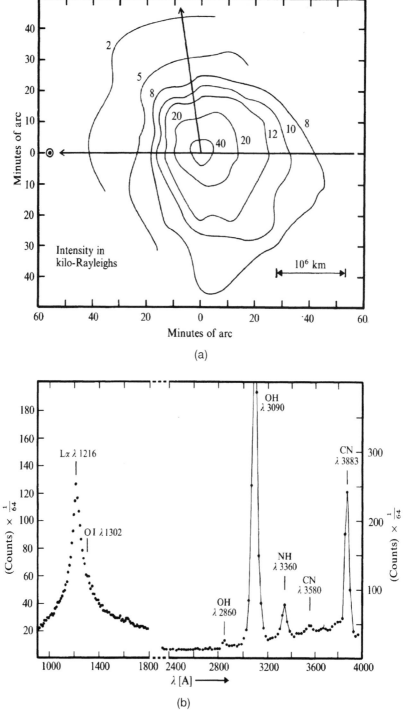

Fig. 5.1. Observations of comet Bennett from the *Orbiting Astronomical Observatory (OAO-2)*. (a) Lyman-α isophotes on April 16, 1970. The arrow points toward the sun. (b) Spectral scans. (Courtesy of A. D. Code, T. E. Hook and C. F. Lillie.)

exchange with solar-wind protons. At 1 AU the photoionization lifetime is $t_{rad} = 1.4 \times 10^7$ s and the solar wind charge-exchange lifetime is $t_{sw} = 3 \times 10^6$ s. The total lifetime of a hydrogen atom, t_H, is given by

$$t_H = \left(t_{rad}^{-1} + t_{sw}^{-1}\right)^{-1}, \tag{5.6}$$

or $t_H = 2.5 \times 10^6$ s. This value is dominated by the solar wind charge-exchange lifetime. At distances other than 1 AU the lifetime can be calculated using the r^{-2} law. In practice t_H is an empirically determined parameter.

Additional assumptions used in the calculations are: Q_H is constant, phase effects are negligible, and resonance scattering in Lyman α is isotropic. Then, the emission from the optically thin parts of the hydrogen cloud is given by

$$4\pi I_c = \left(\frac{\pi e^2}{m_e c}\right) f N F_0. \tag{5.7}$$

The intensity $4\pi I_c$ is usually quoted in rayleighs (10^6 photons $(cm^2 \, s)^{-1}$), the oscillator strength $f = 0.42$, $\pi e^2/m_e c$ are the usual atomic parameters ($e =$ elementary charge, $m_e =$ electron rest mass, $c =$ speed of light in vacuum), N is the column density, and F_0 is the solar Lyman-α flux at 1 AU estimated at 3.2×10^{11} photons $(cm^2 \, s\text{-}Å)^{-1}$. This flux gives a radiation pressure due to Lyman α of $b = 0.57$ cm s^{-2}, which is nearly equal to the solar gravity of 0.6 cm s^{-2}.

Haser's density integral can be evaluated as follows. We introduce a coordinate system with the origin in the nucleus, the sun in the direction of the negative z-axis, and the earth in the direction of the positive x-axis. The space density near the nucleus is given by an equation analogous to (5.1), that is,

$$n_H(x, y, z) = \frac{Q_H e^{-t/t_H}}{4\pi v_H(x^2 + y^2 + z^2)}, \tag{5.8}$$

where t is the time of travel. Away from the nucleus, the atoms are accelerated in the positive z direction. The effect is a density distribution with symmetry around the z-axis rather than the spherical symmetry of (5.8) and (5.1). The trajectories of the hydrogen atoms are parabolas and, for each energy, a point in the hydrogen inner coma can be reached via two separate trajectories. Thus, two components, denoted by $\pm$ in the following equation, must be added to obtain the total density. The result is

$$n_H(x, y, z) = \frac{Q_H}{8\pi v_H} \left\{ \left[a \pm \left(x_0^2 - x^2\right)^{1/2}\right]\left(x_0^2 - x^2\right)^{1/2}\right\}^{-1}$$

$$\times \exp\left\{ -\frac{1}{t_H}\left(\frac{2}{b}\right)^{1/2}\left[a \pm \left(x_0^2 - x^2\right)^{1/2}\right]^{1/2}\right\}, \tag{5.9}$$

where

$$a = z + \frac{v^2}{b},$$ (5.10)

and

$$x_0 = [a^2 + (z^2 + y^2)]^{1/2},$$ (5.11)

Recall that b is the acceleration due to radiation pressure and that v is the initial velocity of the hydrogen atoms. The upper bound to the integration is x_0, which is determined by the maximum value of v considered (see (5.10) and (5.11)); note that x_0 is a function of y and z and forms a rotational parabaloid about the x-axis. The column density is then determined by twice the integral from $x = 0$ to $x = x_0$ or

$$N(y, z) = 2 \int_0^\infty f(v) \int_0^{x_0} n_H(x, y, z) dx \, dv.$$ (5.12)

Here, $f(v)$ is obtained from (5.4) or its equivalent under other assumptions, and $n_H(x, y, z)$ is composed of the two parts of (5.9).

Integrations such as these were carried out by Bertaux *et al.* (1973) and by Keller (1973). The solutions are determined by empirically varying Q_H, v_H, and t_H until a fit to the isophotes is obtained. Sample results are shown in Fig. 5.2. An average hydrogen atom traveling sunward with initial kinetic energy per unit mass $(v_H^2/2)$ against the solar radiation pressure b has a speed that decreases to zero when the work done $Z_{\text{max}}b$ equals the initial kinetic energy. Therefore, we have

$$Z_{\text{max}} = v_H^2/2b$$ (5.13)

Thus, the sunward extent of the hydrogen cloud and the isophotes on the sunward side should be determined primarily by v_H, and empirically this is found to be true. Hence, v_H is first determined from the sunward or forward intensity profile, and then t_H is determined from the antisolar or backward intensity profile. The intensity profile perpendicular to the radius vector can be used as a check on consistency. The fits to the observed profiles are not perfect, but they are certainly adequate to determine values of Q_H, v_H and t_H with reasonable accuracy. The observed intensity profiles require a distribution of initial injection speeds (such as (5.4)) and are not compatible with monokinetic injection, which we would not expect on physical grounds.

The t_H values determined are close to the values expected on theoretical grounds (quoted above). The empirically determined values of Q_H are consistent with values determined by other independent approaches. H. U. Keller (1973) has applied this

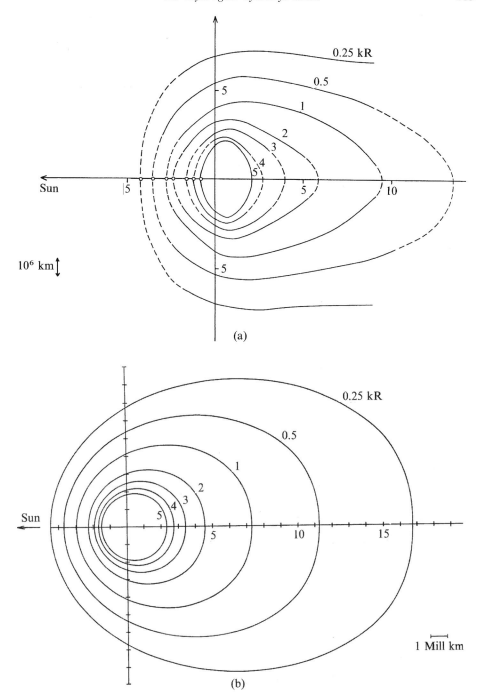

Fig. 5.2. Comparison of observed and computed Lyman-α isophotes for comet Bennett, April 1, 1970. (a) Observed by J.-L. Bertaux and J. Blamont. (b) Calculated. (kR = kilo-Rayleigh). (Courtesy of H. U. Keller, Max-Planck-Institut für Aeronomie.)

Table 5.1 *Hydrogen production rates in comet Bennett*
(C/1969 Y1)

r (AU)	v_H (km s^{-1})	Q_H (atoms s^{-1})	t_H (s)
0.61	7.9	1.2×10^{30}	2.5×10^6
0.62	7.9	1.0	2.5
0.70	7.9	0.92	2.0
0.73	7.9	0.79	2.0
0.78	9.0	0.77	2.5
0.80	7.9	0.69	2.0
0.82	9.0	0.79	2–2.5
0.86	7.9	0.80	2.0

Table 5.2 *Hydrogen production rates in selected comets.*
At or reduced to 1 AU

Comet	Q_H (10^{29} atoms s^{-1})
Kohoutek (C/1973 E1)	0.25–0.4
Tago–Sato–Kosaka (C/1969 T1)	8
2P/Encke	0.03
West (C/1975 V1)	5
1P/Halley (1982 U1)	8 (post-perihelion)

technique to comet Bennett and has found the results shown in Table 5.1. These results are consistent with a variation of Q_H with heliocentric distance as $r^{-1.3}$, with an error in the exponent of about ± 0.5. Other comets show variations in the range $r^{-1.3}$ to r^{-2}. Compare these results with those obtained for other comets as given in Table 5.2.

Additional data concerning the cometary environment can be obtained from the Lyman-α intensity contours:

1. The curvature of the extended atomic-hydrogen tail gives a value for the radiation pressure using the standard syndyne formulas. This method uses only the positions of intensity maxima, and hence it is not subject to the usual uncertainties associated with absolute calibrations. The values found are compatible with the values quoted here, but are somewhat higher.

2. Because the lifetime t_H is determined primarily by the proton flux in the solar wind, its determination provides an opportunity to monitor the proton flux. A possible variation was found: The solar-wind flux may be higher at solar latitudes poleward of $\pm 45°$. As we will see, this conclusion doesn't match the observed *Ulysses* results. The flux is actually lower at the higher latitudes versus the equatorial value (see section 6.2).

The theory of the coma and observational checks indicate that the overall outflow speed from a nucleus should be less than 1 km s^{-1}($\approx$0.5 km s^{-1}). This speed refers to the region relatively near the nucleus where collisions between atoms and molecules are important, that is, a thermalization region. If some dissociation occurs in the collision region, the outflow speed could be increased to 2 or 3 km s^{-1}, but probably no more. Thus, the observed v_H of 8 km s^{-1} is probably related not to details of the flow but to the energy budget of the photodissociation reaction that produces the hydrogen, outside the thermalization region. If H_2O were the primary parent molecule, a speed of 19 km s^{-1} would be expected, and clearly this value was too high. Other candidates for the key parent molecule were OH or other hydrocarbons. Because OH can be produced by the photodissociation of H_2O, it became the likely candidate. In this picture the hydrogen atoms produced by the photodissociation of H_2O are thermalized near the nucleus. The additional hydrogen produced by the subsequent photodissociation of OH would originate outside the thermalization region and hence could reach the outer parts of the hydrogen cloud where the 8 km s^{-1} value for v_H is measured. Observations of the OH intensity in comet Kohoutek have apparently confirmed the validity of this general model.

Observations have been made of several comets in the radio lines of OH at 1665 and 1667 MHz. Both lines were observed in comet West (C/1975 V1) and comet Kohler (C/1977 R1). The line intensity ratio was observed to be consistent with the Local Thermodynamic Equilibrium (LTE) value $I_{1667}/I_{1665} = 0.8$. Both comets exhibited an extended OH cloud, which was 5.5 arcmin in radius for comet West and 7 arcmin for comet Kohler (Despois *et al.* 1981). More detail on the theory of H and OH emission can be found below in section 5.3.1.2. Note that LTE assumes that the distribution of energy among kinetic and internal degrees of freedom is determined by a local temperature.

5.3 Coma

5.3.1 Observed structure, morphology and composition

The coma is composed of neutral molecules and dust particles in a more-or-less spherical volume centered on the nucleus. Molecules in the gaseous coma that have been identified throughout the electromagnetic spectrum are listed in Table 5.3. These molecules are the daughter products of parent molecules that have their origin in the amorphous ices of the nucleus.

Coma sizes, which range up to 10^5 or 10^6 km, typically reach a maximum size when the comet is located at a heliocentric distance of between 1.5 and 2.0 AU. As most comets approach nearer the sun than this distance, their comas are observed to contract. The comas are apparently quite small at large distances from the sun,

Table 5.3 *Chemical species observed by remote spectroscopy in comets*

H	C	O	S	CO_2	HDO	CHO	DCN
HNC	CO	CS	NH	OH	C_2	$^{12}C^{13}C$	
CH	$H_4C_2O_2$	^{13}CN	$H^{13}CN$	OCS	SO_2	S_2	SO
C_3	NH_2	H_2O	H_2	C_2H_2	H_2S	H_2CS	HNCO
CH_4	HCO	CN	HC_3N	Na	NH_3	H_2CO	HCN
CH_3OH	CH_3CN	HC_3N	NH_2CHO	C_2H_6			
C^+	CO^+	CH^+	CN^+	HCO^+	CO_2^+	H_2O^+	H_2S^+
N_2^+	H_3O^+						

that is, at distances greater than 3 or 4 AU. As a matter of terminology, the coma material does not include tail ions passing through.

As we pointed out in Chapter 1, F. W. Bessel carried out visual observations of Halley's Comet at the 1835 apparition. He saw fine structure near the nucleus consisting of jets, rays, fans, cones, and other forms. He concluded that material was ejected sunward from the nucleus, then pushed away from the sun by a repulsive force (Fig. 1.8). Similar structures have been seen in numerous other comets. For instance, Fig. 5.3 shows monochromatic images of comet Hale–Bopp in the continuum at 5270 Å and C_2 images at 5130 Å (Laffont *et al.* 1997). In October 1996, the comet showed clear dust jets extending roughly 3×10^5 km from the nucleus. The gas jets in C_2 were much less visible. Between October 1996 and January 1997 the comet was not observable. But, starting in February and continuing through April 1997, when the comet was once again observable, the monochromatic images were dominated by arcs, which were visible in both the continuum and in C_2 emission. The researchers suggest the dust in the arcs are a secondary source of excited C_2 molecules. Lardière *et al.* (1997) also observed dust shells and jets, as well as CN (Fig. 5.4) jets. They demonstrate that the dust and CN jets are not cospatial.

Thomas *et al.* (1988) found that dust particle intensity decreased as the inverse distance from comet Halley's nucleus, suggesting free outflow. The authors show that the phenomenon can be explained by a model that considers the extended size and nonuniformity of the active region on the surface of the nucleus. The model uses a critical scale length, defined by the opening angle of the jet-like feature and the size of the source. The result is consistent with the observations and provides the basis for studying other mechanisms such as particle fragmentation.

Rodinov *et al.* (1998) found arc-shaped enhancements in monochromatic images showing the emissions due to C_2 and CN from comet Hyakutake (C/1996 B2). Figure 5.5 shows the C_2 arc observed on March 28, 1996 at the Pic du Midi observatory. Similar images were made showing CN emissions. The research team interpret the features as standing shock waves caused by the interaction of supersonic H_2O

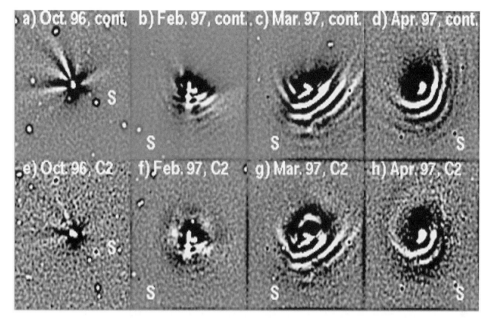

Fig. 5.3. Monochromatic images of comet Hale–Bopp. Upper row are continuum images at $\lambda = 5270$ Å; bottom row are C_2 images at $\lambda = 5130$ Å. North is up and East is to the left. The solar direction is indicated by S. The frames are 200 arcsec square. (Courtesy C. Laffont: From *Earth, Moon, and Planets*, Vol. 78, 1997, pp. 211–17, Jets and arcs in the coma of comet Hale–Bopp from August 1996 to April 1997, by Laffont, Rousselot, Clairemidi, Moreels, and Boice, Figure 1. Copyright © 1997, with kind permission of Kluwer Academic Publishers.)

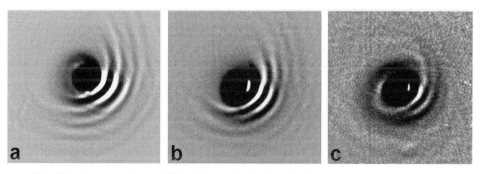

Fig. 5.4. Images of comet Hale–Bopp obtained at the Haute-Provence observatory with a CCD at the 80-cm reflector. (a) Broadband image in V, April 14, 1997. (b) Mid-continuum band image, and (c) CN band image, April 18, 1997. (Courtesy O. Lardière: From *Earth, Moon, and Planets*, Vol. 78, 1997, pp. 205–10, Evolution of dust shells and jets in the inner coma of comet C/1995 O1 (Hale–Bopp), by Lardière, Garro, and Meerlin, Figure 2. Copyright © 1997, with kind permission of Kluwer Academic Publishers.)

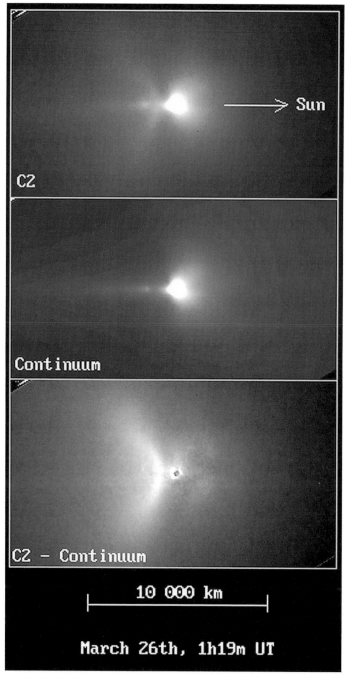

Fig. 5.5. The C_2 arc observed in comet Hyakutake from Pic du Midi on March 26, 1996. The top image was made with the C_2 filter, and the middle image was made in the nearby continuum. The bottom image is the result of subtracting the two upper images. (Courtesy J. F. Crifo: Reprinted from *Icarus*, Vol. 136, by Rodionov, Jorda, Jones, Crifo, Colas, and Lecacheux, Comet Hyakutake gas arcs: First observational evidence of standing shock waves in a cometary coma, pp. 232–67. Copyright 1998, with permission from Elsevier.)

gas jets from the nucleus and an as-yet-unidentified secondary source from the anti-sunward direction.

McBride *et al.* (1997) have analyzed the data collected by the *Giotto Extended Mission (GEM)* as it passed through the coma of comet 26P/Grigg–Skjellerup. Unfortunately, the Halley Multicolor Camera did not survive the Halley encounter. The only optical sensor available at the time of the Grigg–Skjellerup encounter was the Optical Probe Experiment (OPE). The McBride team concluded, based on the OPE data, that at its closest approach, *Giotto* passed within ~100 km of the comet's nucleus on the sunward side of the shadow terminator plane. Several spike-like events were seen in the OPE data, which the team suggests may be jet activity in the innermost coma. In addition, they found one event about 1000 km from the nucleus that suggests a small (10–100 m) nucleus fragment which was producing its own dust coma.

The comet Halley gas production rates determined from both ground-based and spacecraft observations were of order 6–7×10^{29} s^{-1} during the *Giotto* encounter (Huebner *et al.* 1991). The density of most neutrals followed an R^{-2} distribution with distance from the nucleus (Gringauz *et al.* 1986; Krankowsky *et al.* 1986), while the density of ions followed an R^{-1} distribution inside the contact surface and an R^{-2} relation outside (Balsiger *et al.* 1986).

Combi *et al.* (2000) used the SWAN all-sky camera on the *Solar and Heliospheric Observatory (SOHO)* spacecraft to observe the hydrogen Lyman-α coma of comet Hale–Bopp in early 1997, at heliocentric distances from 1.75 AU before perihelion to 1.29 AU after perihelion. Plate 5.1 shows the huge Lyman-α cloud surrounding the comet. An analysis of the data was used to determine the water production rate of the comet. The analysis used a model which takes into account the expansion velocity and kinetic temperatures in the coma. The model image reproduces the distribution of brightness – and the absolute brightness – seen in the observations. The maximum water production rate occurred about 2 weeks after perihelion. Furthermore, the temporal variations agree with other observations analyzed in a similar manner.

The dust composition analyzers on the *VEGA 1* and *2* spacecraft have made the first direct measurements of the physical and chemical properties of the dust in comet Halley. The particles tend to be rich in the light elements H, C, N and O. Figure 5.6 shows the composition of three distinct broad classes of particles, which are now often referred to as CHON based on their composition. Both the *VEGA* and *Giotto* dust detectors show unexpectedly large quantities of particles down to sizes as small as 0.01 μm.

Several papers have been written on the dust size distribution based on the *in situ* measurements made by the *VEGA* and *Giotto* dust detectors. Fulle *et al.* (1995) have developed a dynamical dust emission model, and demonstrated the

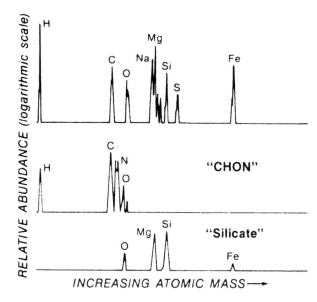

Fig. 5.6. Composition of grains seen near comet Halley by *VEGA* dust analyzer. (Courtesy of J. Kissel: used by permission © *Nature*.)

dependence of the dust fluence on the probability distribution of the dust ejection velocity. They compared their results with the fluences measured by the DIDSY instrument on *Giotto*, and concluded that dust ejection from Halley's nucleus was mainly in the solar direction, with an angular dispersion of about 18°. The differential dust size distribution, $g(s) \sim s^{\alpha}$, is characterized by power index $\alpha = -3.5 \pm 0.2$, which is constant for grain sizes larger than 20 μm. This suggests most of the dust is emitted as large grains. The fluences for very small particles are larger than would be predicted by the power law. The best fitting dust to gas ratio is $\chi = 4 \pm 1$. Fulle *et al.* (1995) point out that their numerical results for the dust size distribution agree with previous DIDSY fitting results and are compatible with dust tail models. However, their analysis shows that the conclusions are sensitive to both the dust size distribution as well as the velocity distribution. The authors suggest that future *in situ* measurements should aim to characterize these distributions.

Horányi and Mendis (1986) pointed out before the Halley encounters that dust particles in the coma will be electrically charged, and will be accelerated by the electric fields in the plasma environment. This may explain the fact that the smallest particles were observed outside the limits predicted by the Finson–Probstein model (see section 5.3.1.5).

The intensive study of comet Halley from the armada of *in situ* observing spacecraft, ground-based observations and observations from earth orbit, revealed a wealth of information about that comet's coma. One of the most interesting new observations is that some gas production seems to arise in "jets" from isolated active

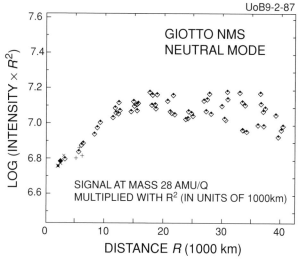

Fig. 5.7. Ion current near 28 *amu*/e deduced from the *Giotto* Neutral Mass Spectrometer, showing a distributed source. The signal is multiplied by R^2, where R is the distance from the nucleus. The fact that the signal is not constant inside about 12 000 km indicates the existence of a source of material. (Courtesy of *Astronomy and Astrophysics*: 1987. Eberhardt, P., Kranowsky, D., Schulte, W., Dolder, U., Lämmerzahl, P., Berthelier, J.J., Woweries, J., Stubbemann, U., Hodges, R.R., Hoffman, J.H., and Illiano, J.M. 1987. The CO and N_2 abundance in comet P/Halley. *Astron. Astrophys.* 187: 481–4.)

regions on the nucleus, as determined by both ground-based observations and the images made with the *Giotto* Halley Multicolor Camera. For instance, A'Hearn *et al.* have observed spiral-shaped jets in images due to emissions from CN and C_2, up to a distance of about 5×10^4 km from the nucleus. Festou (1999) reviews the data on distributed sources from comets including Halley and Hale–Bopp. He points out that the CN and dust coma anisotropies are not spatially correlated. However, the neutral molecule jets may be related to jets of minute CHON particles, which disintegrate and release species such as CN and C_2 (Huebner *et al.* 1991). Infrared observations of comet Halley (Campins and Ryan 1989) have identified crystalline olivine in the silicate spectrum of the comet (see Chapter 8). By contrast, the observations of comet Kohoutek in the same spectral region show a spectrum much like that of comet Halley, but without the crystalline olivine. This may show the differences in the composition of the primordial solar nebula between where the two comets were formed.

There are other indicators of extended sources of gas production which are associated with dust. Eberhardt *et al.* (1987) notes that if CO were all released directly from the nucleus, and moved outward with a constant velocity, then the density of the molecule would decrease as R^{-2}. Figure 5.7 is a plot of the CO intensity measured by the *Giotto* NMS at mass 28 *amu*/e, multiplied by R^2. In the case of constant velocity outflow from the nucleus with no external sources of CO, the $I \times R^2$ curve should be a flat line. However, for $R < 20\,000$ km, the plot clearly shows the existence of a source of CO. Huebner and Boice (1989) have argued that the CO is produced from a form of polymerized formaldehyde (polyoxymethylene or POM). The POM is slowly released from the dust, then dissociated through several steps into formaldehyde (H_2CO). This scenario explains the quantity of H_2CO in comet Halley. Combi *et al.* (1999) discuss an extended source of H_2CO in comet Hale–Bopp. They point out that observations of HCN and CH_3OH are also consistent with extended sources.

Greenberg and Li (1998) observe that coma molecules in addition to CO, for instance C_2, C_3, CN, and H_2CO, arise in part from the organic component in comet dust grains like interstellar core-mantle (Greenberg) grains. In the case of CO, the comet dust grains are heated sufficiently to evaporate most of the volatile fraction of the complex organic refractory molecules, a large fraction of which contain CO groups. The problem is that the predicted maximum CO production rate from the comet dust model is less than the observed distributed CO abundance. Greenberg and Li suggest a solution may be that the dust to gas ratio has been underestimated, or that the extended CO abundance may have been overestimated. To obtain the observed distributed CO production rate from comet dust requires the organic refractory mantles and very high porosity, and also more dust than has been deduced from the space observations.

The coma gases flow away from the nucleus at about 1 km s^{-1}. This rate can be established from the motion of expanding rings or halos, from the speed required to explain the Greenstein effect, and from the theory of coma flows, including the kinetic theory of those gases vaporizing at nuclear temperatures and the speed required to drag dust particles up to the speed required to explain the widths of type II (dust) tails.

The gas–dust environment for the active comet Hale–Bopp is unlike that for other bright comets such as Halley and Hyakutake (C/1996 B2). Combi *et al.* (1997) discuss the consequences of the large dust–gas production rate in comet Hyakutake. These include an increased photochemical heating which leads to higher than usual gas outflow speeds and increased variations with cometocentric and heliocentric distance. The team argues that the gas and dust components of typical comas are tightly coupled, and that the dust serves as an extended source of molecules in many comets. The strong gas–dust interaction in comet Hale–Bopp results in dust particles around 1 µm that are accelerated to nearly the gas speed and millimeter-size particles that are accelerated to atypically high velocities.

A'Hearn *et al.* (1983) reported the detection of molecular sulfur, S_2, in the coma of comet IRAS–Araki–Alcock. No S_2 was detected in comet Halley or any subsequent

comets until comets Hyakutake and Hale–Bopp. A'Hearn *et al.* (2000) and Woodney
et al. (2000) have now found S_2 in comet Hyakutake and OCS, SO, SO_2 and H_2CS
in both comets Hyakutake and Hale–Bopp. Woodney *et al.* (2000) also monitored
the production rates for OCS, CS and H_2S at the National Radio Astronomy
Observatory. Their studies show that the production rate of OCS depended on heliocentric distance as r^{-4}, while the other species' production rates varied as r^{-2}.
They point out that this difference suggests that either there is a radial variation
from the nucleus, or that OCS has an extended source.

5.3.1.1 Structure of the coma – Haser model

Much work on cometary comas consists of the interpretation of observations of
intensity contours of radiation from one molecular constituent. Narrow passband
filters are available that isolate the molecular bands of one molecule, for example, C_2
or CN. The situation is similar to the interpretation of intensities from the hydrogen
cloud (discussed above). The molecules under discussion are assumed to originate
from a spherical volume surrounding the nucleus with a speed determined by flow
conditions near the nucleus, and they are assumed to photodissociate because of
the solar radiation field. The mass of the comet has no significant decelerating
effect on the molecules (outside of the near-nuclear volumes) and hence the flow
is assumed to be at constant speed, v_0. The time scale for the change in density due
to photodissociation is

$$\frac{\mathrm{d}N}{\mathrm{d}t} = -\frac{N}{\tau_0},\tag{5.14}$$

where N is the number density and τ_0 is the time for the number density to fall
to $1/e$ of its initial value. The distance traveled by the average particle, R_0, before
dissociation is

$$R_0 = \tau_0 v_0,\tag{5.15}$$

sometimes called the scale length. Dissociation alone would produce a density
varying as e^{-R/R_0}, and the equation of continuity for a spherical system and constant speed produces an R^{-2} variation. Hence, this simple picture produces a total
variation as

$$N(R) = \left(\frac{R_N}{R}\right)^2 N(R_N)e^{-R/R_0},\tag{5.16}$$

where R_N is the radius of the nucleus. This equation is essentially the same as (5.8),
but the situation in the coma is probably more complicated. The simple molecules
that we observe are thought to originate from photodissociation of relatively complex parent molecules. Equation (5.16) can be generalized in a straightforward

manner to include two decay processes,

$$N(R) = \left(\frac{R_N}{R}\right)^2 N(R_N)\left[e^{-R/R_0} - e^{-R/R_1}\right].$$ (5.17)

Here, R_0 refers to the daughter molecules and R_1 to the parent molecules. In the literature, these processes are often discussed in terms of $\beta_0 = (R_0)^{-1}$, and so on. Equation (5.17) reduces to (5.16) for $R_1 = 0$ or for $\beta_1/\beta_0 = \infty$.

Equation (5.17) must be integrated to obtain the column density along the line of sight that passes the nucleus at the closest distance ρ, and Haser has obtained

$$N(\rho) = 2N(R_m)R_m^2 \frac{\beta_0\beta_1}{\beta_1 - \beta_0} e^{\beta_0 R_m} \frac{1}{\beta_0\rho}[B(\beta_0\rho) - B(\beta_1\rho)].$$ (5.18)

Here, R_m is the distance at which the expression in brackets in (5.17) is a maximum, $N(R_m)$ is the density at that point, and

$$B(z) = \frac{\pi}{2} - \int_0^z K_0(y)dy,$$ (5.19)

where $K_0(y)$ is the modified Bessel function of zero order and the second kind. Thus, the surface brightness $S(x)$ can be expressed in a dimensionless form as

$$S(x) \propto N(\rho) \propto \frac{1}{x}\left[B(x) - B\left(\frac{\beta_1}{\beta_0}\right)x\right],$$ (5.20)

where $\beta_0\rho = x$.

Physically, the preceding discussion can be summarized in terms of the schematic diagram shown in Fig. 5.8, which is a plot of log S versus log ρ. If there is only outflow at constant velocity (no creation or destruction processes), the slope of the brightness curve in this diagram would be -1. If the brightness curve falls slower than -1, the molecules are being created (near the nucleus). If the brightness curve falls faster than -1, the molecules are being destroyed (far from the nucleus).

There is no reason that the Haser model cannot be generalized to cases which involve three species; a parent molecule, a daughter product and a granddaughter product. Or, for that matter, which involve even more generations. O'Dell *et al.* (1988) have applied such a process to observations of the Swan bands of C_2 from comet Halley. The observations show that the normalized surface brightness of C_2 inside a projected nucleocentric distance of 4000 km is constant. There is no two-generation decay process that will explain the flatness; however, a three-generation process where the scale length of the grandparent is of order 15 000 km will explain the observations. The authors argue that there is strong evidence to support the idea that the C_2 molecules are formed at high vibrational temperatures, and require time to equilibrate.

A comparison of the model calculations versus observations of C_2 emission is shown in Fig. 5.9. A reasonably good fit is shown for $\beta_1/\beta_0 = 9$, whereas

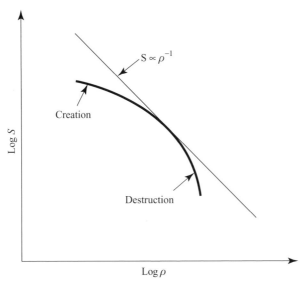

Fig. 5.8. Schematic variation of column density of a species with both creation and destruction mechanisms.

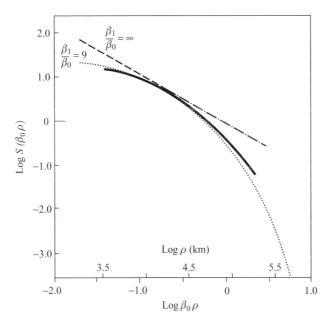

Fig. 5.9. Comparison of theory and observation (solid curve) for the surface brightness variations of the C_2 bands. The dot-dashed curve (which is coincident with the dotted curve at upper left) is the simple $S \alpha \rho^{-1}$ variation. The dashed curve is a fit with destruction processes only. The dotted curve includes both creation and destruction processes. (After C. R. O'Dell and D. E. Osterbrock.)

$\beta_1/\beta_0 = \infty$ does not fit. Equation (5.17) is correct only if the daughter molecules move radially and the parent molecules all decay at about the same distance from the nucleus. The Haser model implicitly assumes the parent molecules are directly emitted from the nucleus, the daughter products are the result of a single photo-dissociation process, the molecules move radially from the nucleus at constant velocity, and the coma is collisionless. Festou (1981a) has generalized the Haser model by introducing the effects of collisions. His revised model – called the vectorial model – takes into account the fact that the dissociation products of the parent molecule are ejected isotropically in the frame of reference of the parent molecule. The derivation of the model equations required an integration over the velocity distribution of the daughter products. Festou (1981b) applied his model to the production rates and lifetimes of the OH radical in comet Kobayashi–Berger–Milon (C/1975 N1).

5.3.1.2 Structure of the coma – recent models

Michael Combi and coworkers have generalized the Haser model, using an average random walk model and the Monte Carlo method. They call their model MCPTM (Monte Carlo particle-trajectory model). The general approach is developed in Combi and Delsemme (1980a) and Combi and Smyth (1988a) and is applied to the Lyman-α coma in Combi and Smyth (1988b). They were able to model effectively the time-dependent outflow speed distribution of H atoms leaving the inner coma, and reproduce the observed Lyman-α isophotes. They have subsequently extended the model to make it physically correct for heavy species (Combi *et al.* 1993). The newly revised model was used to study the distribution of the OH radical in comets based on *IUE* data from comet Halley. They find that near 2.5 AU, production rates should be about 60% less than predicted by the vectorial model.

Smyth *et al.* (1993, 1996) have used their MCPTM model to analyze two sets of ground based Fabry–Perot observations of the coma of comet Halley. The first set was made in the Hα (6563 Å) line of hydrogen, which included line profiles and two-dimensional sky brightness measurements. The other set included line profiles of the [O I] and NH$_2$ lines. The team calculated H$_2$O production rates from the Hα line with almost daily coverage from 50 days before perihelion to 130 days after. The team also took daily measurements with the same instrument in the [O I] λ6300 Å line. Figure 5.10 shows a comparison of water production rates from the team's analysis of the Hα data, the [O I] λ6300 Å data, as well as radio observations at 18 cm of OH emission, and *IUE* observations of OH.

Richter *et al.* (2000) have studied the multiple scattering of Lyman-α radiation in the coma of comet Hyakutake (C/1996 B2) using an improved version of the model described above. The modeled observations were made in April 1996 with the Goddard High Resolution Spectrograph (GHRS) on the *Hubble Space Telescope*,

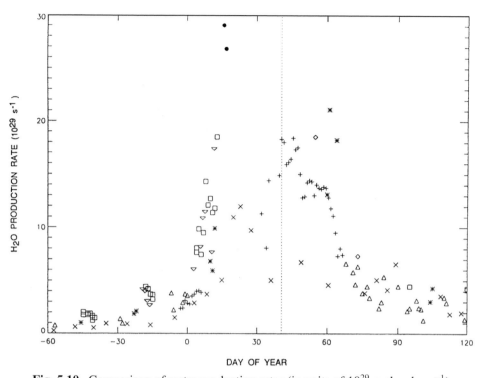

Fig. 5.10. Comparison of water production rates (in units of 10^{29} molecules s^{-1}) for comet Halley, shown for day of the year 1986. The time frame covered by the diagram includes November 1985 to May 1986, with the vertical dotted line indicating perihelion passage. The symbols refer to the following: open triangles, H_2O production rates from *IUE* data for OH emission by Combi, Bos and Smyth (1993); open squares, H_2O production rates from a revised analysis (Smyth *et al.* 1993) of the 6300 Å emission data for atomic oxygen originally published by Magee-Sauer *et al.* (1990); open inverted triangles, H_2O production rates determined from the analysis of Hα emission data by Smyth *et al.* (1993); crosses, H_2O production rates determined from 18 cm OH radio observations by Bockelée-Morvan *et al.* (1990); open diamonds, H_2O production rates determined from rocket observations of H Lyman-α emission by McCoy *et al.* (1992); asterisks, H_2O production rates determined from [O I] 6300 Å emission observations of atomic oxygen by Fink and DiSanti (1990); pluses, H_2O production rates determined from the H Lyman-α observations of Smyth *et al.* (1995); filled circles, average daily H_2O production rates determined by Smyth *et al.* (1995) from [O I] 6300 Å emission observations. (Courtesy W. Smyth: From Smyth, Combi, Roesler, and Scherb, 1995. Observations and analysis of the O(^{1}D) and NH$_2$ line profiles for the coma of comet P/Halley. *Astrophys. J.* 440: 349–60, Figure 11. Water production rates for comet Halley. Reproduced by permission of the AAS.)

which measured Lyman-α line profiles at a number of points in the comet's coma, with spectral resolution corresponding to a velocity resolution of 4 km s^{-1}. The calculation also made use of a new Monte Carlo Radiative Transfer (MCRT) model, which simulates the transfer of many photons emitted by a source (the sun) and randomly walking through the scattering medium. To make the modeling as efficient as possible, photons are injected into the comet from the (GHRS) detector and traced backward to the sun. Four simulations are carried out for each initial photon, as follows:

- The optical depth for scattering, τ, using the probability density p$(\tau) = e^{-\tau}$. The optical depth is converted to a geometrical path length, and the next scattering point is found.
- The photon frequency after it is scattered at the obtained point and the probability that the photon will hit the sun.
- The new direction of the photon based on the phase function.
- The new natural scattered frequency for the photon.

Figure 5.11 shows the normalized Lyman-α emission line profiles resulting from the calculations. The solid lines are the calculated curves and the triangles are the measured data. The abscissa is given in terms of the doppler velocity, with the vertical dashed line being the comet's radial velocity relative to the earth. As is readily apparent, the model represents the observations extremely well.

Radio spectroscopy has contributed significantly to our understanding of comets. Collisional and radiative excitation play roles in the excitation of cometary molecules, depending on where the molecule exists. In the inner coma, densities are sufficiently high for collisional excitation to take place, so that the molecules are in thermal equilibrium. Beyond the inner coma, the solar radiation field dominates the excitation process. In many cases, the radio emissions result from rotational transitions that are a continuation of radiative decays of molecules stimulated by shorter wavelength photons. The details of the excitation by solar photons was proposed many years ago (e.g., Zanstra 1929). New molecules continue to be discovered. For instance, radio studies of the exceptionally active comet Hale–Bopp led to the first discoveries of HC_3N, NH_2CHO, $HCOOH$, SO, SO_2, and $HCOOCH_3$; and the molecules OCS and $HNCO$ were confirmed (Bockelée-Morvan *et al.* 2000).

5.3.1.3 Near-nucleus structure

Observations of comet Halley by the *Giotto* Halley Multicolor Camera show spiral dust jets near the comet's nucleus. The jets are the result of the release of gas from isolated active regions on the surface of the rotating nucleus, which then pushes the small dust particles along. This has the effect of slowing the speed of the gas flow. The terminal speed of the dust is reached within 100 km of the nucleus. We will return to this topic in section 5.3.1.5.

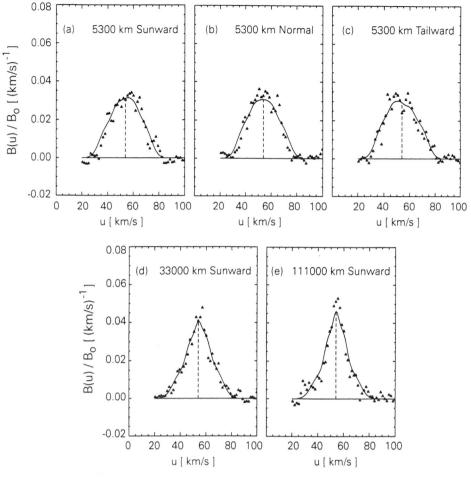

Fig. 5.11. Measured and calculated normalized H Lyman-α profiles in comet Hyakutake as functions of the Doppler velocity relative to the earth at different locations in the plane of the sky. The solid lines are the calculated curves and the triangles represent measured data. The vertical dashed lines indicate the radial velocity of the comet relative to earth. (Courtesy K. Richter: From Richter, Combi, Keller, and Meier. 2000. Multiple scattering of hydrogen Ly α. Radiation in the coma of comet Hyakutake (C/1996 B2). *Astrophys. J.* 531: 599–611. Figure 5, Measured and calculated normalized Ly α profiles in comet Hyakutake. Reproduced by permission of the AAS.)

5.3.1.4 Inner coma

Synthetic spectra

Convincing evidence for the inner coma was presented by Malaise (1970). He analyzed high-dispersion spectra of several comets and computed detailed synthetic spectra. Good agreement could not be obtained with purely radiative excitation;

collisional excitation was required. Besides the atomic parameters of the molecule (here CN) under consideration, the calculation of synthetic spectra requires at least the following quantities:

(1) detailed knowledge of the solar spectrum as seen by the comet;
(2) specification of the comet's radial velocity with respect to the sun (Swings effect);
(3) the expansion velocity of the cometary gas v_e; differential motions in the cometary gas itself due to expansion produces spectral effects (Greenstein effect);
(4) a parameter giving the relative efficiency of collisions and fluorescence in determining the populations of levels in the molecule.

Define the parameter α by

$$\frac{\alpha}{1-\alpha} = \frac{\tau_{fl}}{\tau_{coll}} = Q, \qquad (5.21)$$

where τ_{fl} and τ_{coll} are the characteristic times for establishing a fluorescence or collisional (Boltzmann) equilibrium, respectively, from an arbitrary initial distribution. An α of 1 (or 100%) implies a collisionally dominated regime, and an α of 0 implies a regime dominated by radiation. The collision time is taken as

$$\tau_{coll}^{-1} = 2N\sigma \frac{2\pi kT}{\mu}, \qquad (5.22)$$

where N is the number of particles per unit volume in the model inner coma, σ is the cross section for collisionally exciting the rotational transitions, and $\mu = [m_1 m_2/(m_1 + m_2)]$ is the reduced mass of the colliding particle. The fluorescence time is determined by absorption coefficients and the selection rules for radiative transitions. At 1 AU from the sun $\tau_{fl} \sim 10^2$ s. We have

$$Q = 2.3 \times 10^{12} N\sigma\tau_{fl} \frac{T^{1/2}}{A}, \qquad (5.23)$$

where A is the reduced mass in atomic units, T is the temperature in K, and the collisional excitation cross section is in units of 10^{-6} cm^2. Finally, a parameter describing the photodissociation of the molecule (CN) being studied must be specified; the distance traveled by a molecule before dissociation is called D.

Malaise (1970) has carried out the quite complex calculations of synthetic spectra for CN with the effects of the instrumental profile taken into account. The spectra used were taken at dispersions of 19.5 Å mm^{-1} and 39.0 Å mm^{-1}. The main parameters that determine the shape of the spectra are summarized in Table 5.4. The subscript zero refers to the value of the specific parameter at 10^4 km from the nucleus; T, v_e, and D are constant throughout the inner coma. The Greenstein effect is shown by the values in the column marked v_e. The Swings effect can be

Table 5.4 *Parameters obtained from the synthetic spectra of comets*

Comet	r (AU)	α_0 (%)	T (K)	$N_0 \sigma A^{1/2}$ $\times 10^{-2}$	v_e (km s^{-1})	D (10^5km)
Seki-Lines	0.55	90	600	48	0.80	4.7
Ikeya	0.66	90	580	33	0.74	6.0
Encke	0.69	55	480	5.7	—	—
Honda	0.71	20	550	1.0	—	—
Ikeya	0.74	(75)	500	(10)	0.67	6.4
Encke	0.79	25	350	105	—	—
Burnham	0.99	<2	(350)	<0.1	0.59	2.2
Candy	1.15	20	325	0.5	—	—

Source: After Malaise (1970).

illustrated by the spectra shown in Fig. 5.11. Clearly, models with purely radiative excitation cannot always reproduce the observed spectra. Some collisional effects are definitely present. At 10^4 km from the nucleus the collision frequency is $\sim 10^{-3}$–10^{-2} s^{-1}. Although collisional excitation can be important in the central regions of cometary inner comas, it is important to remember that overall, the influence of collisions is relatively small and that fluorescence is still the dominant excitation process.

When the nature of the exciting particles is specified, their density can be calculated from the known values of $N_0 \sigma A^{1/2}$. Electrons and protons are readily excluded as the exciting particles; the evidence indicates that the exciting particles are the total ensemble of radicals and molecules in the cometary inner coma. The total density is four to five orders of magnitude higher than the densities calculated from observations of visible species, such as CN. (In fact, they are probably too high, due to uncertainties in cross sections.) The total atmospheric densities are in the range 10^7–10^8 cm^{-3} at 10^4 km from the nucleus for a comet at 1 AU or less. Because we have estimates of both total density and outflow speed, the total efflux is roughly known and amounts to between 10^{30} and 2×10^{31} molecules s^{-1}. It is reassuring to note that these total effluxes are close to the values calculated from studies of the hydrogen cloud and from studies of the dust tails of bright comets. Arpigny (1977) has also carried out synthetic spectrum calculations for a number of molecules. He has reviewed the calculations of Malaise and has suggested that they should be redone using a better statistical equilibrium equation and improved molecular parameters, more accurate wavelengths, and including additional lines in the CN spectrum. Arpigny concludes that Malaise's calculations were as good as the state-of-the-art when they were made. However, the importance of CN is such that improved calculations are warranted.

Gas-phase chemistry in the coma

Comet scientists agree that a thorough understanding of the structure and composition of comet nuclei is key to an understanding of how the solar system formed and evolved. It should be clear by now that we simply have not observed directly the physical or chemical makeup of the nucleus. Comet scientists foresee the day when a robotic spacecraft will land on a comet nucleus, scoop up a sample of material and carry out a chemical analysis (see Chapter 11, *Rosetta* mission). Until that day comes, we must be content to infer the nature of the nuclear material from remotely sensed or *in situ* observations.

The state of our understanding of physical processes in comets in 1976 was well summarized by Whipple and Huebner (1976). Around this time Giguere and Huebner began to model the gas-phase chemistry in the collisionally dominated inner coma (Giguere and Huebner 1978, Huebner and Giguere 1980). In their model, they assumed a spherical nucleus with radius $R_0 = 1$ km. Material evaporated at the surface of the nucleus was initially assumed to flow outward at constant velocity. To include both photolytic and chemical reactions, the authors wrote for the number density of species i

$$\frac{\partial n_i}{\partial t} + \nabla \cdot n_i v_i = \sum_j \Re_{ji} - n_i \sum_k \Re_{ik}, \qquad (5.24)$$

where v_i is the outflow velocity, $\Re_{ji}$ is the rate of formation of species i by processes involving species j, and $n_i \Re_{ik}$ is the rate of destruction of species i by processes involving species k.

The authors' code has evolved over time to include more chemical reactions, and to replace the constant outflow velocity with an adiabatic expansion into a vacuum. Table 5.5 is a summary of the photolytic and gas-phase reactions considered in the model, with an example of each. The number density of various species is shown in Fig. 5.12 for an assumed nuclear composition (in units of 10^{13} cm^{-3}) as follows: $H_2O = 2.48$, $CO_2 = 1.46$, $NH_3 = 0.56$ and $CH_4 = 0.56$. This combination is what is to be expected if the nucleus condensed in the vicinity of the giant planets. This, and similar work, has been used to identify the species contributing to observed features found by the *Giotto* mass spectrometers in mass/charge bins (Table 5.6).

5.3.1.5 Coma proper

We now turn to a discussion of the expansion velocities found in cometary comas. According to our definition of the cometary inner coma, the kinetic mean free path is small and the problem can be treated hydrodynamically. Probstein (1968) has considered the problem of the expansion of a two-phase "dusty gas." In essence, the gas sublimates from the nucleus, expands outward, and drags the liberated dust particles along with it. The dust–gas coupling was computed using standard

Table 5.5 *Photolytic and gas-phase reactions*

Reaction	Example
Photodissociation	$h\nu + H_2O \rightarrow H + OH$
Photoionization	$h\nu + CO \rightarrow CO^+ + e$
Photodissociative ionization	$h\nu + CO_2 \rightarrow O + CO^+ + e$
Electron impact dissociation	$e + N_2 \rightarrow N + N + e$
Electron impact ionization	$e + CO \rightarrow CO^+ + e + e$
Electron impact dissociative ionization	$e + CO_2 \rightarrow O + CO^+ + e + e$
Positive ion-atom interchange	$CO^+ + H_2O \rightarrow HCO^+ + OH$
Positive ion charge transfer	$CO^+ + H_2O \rightarrow H_2O^+ + CO$
Electron dissociative recombination	$C_2H^+ + e \rightarrow C2 + H$
Three-body positive ion-neutral association	$C_2H_2^+ + H_2 + M \rightarrow C_2H_4^+ + M$
Neutral rearrangement	$N + CH \rightarrow CN + H$
Three-body neutral rearrangement	$C_2H_2 + H + M \rightarrow C_2H_3 + M$
Radiative electronic state deexcitation	$O(^1D) \rightarrow O(^3P) + h\nu$
Radiative recombination	$e + H^+ \rightarrow H + h\nu$
Radiation stabilized positive ion-neutral association	$C^+ + H \rightarrow CH^+ + h\nu$
Radiation stabilized neutral recombination	$C + C \rightarrow C_2 + h\nu$
Neutral-neutral associative ionization	$CH + O \rightarrow HCO^+ + e$
Neutral impact electronic state quenching	$O(^1D) + CO_2 \rightarrow O(^3P) + CO_2$
Electron impact electronic state quenching	$CO(^1\Sigma) + e \rightarrow CO(^1\Pi) + e$

Source: After Huebner *et al.* (1991).

free-molecular drag coefficients, and the details of the expansion depend on the amount of coupling between the dust and the gas. Generally, the solution contains a sonic or critical point at which the subsonic solution (valid near the nucleus) crosses over to a supersonic solution (valid away from the nucleus). This is the type of transonic flow pattern found in the solar wind and in rocket engines (the de Laval nozzle). We do not discuss the basic physics of such flows here. However, the gas exhaust speed from a de Laval nozzle is $3^{1/2}v_s$, where the sound speed $v_s = (\gamma P/\rho)^{1/2}$. For $T = K$, $\gamma = 1.4$, and mean molecular weight of 20, $3^{1/2}v_s \approx 0.6\,\mathrm{km\,s^{-1}}$. The presence of dust in the gas reduces the terminal speed (see Fig. 5.13). The numerical values quoted here should not be taken too seriously, but clearly these flows can approximately reproduce the injection speeds required by type II tails.

The terminal speed of the dust particles is reached within approximately 20 radii of the nucleus or within some 20 to 100 km. Thus, the terminal speed is reached well within the nuclear region of the coma and, so far as calculations of the structure of dust tails are concerned, the dust can be considered as emanating from a point source.

The terminal speed v_i was found by Probstein to be expressible in the form

$$\frac{v_i}{\left(c_p T_0\right)^{1/2}} = g(M, \beta), \tag{5.25}$$

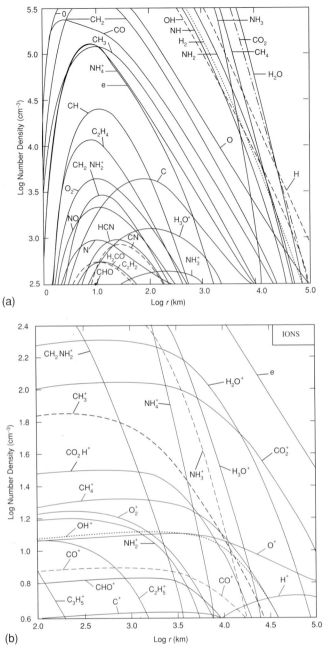

Fig. 5.12. Results of a model of a comet coma which assumes a spherical nucleus of radius = 1 km, albedo = 0.3, and number densities of mother molecules $H_2O = 2.48$, $CO_2 = 1.46$, $NH_3 = 0.56$, $CH_4 = 0.56$, $CO = 0.0$ in units of 10^{13} cm^{-3}. Number density of indicated species are plotted as a function of distance from the center of the nucleus. Frame (b) is a continuation of Frame (a) to lower number densities. (Courtesy of W. Huebner: From Huebner and Giguere. 1980. A model of comet comae. II. Effect of solar photodissociative ionization. *Astrophys. J.* 238: 753–62. Figure 1, Model of number densities in comet comas. Reproduced by permission of the AAS.)

Table 5.6 *Ions and atoms identified by mass spectrometer*

Mass		Mass	
12	C^+	28	CO, N_2?, C_2H_4?, DCN^+
13	CH^+	30	H_2CO
14	CH_2^+, N^+	31	H_3CO^+
15	CH_3^+, NH^+	35	H_3S^+
16	O^+, CH_4^+, NH_2^+	36	C_3^+
17	OH^+, NH_3^+, CH_5^+	37	C_3H^+
18	H_2O^+, NH_4^+, H_2O	39	$C_3H_3^+$
19	H_3O^+, HDO	44	CO_2
23	Na^+		
27	HCN^+		

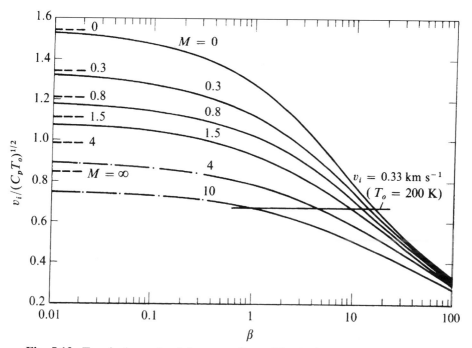

Fig. 5.13. Terminal speeds of dusty gas flows. The dashed lines at the left edge are values for $\beta \geqslant 0$. Note the increase in speed as M (5.26) increases. (Courtesy R. F. Probstein, Massachusetts Institute of Technology.)

where T_0 is the temperature of the gas at the nucleus and c_p is the specific heat of this gas. The parameters M and β are

$$M = \dot{m}_d/\dot{m}_g, \tag{5.26}$$

and

$$\beta = \frac{16}{3}\pi\rho_d dr_0 (c_p T_0)^{1/2}/\dot{m}_g. \tag{5.27}$$

In these equations, $\dot{m}_d$ and $\dot{m}_g$ are the mass flow rates of the dust and gas, respectively; ρ_d is the density of the dust; r_0 is the radius of the nucleus; and d is the diameter of the dust particles. The function $g(M, \beta)$ is given in Fig. 5.13. The parameter β is seen to represent a drag. For an assumed T_0 of 200 K, we have $(c_p T_0)^{1/2} = 0.48$ km s^{-1}. Finson and Probstein (1968) found a range of v_i between 0.2 and 0.4 km s^{-1} by fitting the values required to explain the widths of comet Arend–Roland's dust tail. The value $v_i = 0.33$ km s^{-1} was determined for most of the orbit near perihelion, and the value $v_i/(c_p T_0)^{1/2} = 0.33/0.48 = 0.69$ is plotted in Fig. 5.13. A value of β in the approximate range 0.6 to 25 is indicated. Equation (4.40) can be utilized to calculate a value for $\beta \dot{m}_g$; for $r_0 = 5$ km, $T_0 = 200$ K, $d = 1$ μm, and $\rho_d = 3$ g cm^{-3}, we find $\beta \dot{m}_g \approx 1 \times 10^8$ g s^{-1}. The values of β inferred from the v_i required for comet Arend–Roland then give gas mass loss rates $\dot{m}_g$ of 0.8×10^7 g s^{-1} to 3×10^8 g s^{-1}. For a mean molecular mass of 20, these numbers imply losses in the range 3×10^{29} molecules s^{-1} to 1×10^{31} molecules s^{-1}. The order of magnitude of these results is entirely compatible with the results obtained by other methods. The ratio $\dot{m}_d/\dot{m}_g$ is not well determined (see Fig. 5.13). However, observational estimates for the dust production rates by Liller (1960) for comets Arend–Roland and Mrkos are approximately 10^8 g s^{-1}. Hence, the evidence is consistent with $M \sim 1$ and gas and dust outflow rates $\sim 10^8$ g s^{-1}.

The results discussed above require modification for the case of large dust particles. The usual (micron size) particles that form the ordinary type II tail are not significantly influenced by the gravitational attraction of the nucleus, but the larger particles are. The size of the dust particles enters parameter β through (5.27). For small values of β the ratio of the dust speed to the thermal speed in the gas, v_d/v_g, is nearly constant or independent of β. For large values of β, v_d/v_g is approximately $(9/\beta)^{1/2}$. When the approximation for large values of β are modified to include the gravitational attraction, we have

$$v_d^2 = \frac{9v_g^2}{\beta} - \frac{2GM_c}{r_0}, \tag{5.28}$$

where M_c is the mass of the comet. If the gas mass loss rate $\dot{m}_g$ is known, we can estimate $(\beta/d\rho_d)$ from (5.27). Then, the size of the largest particles that can escape is given by putting $v_d = 0$ in (5.28). Gary and O'Dell (1974) find from studies of comet Kohoutek that the maximum particle size to escape is roughly 10^2 larger than the size that dominates type II tails.

5.3.1.6 Nucleus composition

Comet researchers have tried to follow a logic chain from the composition of comas to the composition of nuclei to the composition of the outer solar nebula. Huebner and Benkhoff (1999b) point out that earlier efforts to follow the chain did not take into account the changing coma composition with heliocentric distance. Comet Hale–Bopp (C/1995 O1), a very active comet, was discovered when it was relatively far from the sun. They used data from that comet to find relative abundances of CO and H_2O over a large range of heliocentric distances, and to infer the abundances in the nucleus. The mixing ratio $(Q(CO)/Q(H_2O))$ as a function of heliocentric distance is plotted in Fig. 5.14. From this information, they ascertain that a mixture, by mass, of 35% amorphous H_2O, 7% CO_2, 13% CO and 45% dust produces a good fit to the data. The dashed line in Fig. 5.14 does not conform to the data, but the fit would be much better if the distributed source of CO were subtracted from the mixing ratio.

5.4 Photometric phenomena

5.4.1 Heliocentric brightness variations

In Chapter 1 we developed the theory for the heliocentric variation of cometary brightnesses based on the adsorption–desorption model of B. U. Levin. Then, in section 3.3.1, we described the observations of heliocentric brightness variations. An interesting conclusion concerning the asymptotic behavior of this variation comes from the picture of cometary phenomena, as outlined in section 5.1.2.

The integrated brightnesses of comets should vary as r^{-2} for a heliocentric distance small enough that essentially all the solar radiant energy incident on the comet nucleus goes into gas production. Following a simple line of argument developed by M. Mumma (private communication), the intensity of light radiated by a kind of atom, molecule, or dust particle is

$$I \sim \sigma FN, \tag{5.29}$$

where I is the intensity emitted in all directions, σ is the cross section for scattering, F is the solar flux in the frequency band of interest, and N is the number of scatterers. (In a situation where more than one species contributes to the intensity, (5.29) will be a sum of similar terms. The generalization of the remainder of the argument is quite simple.) The number of atoms, molecules, or dust particles is clearly

$$N \sim Q\tau, \tag{5.30}$$

where Q is the production rate and τ is the mean or effective (for dust) lifetime. Thus,

$$I \sim (\sigma F)(Q\tau) \sim (\sigma F \tau)_0 r^{-2} Q r^{+2} \sim (\sigma F \tau)_0 \times Q. \tag{5.31}$$

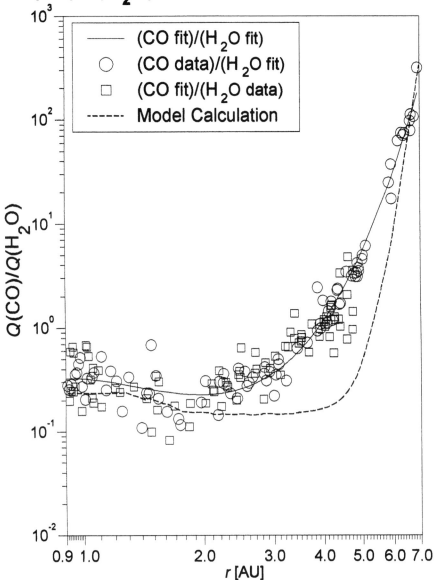

Fig. 5.14. Mixing ratio $Q(CO)/Q(H_2O)$ versus heliocentric distance in comet Hale–Bopp. The heavy dashed curve is the result from model calculations for a mixture of 35% amorphous H_2O, 7% CO_2, 13% CO (half of which is trapped in the amorphous ice) and 45% dust, by mass. The model does not consider the distributed source of CO. (Courtesy, W. Huebner: From *Space Science Reviews*, Vol. 99, 1999, pp. 117–30, From Coma Abundances to Nucleus composition, by Huebner and Benkoff, Figure 7. Copyright © 1999, with kind permission of Kluwer Academic Publishers.)

This reduction assumes (quite reasonably) that the lifetime of species is controlled by solar radiation, which falls off as r^{-2} (this includes radiation pressure for dust particles). Therefore, if the total incident solar radiant energy goes into gas production, we find

$$I \sim (\sigma F \tau Q)_0 R^{-2}, \tag{5.32}$$

or

$$I \propto r^{-2}. \tag{5.33}$$

In addition, the total cometary brightness at large distances from the sun when there is little or no gas production obviously varies as r^{-2}. Hence, the total brightnesses of comets should vary as r^{-2} at both large and small heliocentric distances and should vary as a higher inverse power at intermediate distances where gas production has begun but does not utilize all the radiant energy. Observations of comet Encke (Fig. 5.15) over several apparitions clearly show this variation.

5.5 Summary

Our understanding of comas has increased many fold in the last two decades. This chapter necessarily falls short of a thorough and complete review of the subject. Such a review could easily fill this entire volume. Instead, we have surveyed a number of the most cogent topics, and have cited the extensive literature in the bibliography. The researcher who is interested in more detail can use these citations to carry out a more thorough study of the topic. The review by Combi *et al.* (1997) is a good place to start. These authors stress the importance of considering the gas and dust components of the coma together.

We have talked about the hydrogen–hydroxyl cloud around comets, presenting the work that was carried out in the 1970s. The related topic of the Lyman-α emission from recent comets, and the associated production rates, is presented later in the chapter. This is a topic where models have produced results that fit the observations very well. The models are based on random walk and Monte Carlo simulations of gas motion and radiative transfer.

The *in situ* observations of comet Halley are discussed in the chapter. Of interest was the discovery of large numbers of small CHON grains. The observations have also shown that there are extended sources of species such as CO in Halley. We discuss the argument that the extended source may be the grains. Subsequent observations of comets Hyakutake and Hale–Bopp also show evidence of extended sources of H_2CO and other molecules. The H_2CO breaks down into CO.

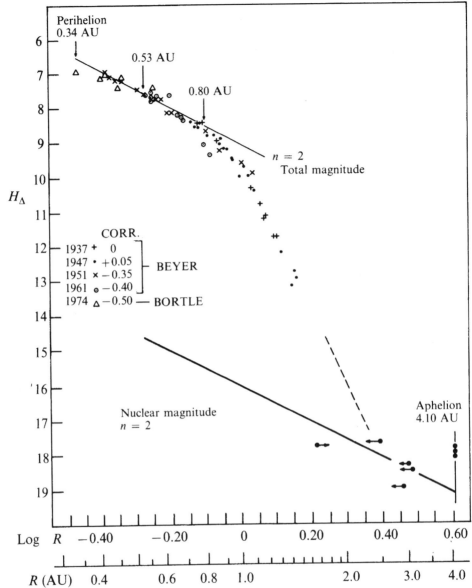

Fig. 5.15. The brightness of comet Encke versus heliocentric distance. The solid curve labeled $n = 2$ corresponds to a r^{-2} variation. (Courtesy of M. Mumma, NASA–Goddard Space Flight Center.)

Jets and arcs of C_2 and CN emission have been observed in comets Hyakutake and Hale–Bopp. Once again, it has been suggested that dust may be the secondary source for these molecules, at least in Hale–Bopp. In the case of the CN, there seems to be no spatial correlation with the dust features.

We briefly discuss gas phase chemistry, and cite some of the results of the models. That work has assisted in the identification of species observed in the *Giotto* mass spectrometer.

The *in situ* studies of comet Halley and remote sensing studies of the two bright comets Hyakutake and Hale–Bopp have provided much new data on cometary comas. There is still much to be learned, and there is little danger that we will exhaust the topic any time soon.

6

Comets and the solar wind

Comets interact with the interplanetary gas flowing away from the sun, the solar wind, and a variety of cometary phenomena are produced including the plasma tails. Historically, the properties of plasma tails were used by Biermann in 1951 to infer the existence of the solar wind, or corpuscular radiation as it was originally called. Thus, the study of the interrelation between comets and the solar wind goes back some 50 years. The interaction involves plasma physics and we provide a basic introduction as an aid to understanding this area of comet research. An excellent plasma physics text is Sturrock (1994).

6.1 Introduction to plasmas

Plasmas are an electrified state of matter, exhibiting collective behavior that is often complex and that can run counter to one's intuition. From spectroscopic observations and *in situ* measurements, we know that the plasmas in comets contain molecular ions such as H_2O^+, CO^+, etc. These are molecules that have lost one electron and thus are positively charged; they make up comet tails along with protons and other atomic ions. The electrons are the negatively charged particles. The charged particles interact with each other and with any electric or magnetic fields present.

A gas is considered a plasma when the following conditions for plasma behavior are satisfied. (1) The Debye length is small compared to the size of the system. The Debye length is the distance to which a charged particle's electric field extends before it is effectively shielded by neighboring charged particles of opposite polarity. (2) A sphere with radius of one Debye length contains many electrons. (3) There is no significant net charge per unit volume, i.e., the plasma is electrically neutral. (4) Plasma oscillations are not strongly damped by collisions.

We evaluate these plasma conditions for environments appropriate to the tail of comet Giacobini–Zinner and the inner coma of comet Halley. Typical parameters are listed in Table 6.1.

Table 6.1 *Typical tail and coma parameters*

	Tail	Coma
Magnetic field, $B(\gamma)$	20	50
Temperature, $T(K)$	10^4	10^3
Density (cm^{-3}), Ions or electrons	50	5000
mass (ions) (AMU)	20	20

The Debye length is:

$$\lambda_0 = 6.9 T^{1/2} N^{-1/2} \ \text{(cm)}, \tag{6.1}$$

where T is the temperature (K) and N is the density (electrons or ions; cm^{-3}).

For the parameters listed, we have λ_0 of 100 cm in the tail and 3.2 cm in the coma. These dimensions are much smaller than any tail or coma diameter. Also, spheres with these dimensions and the densities listed contain many electrons.

Significant violations of charge neutrality would produce large restoring forces acting on the excess charges. As a result, violations of charge neutrality can exist only under very special circumstances such as in electrical insulators. In the cometary environment, particles can move freely to neutralize regions of net charge. The only caveat is that the particles move along the magnetic field lines, see below. Thus, we may safely conclude that charge neutrality holds.

Of course, charge neutrality applies only as a steady-state or time-average condition. Plasma electrons can oscillate collectively around the more massive ions at the plasma frequency:

$$\nu_p = 9 \times 10^3 N^{1/2} \ \text{(Hz)}. \tag{6.2}$$

This value is 63 kHz for the tail and 630 kHz for the coma. The classical electron–ion collision frequency is:

$$\nu_c = 79 N T^{-3/2} \ \text{(Hz)}. \tag{6.3}$$

This value is 4.0×10^{-3} Hz for the tail and 12.5 Hz for the coma. Thus, the frequency of collisions is much lower than the plasma frequency, the plasma oscillations are not damped, and the conditions for plasma behavior in the cometary environment are satisfied. A complication can arise if the plasma coexists with a neutral gas and collisions are important.

The magnetic field in cometary plasma is important and accounts for the organization of the fine structure into linear forms. The ions and electrons can move freely along the magnetic field lines, but not across them. As discussed in section 4.3.2, the Larmor radius r_L for cometary plasmas is much less than observed streamer radii. For the tail conditions listed, r_L for thermal speeds is 0.1 km for electrons and 25 km for ions. These dimensions are small and, in the absence of collisions, means that motion across field lines is effectively inhibited.

An alternate view of the relationship between the plasma and the magnetic field comes from consideration of the magnetic Reynolds number:

$$R_M = 4\pi\sigma l v = l v/\eta, \qquad (6.4)$$

where σ is the conductivity, l is a characteristic dimension, v is the bulk speed, and η is the resistivity. When R_M is very large, the field is said to be *frozen-in*. For this case, the electric field resulting from the motion of the material across the magnetic field lines must vanish. Hence, there is no such motion. Physically, R_M is the ratio of the time required for a magnetic field to diffuse through a distance l ($\tau_{\text{diff}} = 4\pi\sigma l^2$) to the time required for the field to move through a distance l by bulk motions ($\tau_{\text{trans}} = l/v$). For the classical conductivity, $\sigma = 2 \times 10^{-14}T^{3/2}$ (c.g.s.), the magnetic Reynolds number is very large. In real plasmas, the conductivity is often much smaller than the classical value, but the value of R_M is still large and the magnetic field is frozen-in. The field and plasma move together.

The brief introduction to plasmas is completed with the Alfvén speed:

$$V_A = B/(4\pi N\mu m_H)^{1/2}, \qquad (6.5)$$

where μ is the mass of the ions in AMU. For tail conditions, we have 12 km s^{-1} and for the coma, 3 km s^{-1}. This speed can be taken as the characteristic information speed of disturbances in the plasma. A transit time in the coma could be a characteristic dimension divided by the Alfvén speed, say 30 000 km / 3 km s^{-1} = 10 000 seconds or roughly 3 hours. A tail width can be taken as 300 000 km and a transit time would be roughly 7 hours. Of course, for different scales or conditions, these times would be different. Some plasma tails show features changing on shorter time scales. Still, simple calculations of this type can indicate the range of time scales on which we might expect changes in the plasma tail.

6.2 The solar wind

The idea that material could be ejected or flowing from the sun had been around for a long time in the context of geomagnetic or auroral research. Usually the material was considered to be in discrete beams. But, the idea of a flow of material from the sun in all directions at all times is a relatively recent development. The key players in the 1950s were L. Biermann and E. N. Parker.

Biermann (1951) interpreted properties of cometary plasma tails, including their orientation, in terms of a steady outflow from the sun of fully ionized material, the "*solare korpuskularstrahlung*", or solar corpuscular radiation. This research is briefly described in Chapter 4.

Theoretical studies of the outflow were started by Parker (1958, 1963). Despite some early controversy, the hydrodynamic approach featuring a trans-sonic flow, i.e., a sub-sonic flow near the sun and a flow through the sonic point to achieve a supersonic flow away from the sun, is the basis of our current understanding. Many refinements have been added since the early days including the magnetic field, multiple fluids (electrons, protons, alpha particles, and heavier ions), solar rotation, and time-variable phenomona.

The morphology of the magnetic field is important. The shape is determined by the locus of points for material emitted from a specific location on the rotating sun which the magnetic field threads. While the shape in space is curved, azimuthal motion is not required. For purely radial flow, the curvature of the magnetic field for a constant velocity is an Archimedes spiral given by:

$$\tan \psi = \frac{(r - r_0)\Omega \sin \theta}{w}, \tag{6.6}$$

where ψ is the angle between the magnetic field direction and the prolonged radius vector; r is the heliocentric distance; r_0 is the reference distance from which the flow originates; Ω is the angular solar rotation rate; θ is the polar angle; and w is the solar wind speed. Near the earth, $r\Omega$ is approximately 430 km s^{-1}, $r \gg r_0$, and the solar wind speed is roughly 430 km s^{-1}. Thus, ψ averages $\approx 45°$ or $\approx 135°$.

An important feature of the solar wind is the heliospheric current sheet (HCS) (see Fig. 6.1 and 6.2). This feature arises from the global organization of the sun's magnetic field into "hemispheres" of opposite magnetic polarity. Smith (2001) has noted that "... the HCS is unique and represents the magnetic equator of the global heliosphere." This boundary is projected into heliospheric space and is calculated as follows. The line-of-sight photospheric field is measured via the Zeeman effect. A potential field model is used to calculate the field on a *coronal source surface*, often taken as a sphere with radius of 2.5 R_s. The field is projected radially outward into the heliosphere while allowing for solar rotation. These calculations are provided by T. Hoeksema at Stanford University. The calculations tend to smooth out fluctuations with each step. Fairly complex photospheric observations become a smooth HCS in the heliosphere even though some structure and fluctuations are still there. Also, there are some differences in the details of the calculations. Nevertheless, the HCS is a most important feature in the solar wind and it is also important in the solar-wind interaction with comets (section 6.4.3).

Much of the early theoretical work assumed that the solar-wind flow was essentially isotropic. Prior to the launch of the *Ulysses* spacecraft in October 1990, all *in situ* measurements were essentially "equatorial", i.e., within approximately ±20° of the ecliptic or the solar equator (which is inclined to the ecliptic by 7.5°). Of

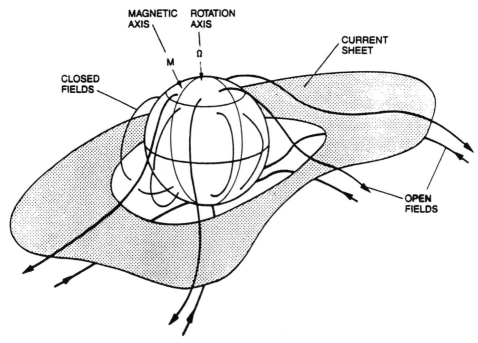

Fig. 6.1. The heliospheric current sheet (HCS) near the sun. The HCS is shown as shaded and curved like the brim of a hat. Both closed and open field lines are shown. (Courtesy of E. J. Smith, NASA–Jet Propulsion Laboratory.)

course, there was evidence that the solar-wind speed from the poles was faster than the equatorial speed. For example, white-light coronal images show a concentration of solar activity (with closed magnetic field lines) in the equatorial latitudes, while the polar regions are generally free of activity and have magnetic field lines open to the heliosphere. High-speed flow from the polar regions were expected from this picture and confirmed by studies of interplanetary scintillations. *Ulysses* (Smith and Marsden 1998), which was launched and put into a highly inclined orbit by a close gravitational encounter with Jupiter, made measurements that showed that the solar wind has a bimodal structure in solar latitude. The equatorial solar wind was slower, denser, and gusty. The polar solar wind was faster, less dense, and steady. The dividing feature separating the equatorial and polar solar wind is the HCS. The HCS is inclined to the solar equator by $10°–20°$ at solar minimum to $70°–80°$ or more at solar maximum. These are average values and we reiterate that the HCS is not really smooth and has considerable structure. Plate 6.1 illustrates the *Ulysses* picture. The inclination value is also the maximum latitude extension of the HCS. Thus, it extends only to low latitudes at solar minimum and close to the poles near solar maximum. Because of the inclination of the HCS (and its structure), a comet or spacecraft at latitudes less than the HCS inclination can be either in the equatorial or polar solar wind. Because the solar wind, excluding transients, rotates with the

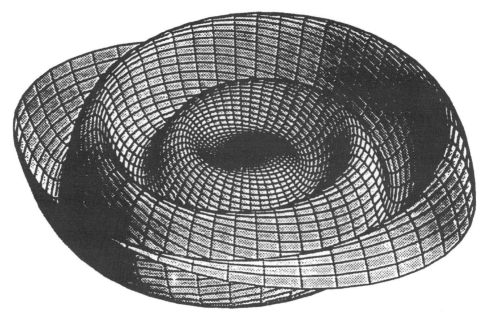

Fig. 6.2. The global configuration of the heliospheric current sheet (HCS). The HCS is transported outward by the solar wind. This transport, the tilt of the HCS, and the solar rotation produce the peak and valley structure which extends throughout the heliosphere. (Courtesy of E. J. Smith, NASA–Jet Propulsion Laboratory. From Jokipii and Thomas. 1981. Effects of drift on the transport of cosmic rays. IV. Modulation by a wavy interplanetary current sheet. *Astrophys. J.* 243: 1115–22, Figure 2 The global configuration of the HCS; reproduced by permission of the AAS.)

sun, a comet or spacecraft successively sees equatorial and polar solar-wind conditions and the HCS separating them. A comet or spacecraft at latitudes higher than the HCS inclination is exclusively in the polar solar wind and does not encounter the HCS. Typical measured solar-wind parameters are given in Table 6.2.

The relatively simple picture just described does not hold near solar maximum. While the HCS makes a small angle with the solar rotation axis (Smith *et al.* 2001), the velocity variation with latitude does not show the simple structure and is essentially chaotic (McComas *et al.* 2002). Apparently, this solar maximum situation persists only for a short time.

In addition to the steady or slowly varying solar wind that co-rotates with the sun, time-varying features such as flare-generated shock waves or coronal mass ejections (CMEs) can also encounter comets. An example of effects produced by transient conditions is the rapid turning of comet Bradfield's plasma tail on February 6, 1980 (Fig. 4.26).

While temperature values are quoted in Table 6.2, the distribution function is neither isotropic nor maxwellian. The speed values are valid for a wide range of

Table 6.2 *Measured average properties of the solar wind*

Property	Equatorial	Polar
Radial speed, w_r (km s^{-1})	450	750
Azimuthal speed, w_ϕ (km s^{-1})	6	3
Polar speed, w_θ	Small	Small
Density, $N_p = N_e$ (cm^{-3})	9	3
Electron temperature, T_e (K)	200 000	100 000
Proton temperature, T_p (K)	50 000	140 000
Magnetic field, $B(\gamma)$	6	6
Variations	Large	Small

heliocentric distance. But, the density values refer to 1.0 AU and need to be scaled by r^{-2} for other distances.

The average orientation of the magnetic field follows the Parker spiral angle (6.6). The field lines away from the equator, on average, are wrapped on a cone with opening angle θ.

6.3 Solar-wind interactions

A comet presents an interesting obstacle to the solar wind. The solar wind does not see the nucleus, and its interaction with the dust and neutral species is limited. The cometary plasma, consisting of ionized molecules and electrons, is a serious obstacle. The ionized molecules cannot cross the magnetic field lines in the solar wind and an interaction takes place.

Comets can be considered "soft" obstacles because of the way the ions get onto the field lines. The neutral molecules are released from the nucleus by sublimation and flow away from the comet at approximately 1 km s^{-1}. The ions are produced by photoionization (an easily modeled process) and the lifetimes are such that this occurs over a wide range of distances. When the ionization occurs, the ions are trapped on the field lines. This is called *mass loading* and, because the ions are essentially at rest with respect to the solar-wind speed of 450 km s^{-1}, the flow is slowed. Near the comet this effect is strong, but it is weak well away from the comet. Thus, the field lines wrap around the comet like a "folding umbrella" (see Fig. 4.20) to produce the general picture due to Alfvén described in sections 1.7.2 and 4.4.2.

6.3.1 Early models

Prior to the *in situ* investigations of comets Giacobini–Zinner and Halley, several models of the comet/solar-wind interaction were produced. One of the first was the

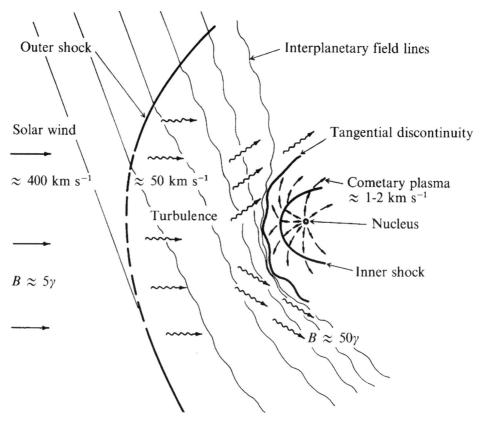

Fig. 6.3. Ideas in the comet/solar-wind interaction *circa* 1980. (Courtesy of J.C. Brandt and D. A. Mendis.)

study by Biermann *et al.* (1967). They assumed axial symmetry and their results apply to the region of the sun–comet line. They also assumed that CO^+ ions were produced by photoionization of an isotropic source of CO.

When the CO^+ ions were produced, they were rapidly accelerated to the flow speed of the initially pure solar-wind plasma. Basically, the equations describing the flow contain a source term to account for the additional mass of the added CO^+ ions; in other words, mass is not conserved. The solar-wind flow is supersonic (and super-Alfvénic) and supersonic flows are very sensitive to the addition of heavy molecules. The result is that a detached or stand-off shock develops when the plasma contains about 1% CO^+ ions by number. The shock was estimated to be approximately 10^6 km sunward from the nucleus. This value was compatible with flow speeds in the coma ~ 1 km s^{-1} and a photoionization lifetime $\sim 10^6$ s. The flow was found to stagnate at roughly 10^5 km from the nucleus (see Fig. 6.3).

Another way to look at the formation of the outer (or detached or stand-off) shock is as follows. The comet is an obstacle in the supersonic and super-Alfvénic solar-wind flow. Thus, the flow cannot receive information about the obstacle ahead because the flow speed exceeds the characteristic information speed. A shock forms which decelerates the flow to a subsonic, sub-Alfvénic regime where the plasma can flow smoothly around the obstacle. In addition to the outer shock, an inner shock in the flow of the cometary plasma away from the nucleus was also expected. Figure 6.1 presents a schematic representation of the pre-*in situ* understanding of the solar-wind interaction. It serves as a guide to the *in situ* measurements that are presented in the next section. We return to current models in section 6.3.3.

6.3.2 Plasma measurements in comets Giacobini–Zinner, Halley, and Borrelly

Despite recent well-observed and well-studied comets, our *in situ* knowledge rests on the spacecraft that encountered comets Giacobini–Zinner and Halley in 1985 and 1986 and comet Borrelly in September 2001. Here we summarize the results for the plasma and magnetic field. Refer to Chapter 3 for an introduction to the instruments carried on the spacecraft.

6.3.2.1 Comet Giacobini–Zinner

The USA's *International Cometary Explorer* (*ICE*) passed through the tail of comet Giacobini–Zinner on September 11, 1985. At the time of the encounter, the total gas production rate was estimated to be approximately 3×10^{28} molecules per second of water. From photographs taken at earlier apparitions (Fig. 6.4), a plasma tail and plasma phenomena were expected. While the comet in 1985 did not show the classic plasma tail (Fig. 6.5), extensive plasma phenomena were measured. See von Rosenvinge *et al.* (1986) and following articles in the same journal.

The magnetic field showed the draped topology and the current sheet as expected in the model by Alfvén discussed in Chapter 4. Sample magnetic field data are shown in Fig. 6.6. The orientation of the field was as expected from the draped field model. The radial component (B_x) on the inbound leg was negative, i.e., the same polarity as the solar wind. Closest approach occurred at 11:02 UT and the magnetic field reversed polarity as the spacecraft passed from one magnetic lobe through the current sheet into the magnetic lobe of opposite polarity. The plasma tail was approximately 10 000 km across and the current sheet was about 1000 km thick.

Far from the comet, the plasma had typical solar-wind values; these were a speed of about 500 km s^{-1}, an electron temperature of 250 000 K, and a density of

Fig. 6.4. Comet Giacobini–Zinner on October 26, 1959. The length of the plasma tail shown is approximately 450 000 km. (Photograph by E. Roemer, University of Arizona; official U.S. Navy photograph.)

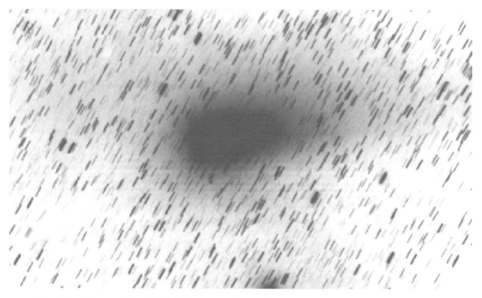

Fig. 6.5. Comet Giacobini–Zinner on September 12, 1985, a day after the spacecraft encounter. The plasma tail shown in Fig. 6.4 is not clearly present. (NOAO photograph by S. R. Majewski, University of Virginia.)

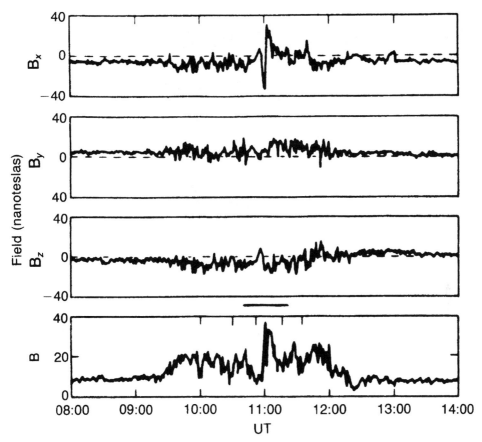

Fig. 6.6. Magnetic field observations obtained as the *International Cometary Observer* (*ICE*) spacecraft traversed the plasma tail of comet Giacobini–Zinner, showing large fluctuations in total magnetic field (*B*) and its components. At the time of closest approach, 11:02 UT, B shows a maximum value and the component in the direction away from the sun (B_x) reverses sign, as expected when the spacecraft passed from one magnetic lobe, through the current sheet, and into the magnetic lobe of opposite polarity. (Courtesy of E. J. Smith, NASA–Jet Propulsion Laboratory.)

5 ions cm^{-3}. See the data for electrons in Fig. 6.7. The general trend was toward higher densities, lower temperatures, and lower flow speeds as the spacecraft moved toward closest approach. These trends reversed after closest approach and these quantities ultimately returned to solar-wind values. Around closest approach (at a distance of 7800 km down the tail from the nucleus), the electron densities exceeded 600 cm^{-3}, the temperature was about 15 000 K, and the flow speed was less than 30 km s^{-1}.

The ions measured were mostly water-group ions, specifically H_2O^+ and H_3O^+. In addition, CO^+ was probably measured. The dominance of water-group ions, of

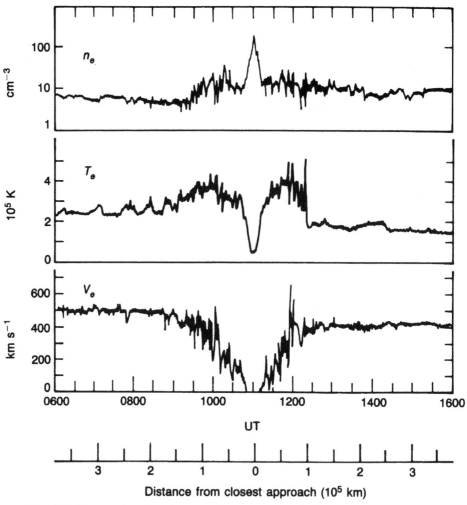

Fig. 6.7. Plasma data for electrons from the *ICE* mission to comet Giacobini–Zinner. Note the trend for the plasma to be dense (n_e), cold (T_e), and slowly moving (V_e) at the time of closest approach. (Courtesy of S. Bame, Los Alamos National Laboratory.)

course, provides support for the dirty-snowball view of cometary nuclei as proposed by Whipple.

Figure 6.8 records the fluxes of high-energy particles near the comet. Note the large range in energies. A reduction of the fluxes was seen when the spacecraft was in the cold, dense plasma tail. Most of the energetic ions can be explained as "pickup" ions. These are ions (mostly water-group) picked up by the solar-wind magnetic field and accelerated by the solar-wind flow. See the additional

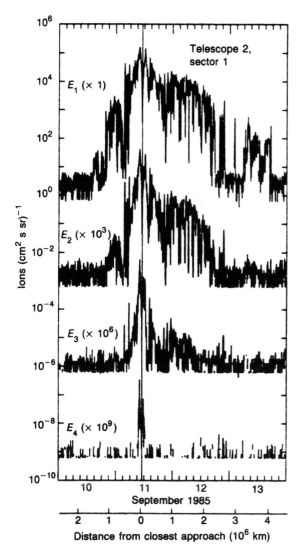

Fig. 6.8. Energetic ions measured at comet Giacobini–Zinner. The midpoints of the energy channels in keV are approximately: $E_1 = 80$; $E_2 = 120$; $E_3 = 170$; and $E_4 = 255$. Note that the fluxes in channels E_2 through E_4 have been scaled. The ions are water and are mostly produced by the pickup process in the solar wind. (Courtesy of R. J. Hynds and A. Balogh, Blackett Laboratory, Imperial College.)

discussion in section 6.3.2.2. The highest energy ions were probably produced by other acceleration processes.

The comet was also surrounded by plasma waves. These waves are created by the pickup process that also produces the energetic ions. Figure 6.9 shows the large spatial extent of the plasma waves.

The bow shock was detected by several experiments at about 130 000 km from the nucleus both on inbound and outbound legs. These features are summarized in Fig. 6.10. Many features of the then current model, based on the pioneering work of Alfvén, were confirmed. The major surprises were the large spatial extent of the

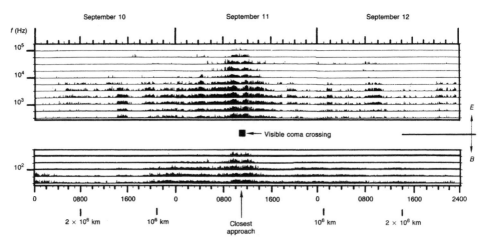

Fig. 6.9. Plasma wave measurements at comet Giacobini–Zinner. The electric field (*E*) measurements (top) are given as peak amplitudes in 11 frequency channels using a logarithmic scale. The magnetic field (B) measurements (bottom) are given as peak amplitudes in five frequency channels using a logarithmic scale. The measurements show a vast region where the pickup process generates plasma waves. (Courtesy of the late F. L. Scarf, TRW Space and Technology Group and E. J. Smith NASA–Jet Propulsion Laboratory.)

high-energy particles and the plasma waves. These measurements are still the only *in situ* data on cometary plasma tails.

6.3.2.2 Comet Halley

Five spacecraft came reasonably close to comet Halley between March 8 and March 14, 1986. The closest approach was by ESA's *Giotto* spacecraft at a distance of about 596 km. The other spacecraft were the then USSR's *VEGA-1* and *VEGA-2* and Japan's *Sakigake* and *Suisei*. Even the USA's *ICE* spacecraft probed the environment far from the comet in late March 1986. The spacecraft missions to comet Halley are summarized in Table 3.3 and Plate 1.2. See Grewing *et al.* (1988).

The physics of the comet/solar-wind interaction is beautifully illustrated by measurements from the ion analyzer on *Giotto* as shown in Plate 6.2a. The flowing plasma is clearly composed of two distinct components. Far from the comet, we find initially only the solar-wind protons at ion energy of 10^3 eV; this corresponds to protons (1 AMU) moving at the solar-wind speed. Then, we begin to see ions at an energy of 3×10^4 eV; these are pickup ions (e.g., 18 AMU for H_2O^+) moving at the solar-wind speed. The pickup ions are accelerated up to a bulk speed equal to the solar-wind speed and this is also the speed at which the

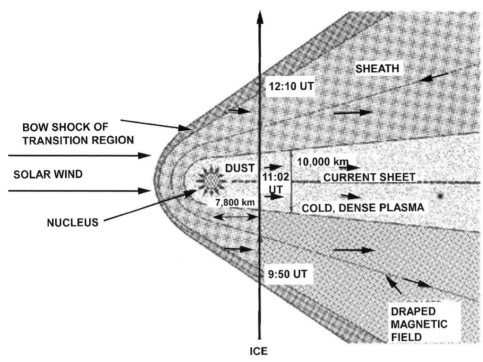

Fig. 6.10. Summary results of the *ICE* encounter with comet Giacobini–Zinner. The UT times along the trajectory refer to September 11, 1985.

ions spiral around the field lines. Thus, when the spiral motion is instantaneously in the same direction as the bulk flow, the ion energy can be four times the average energy of the particles moving at the solar-wind speed. If the spiral motion is not perpendicular to the field lines, the maximum energy is reduced by a factor of $(\sin \psi)^2$, where ψ is the angle between the direction of motion and the field lines. These factors account for the average energy of the pickup ions and the spread in energy.

Again, see Plate 6.2a. The flow speeds steadily decrease. At Point 1, the flow is deflected from the sun–comet line and the width of the distribution increases. This indicates an increase in temperature or random motions. Point 1 has been identified as the bow shock detected at 1 million km from the nucleus with a width of about 100 000 km. At Point 2, another transition point occurs. The flow is further deflected from the sun–comet line, the width of the distribution increases, and pickup ions with lower energy appear. The speeds continue to decrease and at approximately 80 000 km the flowing plasma is depleted. Of course, there is plasma in the inner region, but it is simply not measured by the ion analyzer. The basic behavior occurs in reverse on the outbound leg. The time between Point 2 and Point 1 on the outbound

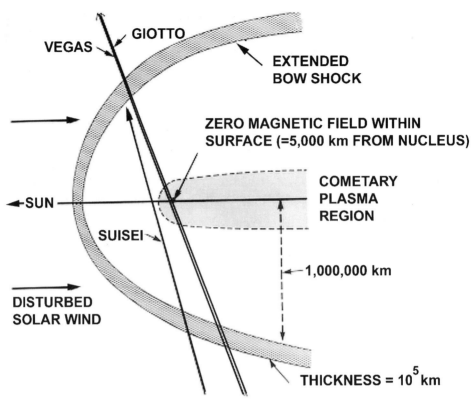

Fig. 6.11. Overview of plasma measurements at comet Halley, drawn approximately to scale. The lines representing the paths of *Giotto* and the *VEGAs* are slightly displaced for clarity. See text for discussion.

leg is smaller because of the trajectory geometry (see Fig. 6.11). Additional plasma data is shown below in connection with model comparisons. See section 6.3.3 and the detailed models by Gombosi *et al.* (1996).

Figure 6.11 shows other plasma and magnetic field features including the bow shock, the draped magnetic field line configuration, a magnetic pileup region, and a central cavity region free of magnetic field. The bow shock was observed at essentially the same distance by experiments on several spacecraft. The magnetic pileup region is the region with magnetic field strength enhanced over the solar-wind value. The pileup region occurred within 135 000 km of the nucleus inbound and 263 000 km outbound. Maximum values for the magnetic field were 57 γ inbound and 65 γ outbound. The central, field-free cavity has a width of about 8500 km. It is formed by the outflowing of pure cometary plasma. The cavity boundary is the *contact surface* that separates (inside) the pure cometary plasma near the nucleus from the mixed cometary and solar-wind plasma (outside). In other words, the outflowing cometary plasma keeps the solar wind with its magnetic field from penetrating inside the contact surface.

The interaction region in which Halley's Comet influences the solar wind is quite large. The consensus figure from the Halley Armada is a value of at least 0.1 AU.

6.3.2.3 Comet Borrelly

The Plasma Experiment for Planetary Exploration (PEPE) on board the *Deep Space 1* spacecraft made ion velocity and mass composition measurements of comet 19P/Borrelly during a flyby on 22 September 2001 (Nordholt *et al.* 2003). Heavy ions were detected when the spacecraft was within approximately 5×10^5 km of the nucleus. Closest approach was at a distance of about 2200 km. The plasma results are shown in Plate 6.2b. The solar-wind protons start at approximately 700 eV/Q and decrease as the flow becomes mass loaded by cometary ions, which show up as the higher energy ions in Plate 6.2b. The situation reverses after closest approach (CA). Note the same general results for comet Halley in Plate 6.2a and discussed in section 6.3.2.2. The composition as determined from closest-approach data is approximately: 63% OH^+; 25% H_2O^+; 2.5% C^+; 2.0% N^+; and 8% CH_3^+. The ions O^+, H_3O^+ and CH^+ were not present at detectable levels.

A major difference from other cometary flybys was a plasma distribution not symmetric about closest approach at 22 h 27 m 29 s (distance approximately equal to 2200 km). The maximum in the water group count rate occurred at 22 h 27 m 19 s (distance approximately equal to 3000 km). This asymmetry is probably due to asymmetrical emission of coma material and model calculations (Hansen *et al.* 2001) confirm this.

6.3.3 Models

Modeling of the comet/solar-wind interaction, in broad outline, is simple. But, in practice, this is a very complex calculation. Many details need to be specified accurately. The boundary conditions need to be physically appropriate. The resolution of the numerical scheme needs to be fine enough to resolve physically important features such as the bow shock. The situation is complicated by the extreme range in scales, from approximately 10 km for the size of the nucleus to roughly 10^7 km for the length of the plasma tail. A realistic model must have a large-enough volume to cover the upstream mass-loading region and the plasma tail.

The model calculations by Gombosi *et al.* (1996) successfully addressed these problems and we follow their discussion here. Also, see Gombosi *et al.* for an extensive list of modeling investigations prior to 1996. The conditions chosen for the model were intended to match comet Halley in March 1986 during the *Giotto* flyby. The calculations model the interaction of an expanding cometary atmosphere, which is being ionized, with the magnetized solar wind. The model is based on single-fluid magnetohydrodynamics (MHD).

The total gas production rate was taken to be 7×10^{29} molecules per second and the molecules moved outward at 1 km s^{-1}. The molecular ions were created by four processes: (1) photoionization by solar ultraviolet radiation; (2) impact ionization by solar-wind electrons; (3) impact ionization by cometary electrons; and (4) charge exchange between solar-wind protons and cometary molecules and atoms. The charge exchange is not a source of net ionization but it produces mass loading because the cometary ions are heavier than protons. These processes produce ions from a neutral source that can be written as:

$$n_n = \frac{Q}{4\pi u_n r^2} \exp\left(-\frac{1}{u_n}\int_{R_n}^{r} \frac{dr'}{\tau(r')}\right), \tag{6.7}$$

where Q is the total gas production rate, u_n is the radial speed in the coma, R_n is the radius of the nucleus, and $\tau(r)$ is the ionization lifetime of the neutral molecules. Equation (6.7) can be used to specify all ionization processes except the impact ionization by cometary electrons. The latter process requires knowledge of an electron temperature profile (which cannot be calculated within the single-fluid MHD framework). This profile was obtained from observations and theoretical work. Recombination is also included in the model.

The MHD equations for the solar-wind flow are solved with a source term of additional mass from ionization of the cometary neutrals as described by (6.7). The effects of ion-neutral friction are also included. The problem of the disparate scales is addressed by using an innovative scheme in which the governing equations are solved on an adaptively refined grid. In other words, a smaller grid was used near the nucleus and a larger grid for large-scale structure. The grid structure is shown in Fig. 6.12. The total number of grid points is 486 000.

Far upstream, the model is a supersonic and super-Alfvénic solar wind. The results of the interaction are illustrated in the following figures. Plate 6.3 shows the bow shock and the magnetic field lines. Note that the incident solar-wind magnetic field is (realistically) inclined to the radial direction by 45°. Plate 6.4 shows the magnetic field lines and structure near the nucleus. The inner shock and the cometary plasma cavity boundary layer are marked. Comparisons with *Giotto* measurements are given in Fig. 6.13a, b, c and d for the magnetic field around the cavity, the plasma density, the plasma speed, and the plasma temperature. While there are some minor discrepancies, the overall agreement with the *in situ* measurements should be described as very good.

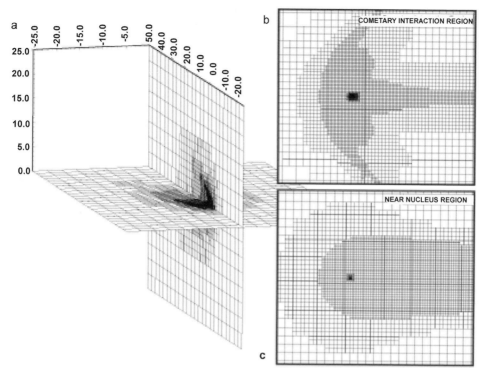

Fig. 6.12. The adaptive grid structure used in the calculations by Gombosi *et al.* (1996). (a) The entire simulation region is shown in a 3D view (50 × 50 × 75 Gm); (b) The cometary interaction region (2.8 × 2.4 Gm); (c) The near nucleus region (37 000 km × 32 000 km). (Courtesy of T. I. Gombosi, University of Michigan, Ann Arbor.)

In summary, the results presented by Gombosi *et al.* (1996) show many features of the interaction and validate the power of advanced computing techniques. The solar wind upstream of the comet is mass-loaded and a bow shock forms at the location measured by *Giotto*, *VEGA-1*, *VEGA-2*, and *Suisei*. The structure in the inner region is reproduced including the peak in plasma density at 10^4 km from the nucleus, the inner shock (now seen to be the boundary of the diamagnetic cavity), and four current systems. Among these current systems is the cross-tail current that separates the lobes of opposite magnetic polarity in the tail. This current lies in a narrow layer perpendicular to the plane containing the interplanetary magnetic field (IMF). Finally, plasma on the nightside of the cavity is ejected into the plasma tail. The model techniques just described have been extended to comet Hale–Bopp by Gombosi *et al.* (1997).

We return to the subject of time-dependent models as part of the disconnection event (DE) discussion below.

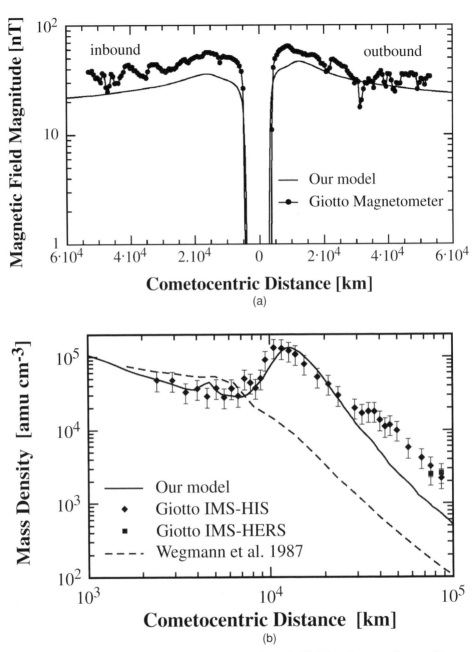

Fig. 6.13. Summary of results from Gombosi *et al.* (1996) and comparisons with the *Giotto* measurements. (a) The modeled magnetic field compared with the measurements along the *Giotto* trajectory; (b) The modeled plasma density compared with the *Giotto* measurements; (c) The modeled plasma speed compared with the *Giotto* measurements on the inbound pass; (d) The modeled plasma temperature of ions compared with *Giotto* measurements on the inbound pass. (Courtesy of T. I. Gombosi, University of Michigan, Ann Arbor.)

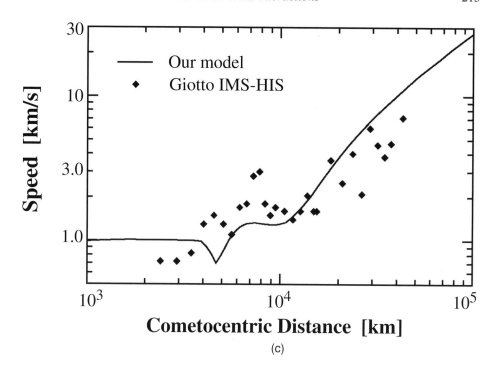

(c)

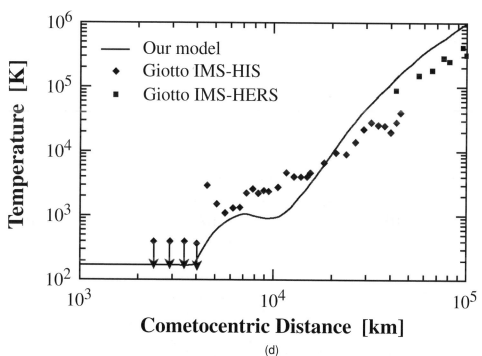

(d)

Fig. 6.13. (*Cont.*)

6.4 Comets as probes of the solar wind

Comets can be regarded as natural probes of the solar wind. They are important adjuncts to direct measurements by infrequent and expensive missions. Interplanetary conditions could be inferred in several ways: (1) solar-wind conditions such as variations can be determined from the appearance and morphology of plasma tails; (2) solar-wind speeds can be derived from the orientation and shape of the plasma tails; (3) disconnection events (DEs) can indicate the location of the heliospheric current sheet (HCS); these data along with orientations and appearance map the latitudinal structure in the solar wind; (4) major brightness variations could be due to solar-wind events such as shocks or CMEs encountering the comet; (5) comets are bright sources of x rays and these are almost surely produced by the interaction with the solar wind.

6.4.1 Solar-wind speeds and aberration angles

The cometary plasma tail acts as a windsock in the solar wind, i.e., it points in the direction of the solar wind as seen by a hypothetical observer riding on the comet. The direction is determined by the components of the solar-wind speed and the orbital motion of the comet. This effect causes plasma tails to point several degrees away from the radius vector in the direction opposite the comet's motion. In Chapter 4, we described Hoffmeister's measurement of this effect and Biermann's interpretation in terms of a continual mass flow from the sun.

A global picture of solar-wind speeds can be derived from a large number of orientation observations. The desired quantity is the position angle on the plane of the sky, measured from north through east. For an assumed solar-wind model, this position angle, θ_c, can be calculated. From the observations, the measured position angle, θ, can be determined. Standard least-squares techniques are used to minimize the quantity

$$\sum_i (\theta - \theta_c)_i^2 \tag{6.8}$$

to determine the model. The results are given in Table 6.3 from a sample of 809 observations (Brandt *et al.* 1972; Brandt and Chapman 1992). The sample covers solar latitudes to approximately $\pm 40°$ to $50°$ and heliocentric distances of about 0.6 to 1.4 AU. The results in Table 6.3 are in agreement with spacecraft for the equatorial region of the solar wind. The sample of observations of comets in the polar region is small. See the next section for examples of aberration angles in the polar solar-wind region.

The wind-sock concept can be applied to specific events such as a tail with an unusual shape or rapid changes in orientation. Fig. 6.14 shows comet Kohoutek on January 20, 1974. A "big bend" is visible in the tail and a change in the polar

Table 6.3 *A global solar-wind model based on plasma tail orientations in the equatorial solar-wind region*

Parameter	Value
Radial speed, w_r (km s^{-1})	400 ± 11
Azimuthal speed, w_ϕ (km s^{-1})[a]	6.7 ± 1.7
Meridional speed, w_θ (km s^{-1})[b]	2.3 ± 1.1
RMS Dispersion (°)	3.74

[a] The azimuthal speed is represented parametrically with smaller speeds away from the solar equator. The value quoted is for the equator.

[b] The meridional speed is assumed to vary as sin 2θ and to have the same flow away or toward the solar equator in both hemispheres. The value quoted is the maximum at latitudes of $\pm45°$ and a small (and poorly determined) flow toward the equator is indicated.

Fig. 6.14. Comet Kohoutek on January 20, 1974 showing the "big bend" feature in the plasma tail. (Joint Observatory for Cometary Research [JOCR] photograph.)

solar-wind speed of 30 km s^{-1} was responsible. This cometary feature was probably the first to be convincingly associated with a specific solar-wind feature (Niedner *et al.* 1978).

Comet Bradfield on February 6, 1980 showed a rapid change in the orientation of the plasma tail (Brandt *et al.* 1980); see Fig. 4.26. The orientation changed by $10°$ in 27.5 minutes for a turning rate of approximately $22°$ per hour. The change was also produced by an excursion in the polar solar-wind speed caused by a solar-flare-related event.

6.4.2 Latitudinal structure and the appearance of comets

The latitudinal structure of the solar wind as mapped by *Ulysses* (Plate 6.1) is clearly reflected in the properties of comets. Here, we follow the discussion of Brandt and Snow (2000). Basically, comets in the equatorial region are exposed to a slower, gustier solar wind while comets in the polar region are exposed to a faster, steadier solar wind. Hence, comets in the polar region should show a smoother, less-disturbed appearance, while comets in the equatorial region should show a relatively structured, more-disturbed appearance. Also, because the HCS does not penetrate into the polar region, we expect DEs (see below) only in the equatorial region.

The change in appearance of comet Hyakutake as it moved from the equatorial to the polar region is documented in Fig. 6.15. The gusty equatorial solar wind produces a disturbed appearance below ecliptic latitudes of about $30°$N ($24°$N heliographic) while the steady polar solar wind does not and the comet appears smooth.

The relative stability of solar-wind conditions in the polar region, particularly the absence of the HCS, may have made possible the detection of comet Hyakutake's plasma tail 550 million km or 3.8 AU from its nucleus. The tail signature appears in the magnetic field data (Jones *et al.* 2000) and in ion composition measurements (Gloeckler *et al.* 2000) obtained from experiments on *Ulysses*. Comet Hyakutake was in the polar region for the entire period of interest and the detection at these immense distances was remarkable.

The change in speeds can be documented using aberration angles as described in the last section. The plasma tail would point very close to the anti-solar direction for a very high solar-wind speed. It would depart from the anti-solar direction somewhat for a speed of 750 km s^{-1} and a larger amount for a speed of 450 km s^{-1}. Fig. 6.16 shows the position angles for these cases plotted for comet 122P/de Vico as it passed from the polar region into the equatorial region and back into the polar region in 1995 (Brandt *et al.* 1997). The speed changes are clearly shown. Also, the spread seems to be higher in the equatorial region.

Investigations into the structure in different solar-wind regions are usually hampered by the lack of polar region or trans-regional comets. Three recent

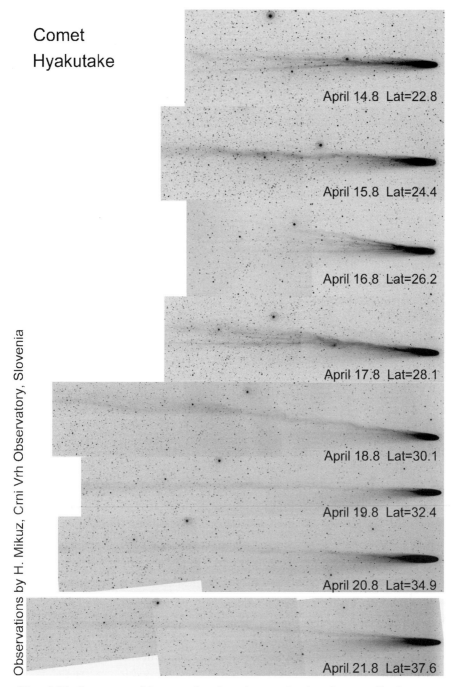

Comet
Hyakutake

April 14.8 Lat=22.8

April 15.8 Lat=24.4

April 16.8 Lat=26.2

April 17.8 Lat=28.1

April 18.8 Lat=30.1

April 19.8 Lat=32.4

April 20.8 Lat=34.9

April 21.8 Lat=37.6

Observations by H. Mikuz, Crni Vrh Observatory, Slovenia

Fig. 6.15. Sequence of images showing changes in the plasma tail of comet Hyakutake in April 1996. The dates and latitudes are shown. The comet has a relatively disturbed appearance in the equatorial region and a relatively smooth appearance in the polar region. The boundary was located near 30° N (ecliptic) or 24° heliographic. (Images courtesy of H. Mikuz, Crni Vrh Observatory, Slovenia; see Brandt and Snow 2000; Snow *et al.* 2004.)

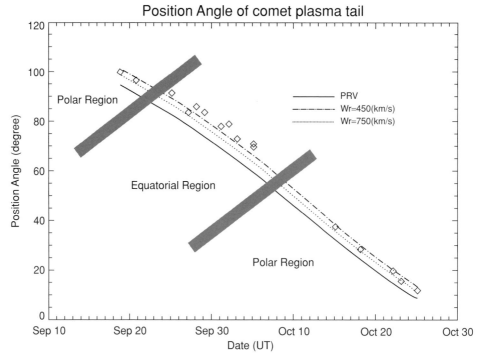

Fig. 6.16. Comparison of plasma-tail orientations (via position angles) and solar-wind speeds for comet de Vico in 1995. The boundaries between the polar and equatorial regions are marked by the gray bands at 20° (ecliptic, north and south). The observations fall near the 750 km s^{-1} curve and have a small scatter in the polar region. The observations fall near the 450 km s^{-1} curve and have a larger scatter in the equatorial region. (Reprinted from *Planetary and Space Science*, Vol. 45, by Brandt, Yi, Petersen, and Snow, 1997. Comet de Vico (122P) and latitude variations of plasma phenomena, pp. 813–19. Copyright 1997, with permission from Elsevier.)

comets – 122P/de Vico in 1995, Hyakutake in 1996, and Hale–Bopp in 1997 – have been trans-regional and have contributed to our understanding.

   One element of the appearance of comets, the visual brightness, has been suggested as possibly providing useful information on the solar wind. Comets encountering an interplanetary shock or undergoing a disconnection event could reasonably be expected to have an increased brightness. Unfortunately, this idea has not been useful. A major part of the problem is that determining accurate and reliable brightnesses of plasma structures is difficult.

6.4.3 Disconnection events (DEs)

DEs can be very spectacular. Historical examples were given in Chapter 4, and Fig. 6.17 shows a recent example in comet Hyakutake. Basically, the entire plasma

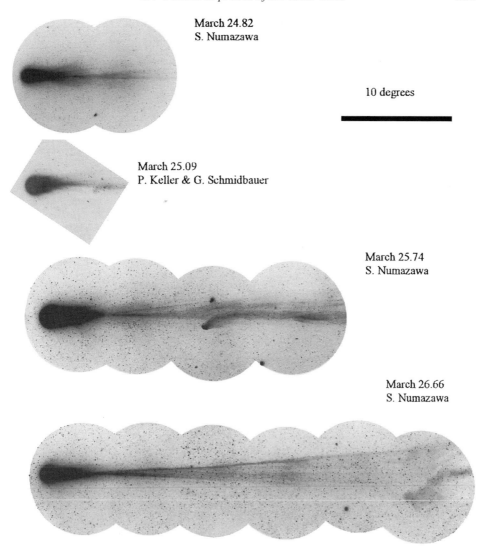

March 24.82
S. Numazawa

10 degrees

March 25.09
P. Keller & G. Schmidbauer

March 25.74
S. Numazawa

March 26.66
S. Numazawa

Fig. 6.17. Spectacular DE in comet Hyakutake. Top to bottom, the images in 1996 are: (1) March 24.82; (2) March 25.09 (Courtesy of P. Keller & G. Schmidbauer, Ulysses Comet Watch); (3) March 25.74; (4) March 26.66. The angular scale of $10°$ is indicated. (Images (1), (3), and (4) appeared on the cover of *Sky and Telescope* for July 1996. These images courtesy of *Sky and Telescope* and S. Numazawa, Japan.) (Image sequence courtesy of the Ulysses Comet Watch; see Snow *et al.* 2004.)

tail disconnects and then reforms with a fairly regular morphological sequence, as illustrated in Fig. 4.23. While many details about DEs were known to Barnard and contemporaries some 100 years ago, the early work essentially had been forgotten. Since their rediscovery by Niedner and Brandt in 1978, they have been the subject of many investigations.

The generally accepted view is that DEs are produced by magnetic reconnection on the sunward side of comets when they cross the heliospheric current sheet (HCS). When the HCS is crossed, magnetic field of opposite polarity is pressed into the comet and reconnection occurs. The basic ideas were contained in the Niedner and Brandt (1978) paper as illustrated in Fig. 6.18. While the generally accepted view is presented here, questions need to be settled in two areas. (1) The location of the HCS relative to the comet is not always known with certainty. Farnham and Meech (1994), for example, have expressed concerns that the HCS is not always near the comet at the time of DEs. (2) Wegmann (1998, 2000) has questioned the MHD model calculations by Yi *et al.* (1996, 1998) that reproduced DEs at comet crossings of the HCS. The basic issue concerns the correct value of the resistivity in real plasmas. See also Wegmann 1995.) While we address these issues in the discussion, they will only be settled over a period of time.

Unfortunately, comets are not usually near spacecraft. Hence, inferring the solar-wind conditions for a comet at the time of a DE involves using measured solar-wind parameters from another location, often from satellites in earth orbit. Assumptions must be made to carry out this calculation. Recall that solar wind from a fixed location on the sun forms a spiral pattern due to the sun's rotation. For reasonably steady conditions, an observer almost anywhere along this spiral sees the same solar wind. Also, differences in latitude are usually ignored. Calculating the time shifts to identify the correct solar wind for comparison is called *co-rotation*. The time delay can be written as:

$$\Delta t = \frac{(l - l_\mathrm{E})}{\Omega} + \frac{(r - r_\mathrm{E})}{w}, \tag{6.9}$$

where l is the longitude, Ω is the siderial pattern speed for solar rotation, r is the heliocentric distance, w is the radial solar-wind speed, and we have chosen the earth (E) as the reference location. Thus, the solar wind measured by an earth-orbiting satellite at time t is calculated to encounter the comet at time $t + \Delta t$. This basic method has been used to seek associations of DEs with solar-wind features such as the HCS, high-speed streams, and density enhancements.

The procedure can work well for individual DEs and an example is presented below. To minimize the effects of uncertainties in individual events, Brandt *et al.* (1999) compared all 19 DEs in comet Halley against the solar wind. The 19 DEs were found to be associated with crossings of the HCS and no other property of the solar wind showed a one-to-one association with DEs.

Two specific DEs deserve special mention because of the proximity of spacecraft. The first DE occurred in comet Halley on March 8.4, 1986. This was during the time of the passage of the Halley Armada. The HCS presence and timing could be verified for this DE. The reversal of the polarity of the magnetic field in the comet's

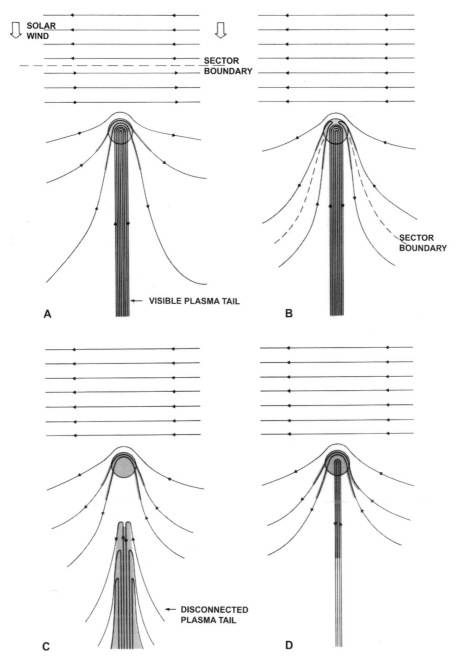

Fig. 6.18. Schematic of the sunward, magnetic reconnection model of disconnection events. (a) The sector boundary (or HCS) separating regions of opposite
magnetic pol-arity in the solar wind approaches a comet; (b) Fields of opposite polarity are pressed together in the comet causing the severing of field lines by reconnection. (c) The disconnected tail moves away and a new tail develops. (d) The old
tail is gone and the new tail continues to develop. This sequence repeats at each
new HCS (or sector boundary) crossing. (Courtesy of M. B. Niedner, Jr., NASA–
Goddard Space Flight Center, and J. C. Brandt, University of New Mexico.)

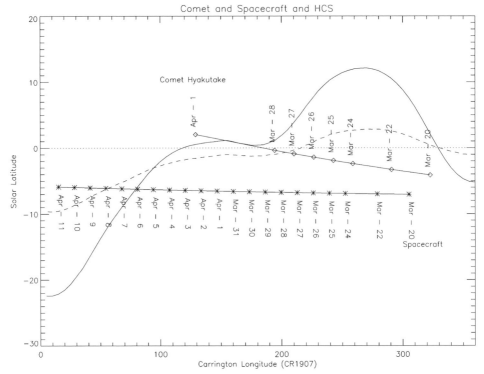

Fig. 6.19. Carrington–Niedner diagram for the DE of March 24.6, 1996 in comet Hyakutake. This diagram is constructed by placing each feature – comet Hyakutake, spacecraft, and the HCS – on a Parker spiral and plotting its Carrington longitude on the coronal source surface. The HCS is plotted for two different methods of calculation (solid curve and dashed curve); see text for discussion. (Courtesy of M. Snow, University of Colorado; see Snow *et al.* 2004.)

ionosphere (as expected on the HCS picture) was verified by *VEGA-1* (before the DE) and *VEGA-2* (after the DE). This is a very important DE and all measurements are consistent with the HCS/magnetic reconnection model (Niedner and Schwingenshuh 1987).

The second is the spectacular DE in comet Hyakutake shown in Fig. 6.17. The DE was well observed, the time of disconnection was well determined at March 24.6, 1996, and the comet was within about 0.1 AU of earth. Thus, it was likely that essentially the same solar wind was seen by the comet as by the earth-orbiting satellites *IMP-8* and *WIND*. This DE was close to the calculated positions of the HCS as shown in Fig. 6.19 (Snow *et al.* 2004). On March 24–25, 1996, the comet was approximately 5° to 14° below the HCS; the 5° is for the (preferred) "radial" calculation of the HCS and the 14° is for the "classic" calculation. These are typical values for the association of DEs and the HCS (Brandt *et al.* 1999).

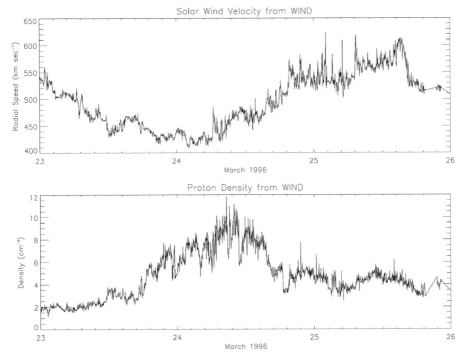

Fig. 6.20. Solar-wind velocity and proton density from the *WIND* spacecraft for the time period around the March 24.6, 1996, DE in comet Hyakutake. (Courtesy of M. Snow, University of Colorado; adapted from Snow *et al.* 2004.)

The two methods of computation for the HCS are included in Fig. 6.19 to illustrate the uncertainty in the computed location of the HCS and as a caution against taking the computed location too seriously. Each computation starts with the measured photospheric field, uses a potential field model to determine the field on a coronal source surface, typically at 2.5 R_s, and then projects the field radially outward into the heliosphere. The computation in Fig. 6.19 can be verified from the expected crossing of the HCS by the spacecraft approximately between April 6.0 and 8.0. This crossing was measured by *WIND* and *IMP-8* on April 6.3.

The spacecraft measurements can be used to check for other solar-wind features that could have produced the DE. Figure 6.20 shows the solar-wind radial speed and density for March 22, 23, 24, and 25, 1996. The speed is slowly varying and is approximately 450 km s^{-1} at the time of the DE. The density is low in this time interval and does not exceed 10 cm^{-3} in this interval. Nothing in these data indicates a likely alternative candidate feature responsible for the comet Hyakutake DE.

The MHD simulation of a comet crossing the HCS was carried out by Yi *et al.* (1996, 1998). The approach was basically similar to the steady-state model of Gombosi *et al.* (1996) (section 6.3.3), but with simplifications to allow treatment

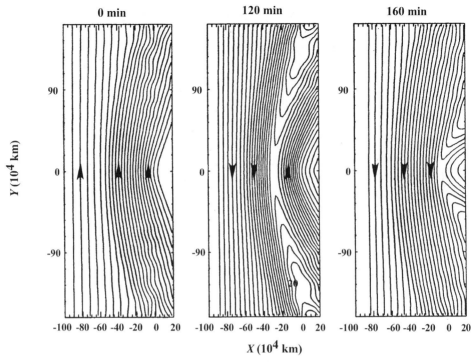

Fig. 6.21. Evolution of the magnetic field line, configuration in the *XY* plane (the plane containing the solar-wind magnetic field or equatorial plane) after the field direction is reversed. The magnetic polarity is marked and the nucleus is located at $(0, 0)$. Reconnected field lines are clearly shown in the panel for 120 min. (Courtesy of Y. Yi, Chungnam National University, South Korea. See Yi *et al*. 1996.)

of the time variations. For example, in Yi *et al*. the solar-wind magnetic field is perpendicular to the flow direction (thus allowing a quarter box simulation), the only ionization process is photoionization, and a variable (but not adaptive) grid is used. No attempt was made to minimize the numerical resistivity. In fact, the current in the compressed HCS was modeled by an approximation because the thickness of the compressed HCS was smaller than the smallest grid size in the model.

After achieving a steady-state solution, the polarity of the incident solar-wind magnetic field is reversed at $t = 0.00$ hours. The results are shown in Plates 6.5 and 6.6 which show the structure after the IMF is reversed. In both the planes containing the IMF and perpendicular to the IMF, ray formation and the DE are clearly shown. Evidence for magnetic reconnection on the sunward side (in an *x*-type neutral point) is clearly seen in Fig. 6.21. Additional evidence for reconnection is the flow of plasma across the current sheet and the increase in temperature across the current sheet. The evidence for magnetic reconnection seems to be overwhelming.

Figure 6.22 shows a three-dimensional contour surface of the region with ion density over 200 AMU cm^{-3} or about 11 H_2O^+ cm^{-3} after the IMF direction

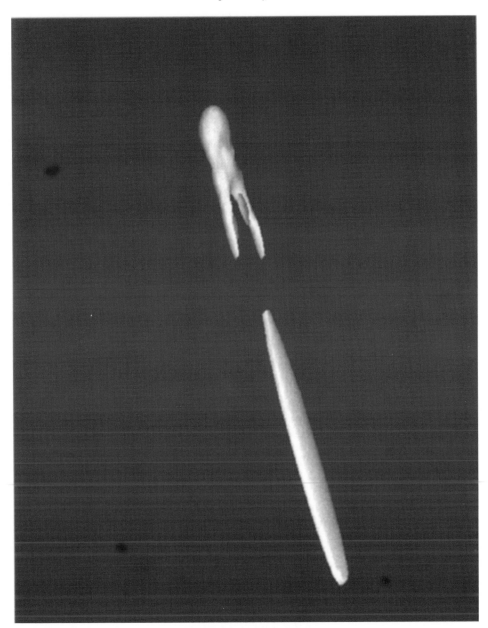

Fig. 6.22. The three-dimensional contour surface of the region with ion density greater than 200 m_p($\approx$11 H_2O) cm^{-3} after the polarity of the solar-wind magnetic field is reversed. (Courtesy of Y. Yi, Chungnam National University, South Korea. See Yi *et al.* 1996.)

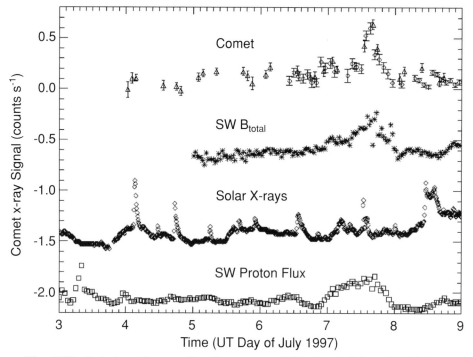

Fig. 6.23. Variation of x-ray flux from comet P/Encke (0.090 to 0.75 keV) showing the variability. The solar-wind magnetic field (SW B_{total}), solar x rays (from *SOHO*), and the solar-wind proton flux are also given for comparison. See Lisse *et al.* (1999). (Courtesy of C. M. Lisse, University of Maryland, College Park: Reprinted from *Icarus*, Vol. 141, by Lisse, Christian, Dennerl, Englhauser, Trümper, Desch, Marshall, Petre, and Snowden, 1999. X-ray and extreme ultraviolet emission from comet P/Encke 1997, pp. 316–30, Copyright 1999, with permission from Elsevier.)

is reversed. The expected structure of the DE is clearly shown. The combined evidence of the physical association of DEs and the HCS plus the results of the MHD simulation modeling the comet crossing the HCS give considerable support to the picture presented here. Continued research on the circumstances of individual DEs plus advances in modeling techniques should clarify any remaining uncertainty.

6.4.4 Cometary x rays

The unexpected discovery of bright x rays from comet Hyakutake (Lisse *et al.* 1996) has opened a new and promising chapter in the field of comet/solar-wind interactions. The emission was concentrated on the sunward side of the comet and was seen in the 0.09 to 2.0 KeV energy range. The x-ray flux was highly variable in comet Hyakutake and also in comet Encke (Lisse *et al.* 1999) as shown in Fig. 6.23. An x-ray image of comet LINEAR (C/1999 S4) is shown in Plate 6.7.

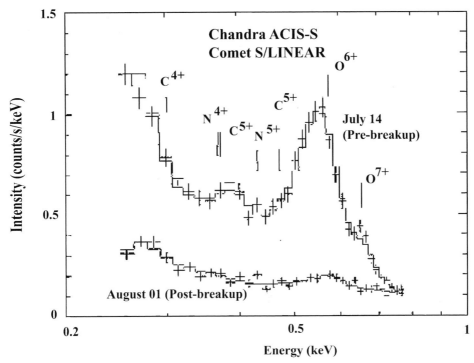

Fig. 6.24. *Chandra X-Ray Observatory* spectrum of comet LINEAR (C/1999 S4) and comparison with a model consisting of lines (broadened by the instrumental width) from charge-exchanged ions and a continuum from thermal brehmsstrahlung. The agreement is excellent. See Lisse *et al.* (2001). (Courtesy of C. M. Lisse, University of Maryland, College Park: Reprinted with permission from *Science*, Vol. 292, pp. 1343–8, Charge-exchange-induced X-ray emission from comet C/1999 S4 (LINEAR), by Lisse, Christian, Dennerl, Meech, Petre, Weaver, and Wolk. Copyright 2001 American Association for the Advancement of Science.)

High-speed collisions between interplanetary dust particles and dust particles released by the comet were the original motivation for looking at comet Hyakutake (e.g., Ibadov 1996). This comet came very close to earth and the circumstances were believed favorable for a detection. The observed brightness was much too high to be from dust–dust collisions, and early on some form of solar-wind interaction seemed likely. Dennerl *et al.* (1997) searched the *ROSAT* database and found several more comets in x rays. They concluded that their "findings establish comets as a class of x-ray sources." Thus, x-ray emission from comets must be regarded as a universal phenomenon.

Subsequent analysis involving 15 comets showed that the emission was confined to the coma volume between the nucleus and the sun. No correlation was found between the x-ray emission and dust or plasma tails or the sun's x-ray flux. The spectrum was soft with $kT \sim 0.3$ KeV. Many mechanisms involving the solar wind have been proposed including: charge exchange of solar-wind heavy ions with cometary neutrals; electron–neutral bremsstrahlung; electron–dust bremsstrahlung; electron–ion bremsstrahlung; instabilities producing lower hybrid waves; and mini-flares produced by the capture and dissipation of small current sheets in the solar wind (Lisse *et al.* 1997).

Lisse *et al.* (2001) have reviewed the evidence on x rays from 15 comets and conclude that only two remain plausible. First, is the proposal by Cravens (1997, 2002); heavy minor species in the solar-wind charge exchange with neutrals in the coma and produce ions in an excited state, e.g.,

$$O^{6+} + M \rightarrow O^{5+*} + M^{+}. \tag{6.10}$$

In this example, a six-times ionized oxygen ion charge exchanges with a neutral molecule represented by M. This can be any of a variety of neutrals in the coma such as H_2O, OH, O, H, etc. The products are a new ion, M^{+}, and a five-times ionized oxygen ion. Some of these oxygen ions can be in an excited state, denoted by the * in (6.10). These excited ions spontaneously decay and emit photons in the extreme ultraviolet/x-ray region of the spectrum.

Second is the electron–neutral thermal bremsstrahlung proposal of Northrop *et al.* (1997). This process involves energetic solar-wind electrons colliding with a cometary neutral and a photon is emitted.

The charge-exchange mechanism produces lines and the electron–neutral thermal bremsstrahlung produces a continuum. Both mechanisms use coma neutrals as the target.

Lisse *et al.* (2001) obtained observations of comet C/LINEAR 1999 S4 using the *Chandra X-Ray Observatory*. Line emission was detected as shown in Fig. 6.24. The fit to the observations contains a six-line charge exchange model plus a thermal bremsstrahlung contribution. The clear peak at 570 eV is caused by charge exchange to O^{+5}. The agreement with the observations is excellent, but the contribution from thermal bremsstrahlung could decrease as spectral resolution improves.

The study of x rays from comets is relatively new and further developments are to be expected. The x rays provide another cometary phenomenon caused by the solar-wind interaction.

6.5 Summary

The solar-wind/comet interaction is a challenging plasma physics problem. It produces the plasma tail and occasionally a disconnection event (DE). *In situ* measurements at three comets and detailed modeling demonstrate the complexity involved. The discovery of comets as x-ray sources suggests that additional discoveries should be expected.

7

The nucleus

The nucleus is the origin of all cometary activity. Without the nucleus, the spectacular tails and other phenomena simply would not exist. Unfortunately, we have detailed knowledge of only two nuclei: comet Halley observed in 1986 and comet Borrelly observed in 2001. While the images of these nuclei are wonderful, they are a mixed blessing because we reasonably expect comet nuclei to show a broad range of characteristics. We must constantly be on guard against interpreting the results from these two comets as characteristics of all comets. Still, the observations of comet Halley and comet Borrelly are invaluable and we begin with them to set the stage for our discussion of cometary nuclei.

7.1 The nucleus of comet Halley

The *Giotto* mission to Halley's Comet (described in Chapter 3) passed within approximately 596 km of the nucleus on the sunward side. The spacecraft carried the Halley Multicolor Camera (HMC), which obtained excellent images on the inbound leg of its trajectory. Unfortunately, the HMC system was damaged by dust impacts around the time of closest approach. Specifically, an analysis of the data indicated that the HMC baffle (a sunshade designed to reduce the amount of stray sunlight) probably collapsed and covered the HMC entrance aperture. Figure 7.1 shows a sequence of images as the spacecraft approached the nucleus. Figure 7.2 shows a composite of 68 HMC images of the nucleus, a schematic diagram identifying many of the features, and enlarged images of these features. Considerable detail is shown.

The *VEGA* missions to Halley's Comet (described in Chapter 3) also imaged the nucleus. They passed the nucleus at considerably larger distances than *Giotto* and their images were noisy due to some technical difficulties. Nevertheless, these images were valuable. Figure 7.3 shows a *VEGA-1* image taken near closest approach at a distance of 8904 km and Fig. 7.4 shows a *VEGA-2* image taken near closest approach at a distance of 8031 km.

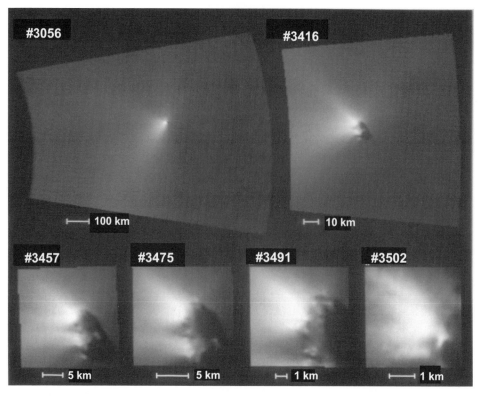

Fig. 7.1. Time sequence of images from the Halley Multicolor Camera (HMC) as the *Giotto* spacecraft approached the nucleus of comet Halley. The scale on the nucleus is indicated by the bars. (Courtesy of H. U. Keller, Max-Planck-Institute für Aeronomie, Katlenburg-Lindau, Germany. Copyright © MPAE.)

What was found? And, what was expected? For many, the images contained surprises. See Keller (1990) for a review of imaging results. Refer to the ideas on the nuclei of comets outlined in Chapter 1. The nucleus was not spherical or even close. Also, the surface has many local features that can be labeled valleys, hills, craters, etc. Figure 7.5 gives a good idea of the large-scale shape and dimensions. The long axis was approximately 16 km and the short axes were approximately 8 km. These values were much larger than the radius of 2.5 km expected for a spherical nucleus. The culprit was the albedo, the fraction of light reflected from the surface. Estimates for the albedo varied, but some were as high as 0.6. The actual value was 0.04. Thus, the nucleus was much larger and darker than expected.

The gas and dust emission was not uniform over the entire surface, but occurred only in discrete jets, and these were active only on the sunlight side. The fraction of the surface covered with active jets was ∼10%. The jets "turned on" as their portion of the nucleus rotated into sunlight and "turned off" as their portion rotated onto

(a)

Fig. 7.2. (a) Nucleus of Halley's Comet as shown in a 68-image composite taken by the HMC on March 14, 1986. (b) Schematic diagram identifying features on the nucleus. (c) Images of individual features. (Courtesy of H. U. Keller, Max-Planck-Institute für Aeronomie, Katlenburg-Lindau, Germany. Copyright © MPAE.)

the night side. The emission is toward the general direction of the sun. The rapid turn-on and turn-off of the jets implied a thin dust crust over the jets and a thick crust elsewhere. Erosion of the pits from which the jets emanate could produce Brownlee-type particles, from either the pit surface or the pit edges.

An important measurement of the surface temperature for comet Halley was made by the infrared instrument (IKS) on *VEGA-1* (Emerich *et al.* 1987) at a wavelength of around 10 μm. The IR fluxes measured were much larger than expected for a

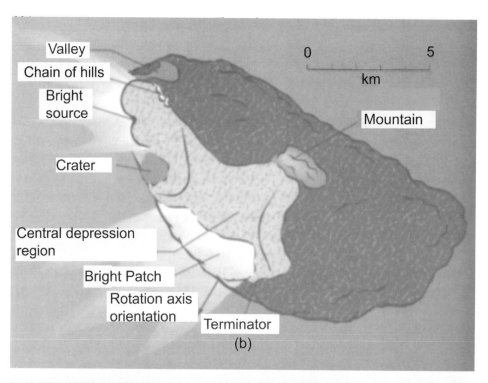

(b)

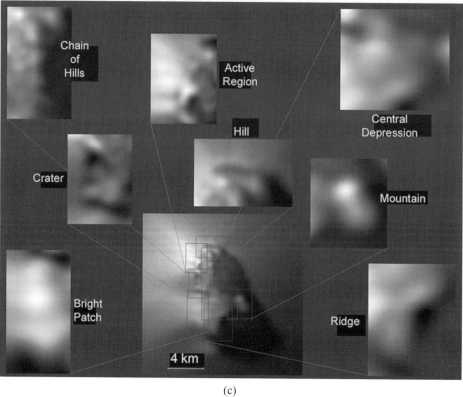

(c)

Fig. 7.2. (*Cont.*)

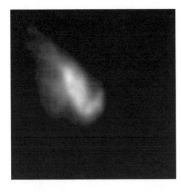

Fig. 7.3. *VEGA-1* image of comet Halley's nucleus taken near closest approach. Range was 8904 km. (International Halley Watch 1995. Comet Halley Archive, CD-ROM Vol. 26. USA-NASA-IHW-HAL-0026, Small Bodies Node, University of Maryland, College Park, MD.)

Fig. 7.4. *VEGA-2* image of comet Halley's nucleus taken near closest approach. Range was 8031 km. (International Halley Watch 1995. Comet Halley Archive, CD-ROM Vol. 26. USA-NASA-IHW-HAL-0026, Small Bodies Node, University of Maryland, College Park, MD.)

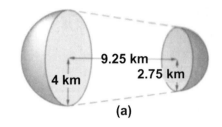

(a)

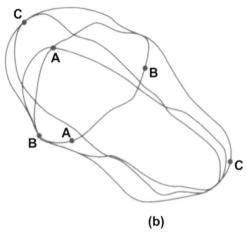

(b)

Fig. 7.5. Representations of comet Halley's nucleus after drawings by R. Z. Sagdeev *et al.* (1987). (a) A simple model with dimensions marked. (b) A more realistic model with the following approximate dimensions: A–A, 7.5 km; B–B, 8.2 km; and C–C, 16.0 km.

layer of sublimating ices. The measurements gave an average value of 320 K and a possible maximum value of 400 K.

The significance of these values can be seen by a comparison with the temperature for the surface of a black body:

$$T_s = \frac{277 \left(\frac{1-A}{\varepsilon}\right)^{1/4}}{\sqrt{r}} \text{(K)}, \qquad (7.1)$$

where A is the albedo, ε is the emissivity ($=1$ for a black body), and the heliocentric distance r is in AU. This equation is valid for the case of a rapidly rotating body that absorbs solar radiation over an area πa^2 and radiates over $4\pi a^2$. If the body is dark, $A \ll 1$, and the albedo term can be neglected. The *VEGA-1* encounter took place at 0.79 AU from the sun and the temperature from (7.1) would be about 312 K. For a slowly rotating body, the temperature would be $2^{1/4}$ (or 19%) higher for a value of 371 K.

Thus, the surface of Halley's Comet has a temperature characteristic of a dark, dusty surface much higher than the expected 215 K temperature for sublimating water ice. These measurements are a critical fact to be explained by any physical model of ice sublimation from the surface layers of comets.

The interior temperature in Halley's nucleus may have been determined by infrared measurements at 2.65 μm from the *Kuiper Airborne Observatory*. The spectrum was obtained by Mumma *et al.* (1987). (Also see Fig. 7.26 and discussion for comet Hale–Bopp). In ortho water, the nuclei of the two H atoms spin in the same direction, while in para water, they spin in the opposite direction. The ratio of ortho to para hydrogen (OPR) depends on the temperature at the time the water molecules formed. This temperature may also be the temperature of the interior. The two states of water are almost independent species that can be changed from one to the other only by chemical reactions. Thus, the interior ices can be sublimated and flow through the crust into the coma without altering the OPR. The measurements indicated a temperature of 50 K or less.

The rotation of the nucleus was originally thought to be constrained from the period of observed brightness variation and the snapshots (in time) taken by the *VEGA*s and *Giotto*. An accurate description of the rotation is rather complex and the rotation axis does not remain fixed in the nucleus. This subject is discussed further below in connection with the interior structure of comets.

The images of the Halley nucleus placed the final nail in the coffin of the sandbank model of the nucleus as proposed by Lyttleton (1953); see Chapter 1. The sand-bank model could not explain the survival of comets after many passes through the inner solar system or the non-gravitational forces. Any remaining doubt should have been dispelled by the images of the nucleus of Halley's Comet.

7.2 The nucleus of comet Borrelly

The *Deep Space 1* mission passed by comet Borrelly (19P) on September 22, 2001. The closest approach was at a distance of 2171 ± 10 km. (See section 11.1 for a brief discussion of the mission history.) The Miniature Integrated Camera and Spectrometer (MICAS) on the spacecraft provided superb images and spectral information (Soderblom *et al.* 2002).

The distant view (Fig. 7.6) shows several jets. The active areas on the nucleus are approximately 10% or less of the surface area. The coma shows dust features consisting of fans and highly collimated jets. At the time of encounter, the images showed a prominent dust jet containing at least three smaller jets. This feature emanated from a broad basis near the middle of the nucleus.

Fig. 7.6. Image of comet Borrelly taken by *Deep Space 1* on September 22, 2001. The image shows several jets. (Courtesy of NASA/JPL.)

Figure 7.7 shows the close-up image of the nucleus. The high-resolution images range from 47 to 58 meters per pixel. The terrain is complex, but does not show impact craters larger than approximately 200 meters in diameter. The long axis of the nucleus is 8 km, a value in good agreement with the value obtained by Lamy *et al.* (1998b) from *Hubble Space Telescope* observations (see below). The albedo over the surface varies from 0.01 to 0.03. The short-wavelength infrared spectra indicate a hot, dry surface. The temperatures are in the range 300 to 345 K and are reasonably close to values for a slowly rotating body with low albedo and an emissivity of 0.9 (at a heliocentric distance of about 1.4 AU); see (7.1). The spectra do not show evidence for water ice or hydrated minerals.

The excellent spatial resolution of the images allowed detailed studies of the terrain as shown in Fig. 7.8. The "smooth" terrain (indicated) is found mostly on

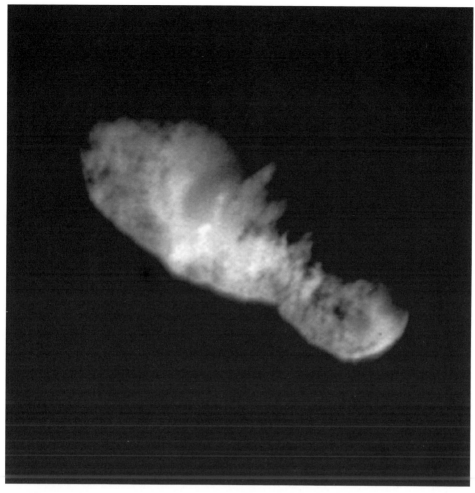

Fig. 7.7. Close-up image of the nucleus of comet Borrelly. See text and Fig. 7.8 for discussion. (Courtesy of NASA/JPL.)

the central part of the nucleus and includes several mesa-like features. The smooth terrain has a higher than average albedo and contains the source regions for the jets. The "mottled" terrain is generally inactive and contains bumps, pits, troughs, and ridges. Strong albedo variations are present, but the overall albedo is lower than average. As shown in Fig. 7.8, the information available allows the investigation of complex geological relationships.

7.3 The nature of the nucleus

This section summarizes the facts known about cometary nuclei in addition to the nuclei of comet Halley and comet Borrelly.

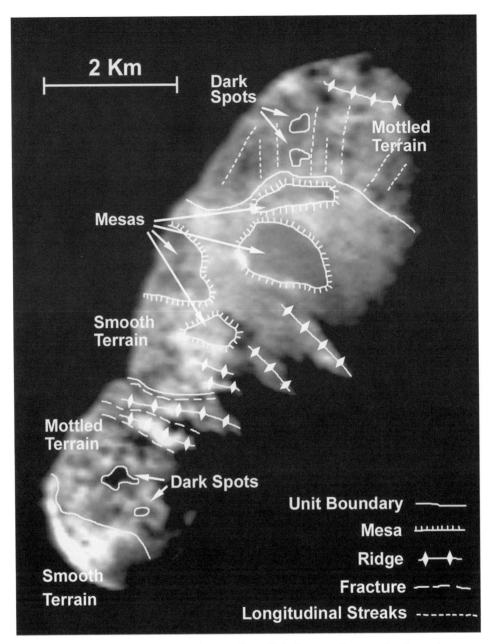

Fig. 7.8. Morphological map of comet Borrelly's features. See text for discussion. (Courtesy of NASA/JPL/Dan Britt, University of Tennessee.)

7.3.1 Sizes

Accurate sizes are difficult to determine. Occasionally, fortunate circumstances such as a comet passing close to earth can provide a measurement opportunity. But, most of the time astronomers are left to infer the size from brightness observations. Simply, a brightness gives a measure of a characteristic size (a) squared times the albedo (A) or $a^2 A$. The albedo must be assumed or determined to calculate the characteristic size. See Chapter 3 for specific definitions for albedos.

In 1927, comet Pons–Winnecke passed within 6 million kilometers of earth, but the nucleus could not be resolved. Had the nucleus been as large as 5 km across, it would have been resolved. The 5 km figure is then an upper limit.

Occasionally, a comet can pass directly between the earth and the sun. The silhouette, in principle, could give information on the size of the nucleus. The Great Comet of 1882 passed between the earth and the sun, but no effects could be seen. Certainly the interpretation was difficult, but an upper limit of 70 km across was reported.

More direct observations have the problem of separating the light from the nucleus from the surrounding coma radiation. Roemer (1966) tackled this problem by using long focal-length observations that, in principle, suppressed coma radiation. Roemer assumed albedos of 0.02 and 0.7 to cover the possible extremes in the solar system. We now consider 0.02 to be closer to the true value based on experience with Halley's Comet. Radii for 29 comets were determined with values between 0.8 and 65 km.

The planetary camera on the *Hubble Space Telescope* has been used to provide extensive data on the sizes of nuclei. This approach has been pioneered by P. Lamy and his collaborators. The high spatial resolution available allows the nucleus to be observed essentially as a delta function in the midst of a slowly varying coma contribution. Lamy and Toth (1995) wrote the brightness as a function of the apparent distance from the center (ρ) as:

$$B(\rho) = a\{1/\rho + b\delta(\rho)\}. \tag{7.2}$$

Here, B is the brightness; $1/\rho$ is the expected variation of the coma brightness for a constant source and an inverse square dependence of space density; $\delta(\rho)$ is the delta function containing the light from the nucleus; and a and b are constants. A fit to the data produced a radius for comet 4P/Faye of 1.8 km for an assumed albedo of 0.04. The sizes calculated from individual observations showed a fairly small variation and the nucleus was taken to be rather spherical. Roemer's value scaled to an albedo of 0.04 is 3.8 km. While these values are substantially different, they are still reasonably close. Note that the investigation of comet Faye was carried out when the *Hubble Space Telescope* still suffered from spherical aberration.

An investigation without the spherical aberration was applied to comet Borrelly. The basic approach was similar to (7.2) multiplied by the instrumental point-spread function (PSF). A complication was the need to model the asymmetric coma. Lamy

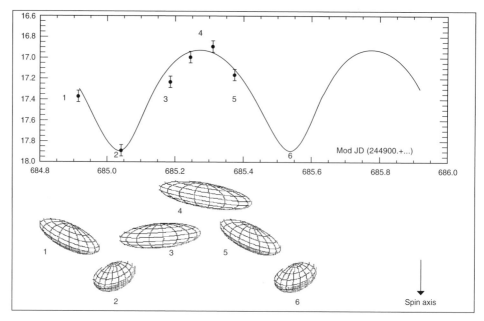

Fig. 7.9. The light curve for comet Borrelly. The fitted curve is shown along with the model cross-sections at different phases. (Courtesy of *Astronomy and Astrophysics*: Lamy, P. L., Toth, I., and Weaver, H. A., 1998. Hubble Space Telescope observatons of the nucleus and inner coma of comet 19P/1904 Y2 (Borrelly). *Astron. Astrophys.* 337: 945–54.)

et al. (1998a) unambiguously detected the nucleus. The light curve (Fig. 7.9) showed substantial variations and they were able to derive semi-axes of 4.4 ± 0.3 km and 1.8 ± 0.15 km for this "spheroidal" body, all by assuming an albedo of 0.04. The surface area can be calculated from these dimensions (by assuming one long and two short axes) and compared with the active area determined by A'Hearn *et al.* (1995), see below. The comet has an active area of 8% making it a moderately active comet.

The *Deep Space 1* imaging of 19P/Borrelly provides an excellent opportunity to check the basic approach just outlined. The long axis value from *Deep Space 1* is quoted as 8 km, which would be somewhat smaller ($\approx 10\%$) than the results from Lamy *et al.* (1998b). When the quoted errors are considered, the differences are inconsequential. Also, Lamy *et al.*'s result is based on an assumed albedo of 0.04; they noted that a change in the assumed albedo of 0.005 would change the dimensions by 7%. All in all, the agreement is excellent and can be regarded as verifying the basic approach.

The size of the nucleus of comet Hale–Bopp was estimated with a similar approach by Weaver *et al.* (1997) and Weaver and Lamy (1997). The *HST* observations are shown in Fig. 7.10. The profiles are given for two different slices depending

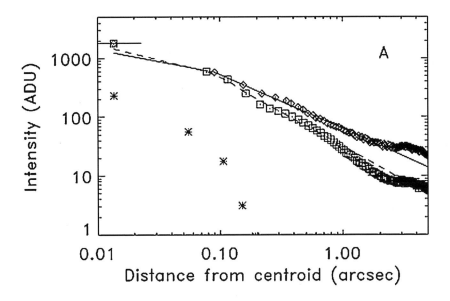

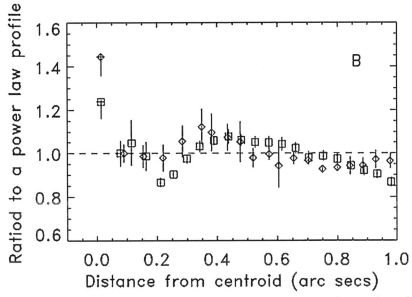

Fig. 7.10. *Hubble Space Telescope* observations of the near-nucleus region of comet Hale–Bopp on October 23, 1995. (A) Intensity profiles along two different directions are shown along with the power-law fits. The asterisks show the point-spread function (PSF) scaled from a nucleus having a diameter of 27 km and a geometric albedo of 0.04. (B) The observed intensities are divided by the power-law fits to show more clearly the signature of the nucleus. The two data points near zero distance imply diameters of 27 km and 42 km. Courtesy of H. Weaver. (Reprinted with permission from *Science*, Vol. 275, 28 March 1997, pp. 1900–4. The activity and size of the nucleus of comet Hale–Bopp (C/1995 O1), by Weaver, Feldman, A'Hearn, Arpigny, Brandt, Festou, Haken, McPhate, Stern, and Tozzi, Figure 2. Copyright 1997. American Association for the Advancement of Science.)

on whether a weak but persistent jet is included. Extrapolating the fitted power law curves and dividing by the power law curve gives the two points nearest to the center. For the standard 0.04 albedo, these correspond to a nucleus with radii of 27 to 42 km. These values are substantially greater than the values for Halley's Comet.

An important application of this general approach was carried out by Lamy *et al.* (1998a) for comet 46P/Wirtanen, the target of the *Rosetta* mission (section 11.1.4). *Hubble Space Telescope* observations were used to separate the signal from the nucleus from that of the coma. The nucleus signal would be produced by a spherical body with an albedo of 0.04 if its radius were 600 ± 20 meters. An earlier estimate of 700 to 800 meters was obtained by Boehnhardt *et al.* (1997) from observations of the comet while it was in a star-rich field. More recently, Boehnhardt *et al.* (2002) obtained observations at the Very Large Telescope Observatory in Chile and determined a mean radius of 550 ± 40 meters. Note that the error bars of the last two values overlap. While comet Wirtanen has been observed often since its discovery in 1948, it was not well-characterized until recently. The current effort, started with a coordinated observing campaign in 1996/1997, was the result of the comet being selected as the target of the *Rosetta* mission. Table 7.1 lists some optically determined nuclear radii of comets.

Radar measurements can usefully probe comets. But they must come quite close to earth because the strength of the radar echoes varies as the surface area (S) times the range (Δ) to the minus fourth power or $S\Delta^{-4}$. Comet IRAS–Araki–Alcock, in May of 1983, was studied at a range of 0.031 AU and a wavelength of 12.6 cm by Harmon *et al.* (1989). The return spectrum showed two distinct components. The narrow-band component corresponded to the nucleus and a broad-band component was interpreted as coming from a particle cloud. The size of the nucleus depends on the dielectric constant of the surface. Figure 7.11 shows a plot of nucleus size for various materials. A rock surface is unlikely and the radius probably falls in the range 2 to 8 km. Thermal radiation measurements by Hanner *et al.* (1985) imply a radius of about 5 km. This would correspond to a deep layer of packed snow.

The particle cloud is confined to within 1000 km of the nucleus and is predominantly on the sunward side. The size of the particles is in the order of centimeters.

Observing conditions for radar probing of comet Halley were poor during the last apparition. Still, weak echoes were detected at 12.6 cm in late 1985 (Campbell *et al.* 1989). The echo spectrum was dominated by a broad-band feature whose nature was inconsistent with an echo from the nucleus. The echo was interpreted as backscatter from grains larger than 2 cm in radius ejected from the surface. These grains are presumably similar to those found from the radar study of comet IRAS–Araki–Alcock. The echoes from comet Halley were obtained at the large range of 0.63 AU.

Table 7.1 *Optically determined nuclear radii*

Comet	Effective radius or dimensions (km)	Reference
4P/Faye	1.8	Lamy *et al.* (2000)
10P/Tempel 2	4.6	Lamy *et al.* (2000)
17P/Holmes	1.7	Lamy *et al.* (2000)
19P/Borrelly	4.4 × 1.8	Lamy *et al.* (1998b)
37P/Forbes	0.8	Lamy *et al.* (2000)
44P/Reinmuth 2	1.6	Lamy *et al.* (2000)
50P/Arend	0.95	Lamy *et al.* (2000)
59P/Kearns–Kwee	0.8	Lamy *et al.* (2000)
63P/Wild 1	1.5	Lamy *et al.* (2000)
71P/Clark	0.7	Lamy *et al.* (2000)
84P/Giclas	0.9	Lamy *et al.* (2000)
106P/Schuster	0.9	Lamy *et al.* (2000)
112P/Urata–Niijima	0.9	Lamy *et al.* (2000)
114P/Wiseman–Skiff	0.8	Lamy *et al.* (2000)
C/1995 O1 Hale–Bopp	27 to 42	Weaver *et al.* (1997)

The albedo is usually assumed to be 0.04.

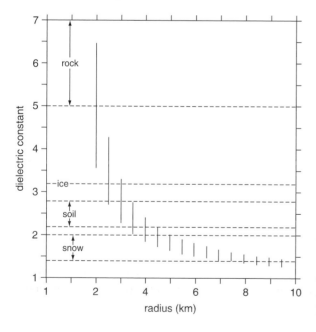

Fig. 7.11. A plot of radius vs. dielectric constant for the radar observations of comet IRAS–Araki–Alcock (1983d). A deep layer of packed snow and a radius of about 5 km are implied; see text. Courtesy of J. K. Harmon. (From Harmon, Campbell, Hine, Shapiro, and Marsden, 1989. Radar Observations of comet IRAS–Araki–Alcock 1983d. *Astrophys. J.*, 338: 107–93, Figure 4, Radius vs. dielectric constant for radar observations of comet IRAS–Araki–Alcock. Reproduced by permission of the AAS.)

Sample radar observations of comet Hyakutake at a range of 0.10 AU are shown in Fig. 7.12 (Harmon *et al.* 1997). The sharp peak is the echo from the nucleus; the radius was estimated to be 1.5 to 2.0 km. The broad-band feature was again attributed to large grains ejected from the nucleus. Comets Encke and Grigg–Skjellerup were also detected by radar, but an estimate for nuclear size was not possible.

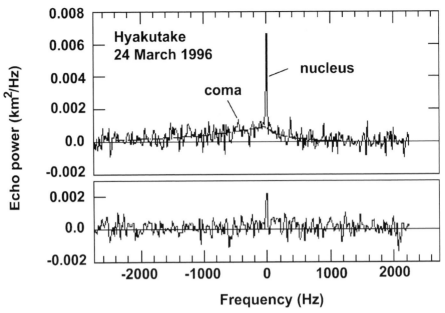

Fig. 7.12. Radar observations of comet Hyakutake (C/1996 B2) showing the detection of the nucleus and coma. The labels indicate the opposite-sense (OC) and same-sense (SC) circular polarization echos from a circularly polarized transmitted wave; see Harmon *et al.* (1997). (Courtesy of J. K. Harmon, National Astronomy and Ionosphere Center, Arecibo Observatory, Puerto Rico.)

Harmon *et al.* (1999) have reviewed the radar observations of comets. They note that the "...highest priority of future radar observations will be to obtain delay-doppler images of a nucleus, which would give direct size and shape estimates as well as a more reliable albedo." They also list future apparitions of short-period comets that offer excellent opportunities for radar studies.

Sizes and albedos of bare nuclei can be determined by using the "standard method" developed for asteroids and satellites. A non-icy surface reflects light in an amount proportional to $p_\lambda S F_\lambda$ (p_λ is the relevant geometric albedo, S is the projected surface area, and F_λ is the solar flux in the wavelength of interest) and emits thermal black-body radiation at a rate proportional to $(1 - A_{BB})SF \propto S\varepsilon T^4$ (where A_{BB} is the bolometric Bond albedo, and ε is the emissivity). If all the reflected and re-emitted radiation were measured, the calculations would be relatively simple. This is not the case, and the calculations are simple in principle and complicated in practice. A major complication is the phase integral (see Chapter 3) formally given by

$$A_B = p_\lambda \times q_\lambda, \tag{7.3}$$

where A_B is the wavelength-dependent Bond albedo, p_λ is the geometric albedo, and q_λ is the phase integral. Also, the infrared beaming factor is a source of uncertainty; the infrared emission is not isotropic, but rather is peaked at opposition. These determinations require visual observations (where most of the light is reflected) and infrared observations (where the thermal radiation is emitted). According to Wien's Law, dark bodies near 1 AU radiate around 10 μm, and those near 4 AU radiate around 20 μm.

The emissivity ε is taken to be 0.9. The black-body nature of the infrared can be verified observationally. Then, the visual magnitude and the infrared measurement of the thermal flux determine the projected surface area and albedo. Earlier descriptions of this technique were given by Morrison (1973) and Jones and Morrison (1974). A "standard" approach is due to Lebofsky *et al.* (1986), where they used occultation diameters of the asteroids 1 Ceres and 2 Pallas to refine the technique; see also Lebofsky and Spencer (1989). An early cometary application of this method was the determination of the diameter of comet Schwassmann–Wachmann 1 at 40 ± 5 km (Cruikshank and Brown 1983).

In fact, there are at least two "standard models" in the literature. The *standard thermal model* (STM) applies to non-rotating objects in equilibrium with solar radiation. The STM produces the highest nuclear temperature. The *isothermal latitude model* (ILM) (Spencer *et al.* 1989) applies to rapidly rotating objects and produces the lowest nuclear temperature. Campins *et al.* (1995) applied both variations to 4015 Wilson–Harrington. For the STM, they found an effective radius of 1.3 ± 0.16 km and a geometric albedo of 0.10 ± 0.02. For the ILM, they found an effective radius of 2.0 ± 0.25 km and a geometric albedo of 0.05 ± 0.01. Campins *et al.* conclude that the ILM is the better description of 4015 Wilson–Harrington.

We mention two other applications of this method to comets of special interest. A'Hearn *et al.* (1989) studied comet 10P/Tempel 2, a perennial target for cometary missions. The optical and thermal light curves are in phase with a period of 8.9 hours. The results are: the shape is a prolate spheroid with dimensions of $16.3 \times 8.6 \times 8.6$ km; the albedo $p_V = 0.022$. Based on the gas production rate, the active fraction of the surface is less than 3%. Table 7.2 lists some nuclear radii based on the standard method.

Hanner *et al.* (1987) studied comet Sugano–Saigusa–Fujikawa (C/1983 J1) using almost the standard model. They found that the radius had a maximum value of about 370 meters; the maximum value comes about from assuming that the entire signal came from the nucleus. The radius value is compatible with the result from radar ranging and the geometric albedo was consistent with 0.03. Despite its small size, the comet had substantial gas production rates and Hanner *et al.* consider it to be "A Small, Puzzling Comet."

Table 7.2 *Nuclear radii from the standard method*

Comet	Radius (km)	Albedo	Reference
2P/Encke	4.8	0.046	Fernández (1999)
10P/Tempel 2	5.9[a]	0.022	A'Hearn *et al.* (1989)
29P/Schwassmann–Wachmann 1	40	0.13	Cruikshank and Brown (1983)
55P/Tempel–Tuttle	3.5	0.06	Fernández (1999)
81P/Wild 2	<6.0	>0.018	Fernández (1999)
107P/4015 Wilson–Harrington	2.0	0.05	Campins *et al.* (1995)
C/1983 J1 Sugano–Saigusa–Fujikawa	≤0.37	≥0.03	Hanner *et al.* (1987)
C/1995 O1 Hale–Bopp	50	0.045	Fernández (1999)

[a] The nucleus is approximately a prolate spheroid; see text.

Another approach to maximizing the ratio of nuclear : coma contribution to an image is to observe the comet at large heliocentric distances, usually with CCDs. Of course, the comet must still be bright enough to observe. Examples are comet 10P/Tempel 2 (Jewitt and Luu 1989) and 31P/Schwassmann–Wachmann 2 (Luu and Jewitt 1992). Comet 10P/Tempel 2 was observed as far out as 4 AU. The bare nucleus appeared to be visible beyond 2.5 AU. The light curve leads to the conclusion that the nucleus is prolate with axes of $1.9 : 1 : 1$. Adopting a geometric albedo of 0.024 (final value of 0.022) from A'Hearn *et al.* (1989) gives semi-axes of $8 : 4 : 4$ km. Comet 31P/Schwassmann–Wachmann 2 was observed at 4.6 AU. An upper limit to the radius of 3.1 km was derived assuming a geometrical albedo of 0.04.

The sizes and size distribution can be estimated from cratering on the Galilean satellites of Jupiter. Of course, the assumption is made that the crater-producing comets near Jupiter are a useful surrogate for the global comet population. With this caveat, the crater size distribution at the poles of Ganymede has the following interpretation (Shoemaker and Wolfe 1982). The differential size distribution of $f(R)dr \propto r^{-3}dr$ holds down to about 2 km. The usual scaling law is that craters about 15 to 20 times the size of the impacting bodies are produced. Thus, the r^{-3} differential distribution could hold down to the range of 100 to 200 meters. The number of comets in a range dr can be written as:

$$n(r)dr \propto r^{-3}dr, \tag{7.4}$$

and the number greater than a given radius becomes:

$$N(r) \propto r^{-2}. \tag{7.5}$$

While bearing the assumptions and uncertainties in mind, this distribution is useful in considering the size distribution of the comet population. Of course, there

Plate 1.1. A facsimile reproduction of the section of the Bayeux Tapestry showing Halley's Comet in the sky above King Harold. (Courtesy of Delphine Delsemme.)

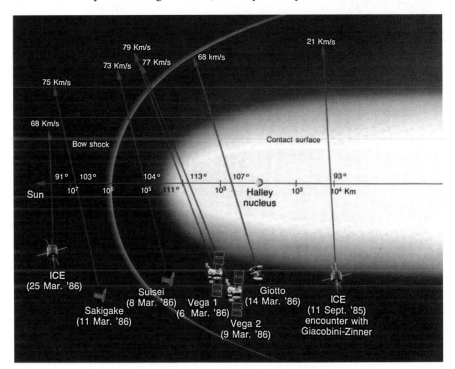

Plate 1.2. Diagrammatic summary of the spacecraft encounters with comets Halley and Giacobini–Zinner. The summary shows the dates and distances of closest approaches. The distance scale is in increments of a factor of ten. (NASA.)

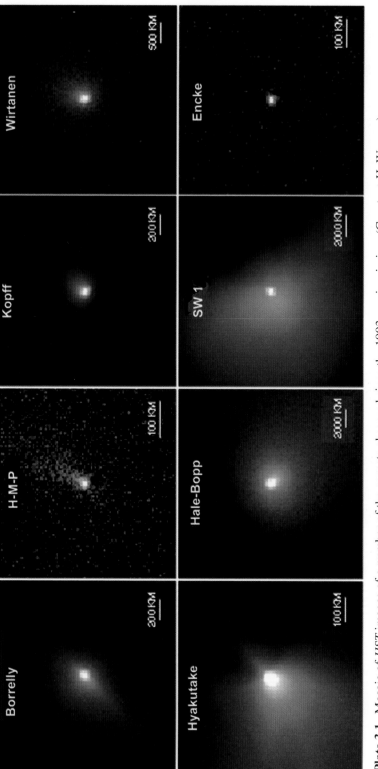

Plate 3.1. Mosaic of *HST* images of a number of the comets observed since the 1993 repair mission. (Courtesy H. Weaver.)

Plate 3.2. The Giotto fresco showing Halley's comet. © Scala/Art Resource, NY.

Plate 4.1. Dramatic image of comet Hale–Bopp on April 8, 1997, showing the blue plasma tail and the whitish dust tail. (Courtesy of H. Mikuz, Crni Vrh Observatory, Slovenia.)

Plate 4.2. *Infrared Astronomical Satellite* (*IRAS*) false-color image constructed from 12 μm, 60 μm, and 100 μm scans. The dust trail is the thin blue line stretching from the comet's head at upper left to the lower right. (Courtesy of Mark Sykes, University of Arizona.)

Plate 4.3. Comet Ikeya–Seki (1965f) as photographed from Kitt Peak at dawn on October 29, 1965. The yellow color is real, and is due to strong sodium emission. (Courtesy of Roger Lynds, National Optical Astronomy Observatory.)

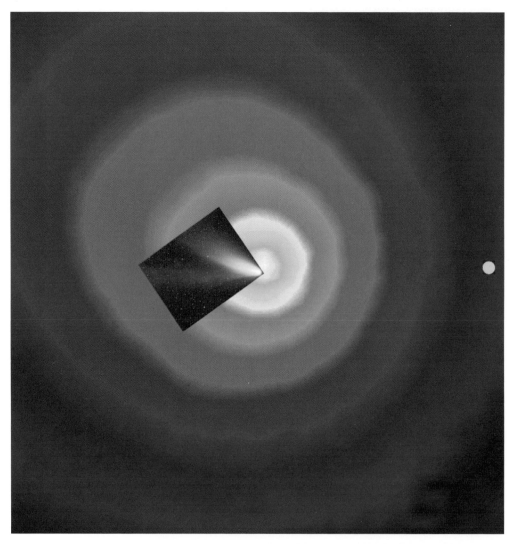

Plate 5.1. Hydrogen Lyman-α image of the coma and hydrogen cloud of comet Hale–Bopp together with a comparison with the visible ion and dust tails. The field of view is approximately 40° on a side and the image was constructed from a *SOHO* SWAN image taken on April 1, 1997, the day of perihelion. The images are to scale and the Lyman-α contours are shown in shades of blue. The small yellow disk shows the angular size of the sun and the solar direction. The visual photograph is by Dennis di Cicco and *Sky & Telescope*. (Courtesy of M. Combi, University of Michigan; see Combi *et al.* 2000.)

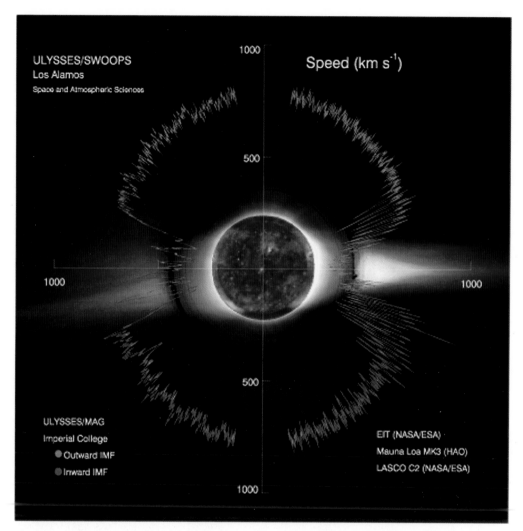

Plate 6.1. Polar plot of *Ulysses* solar-wind speeds and additional images and information as noted. The solar-wind speed of 750 km s^{-1} with small variations at polar latitudes and the solar-wind speed of 450 km s^{-1} with large variations at equatorial latitudes is clearly shown. See http://swoops.lanl.gov and McComas *et al.* 1998. (Courtesy of D. J. McComas, Southwest Research Institute.)

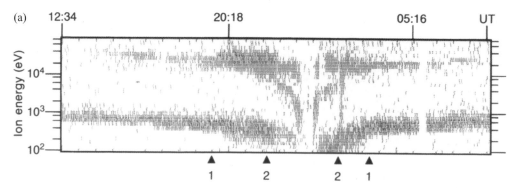

(a)

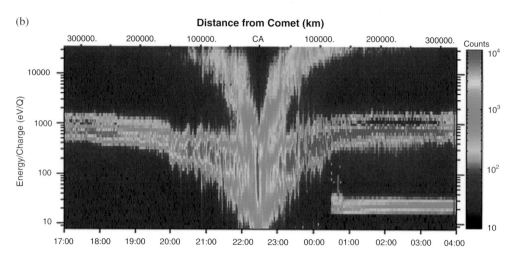

Plate 6.2. (a) Plasma results at comet Halley obtained by the ion analyzer on *Giotto*. The times refer to March 13 and 14, 1986. The data near 10^3 eV are from protons in the solar wind. The fluxes are color-coded, with red indicating the highest and dark blue the lowest. See section 6.3.2.2 for a detailed discussion. (Courtesy of the late A. Johnstone and A. Coates, Mullard Space Science Laboratory, University College London.)

(b)

Plate 6.2. (b) Plasma results at comet Borrelly obtained by the ion analyzer on the *Deep Space 1* mission. The times refer to September 22 and 23, 2001. The bar at lower right is produced by xenon ions from the spacecraft thruster. See section 6.3.2.3 for discussion. See Nordholt *et al.* 2002. (Courtesy of Los Alamos National Laboratory.)

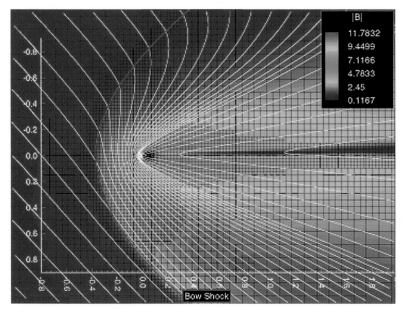

Plate 6.3. The large-scale view of model results. Approximately 2×3 Gm are shown. The bow shock is marked and the white lines mark the magnetic field lines in the equatorial plane (the plane of the solar wind magnetic field). The field strength is given in units of the solar-wind field, 4.81 nT. See section 6.3.3 and Gombosi *et al.* (1996). (Courtesy of T. I. Gombosi, University of Michigan, Ann Arbor.)

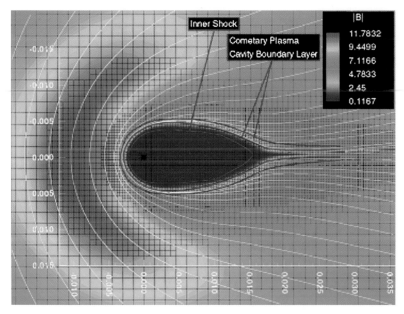

Plate 6.4. The near-nucleus view of model results. Approximately $50\,000 \times 30\,000$ km is shown. The inner shock and the diamagnetic cavity are marked. The white lines mark the magnetic field lines in the equatorial plane. The field strength is given in units of the solar-wind field, 4.81 nT. See section 6.3.3 and Gombosi *et al.* (1996). (Courtesy of T. I. Gombosi, University of Michigan, Ann Arbor.)

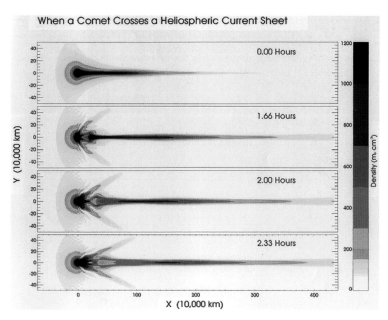

Plate 6.5. MHD simulation of a disconnection event (DE) shown in the *XY* (or equatorial) plane, the plane containing the solar wind magnetic field or IMF. The times after the polarity of the solar wind magnetic field is reversed are given. See Yi *et al.* (1996). (Courtesy of Y. Yi, Chungnam National University, South Korea.)

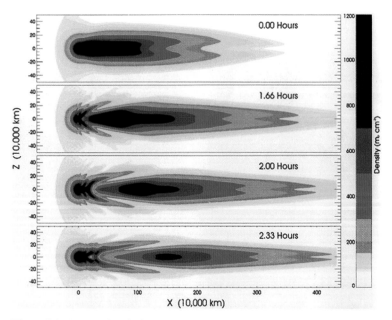

Plate 6.6. MHD simulation of a DE shown in the *XZ* plane, the plane perpendicular to the solar wind magnetic field or IMF. The times after the polarity of the IMF is reversed are given. See Yi *et al.* (1996). (Courtesy of Y. Yi, Chungnam National University, South Korea.)

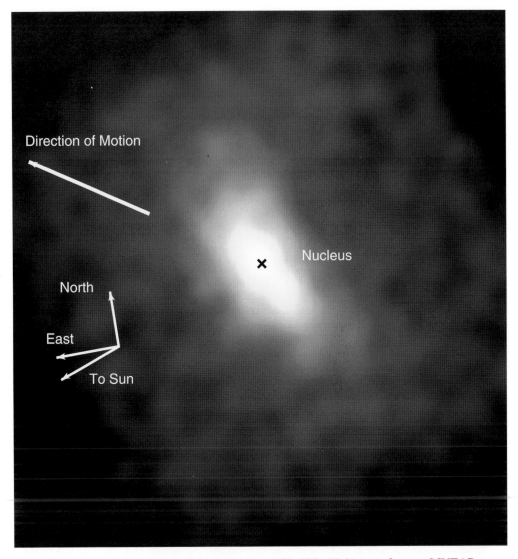

Plate 6.7. False-color representation of an x-ray (200–800 eV) image of comet LINEAR (C/1999 S4) obtained on July 14, 2000 by the Chandra X-Ray Observatory. See Lisse *et al.* (2001). (Courtesy of C. M. Lisse, University of Maryland, College Park, and S. J. Wolk, Chandra X-Ray Center, Harvard-Smithsonian Center for Astrophysics.)

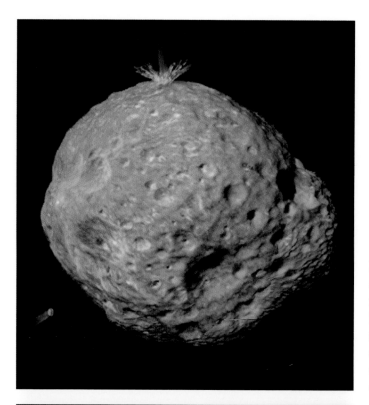

Plate 8.1. (a) Schematic of stages in the fragmentation history of a moderately large asteroid. It is originally composed of strong rock and is

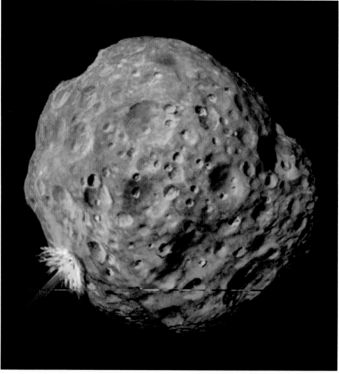

(b) cratered by impacts. An energetic impact

(c) totally disrupts the asteroid. Most of the fragments fail to reach escape velocity and the body reassembles.

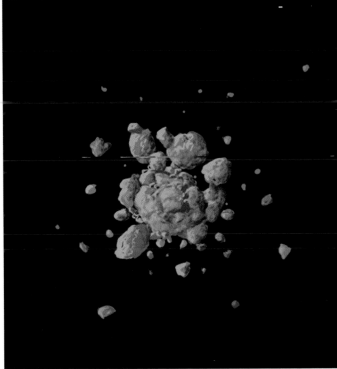

(d) Later impacts continue to fragment the body, and it is converted to a gravitationally bound pile of boulders.

(e) Finally, a gigantic collision totally disrupts the asteroid

(f) and its remnants become scattered through space to form an asteroid family. (Courtesy of Don Davis/*Sky & Telescope*.)

Plate 9.1. The Eagle Nebula (M16) is a region of star formation in our galaxy. (Courtesy of NASA/STScI/AURA.)

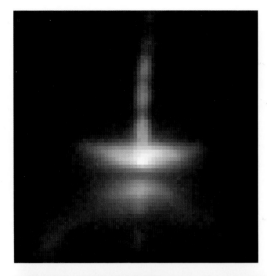

Plate 9.2. Herbig–Haro 30, a young stellar object. (*Hubble Space Telescope* (AURA/STScI/NASA).)

Plate 9.3. Tarantula Nebula complex in the Large Magellanic Cloud. The star cluster, Hodge 301 is seen in the lower right-hand corner of the image. (Hubble Heritage Team (AURA/STScI/NASA).)

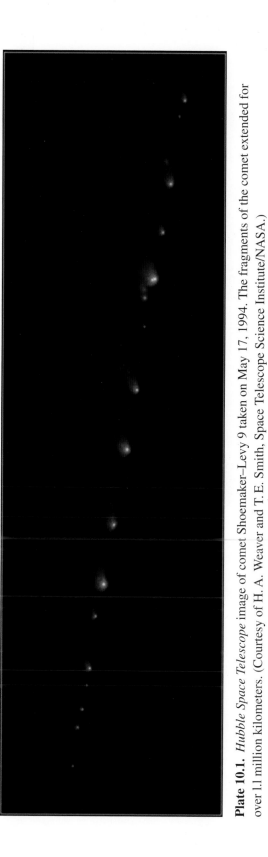

Plate 10.1. *Hubble Space Telescope* image of comet Shoemaker–Levy 9 taken on May 17, 1994. The fragments of the comet extended for over 1.1 million kilometers. (Courtesy of H. A. Weaver and T. E. Smith, Space Telescope Science Institute/NASA.)

Plate 10.2. *Hubble Space Telescope* (WF/PC-2) images of the G fragment impact site. See Fig. 10.7 for a schematic. The images from lower left to upper right show: (1) The impact plume at July 18, 1994, 07:38 UT, about 5 minutes after the impact. (2) The fresh impact site at 09:19 UT, about 1.5 hours after the impact. (3) The modified impact site at July 21, 1994, 06:22 UT, about 3 days after the G impact. An additional impact is seen. (4) Further modified impact site at July 23, 1994, 09:08 UT, about 5 days after the G impact. The S impact has also occurred near the G impact site. (Courtesy of R. Evans, J. T. Trauger, H. Hammel, and the *HST* Comet Science Team/NASA.)

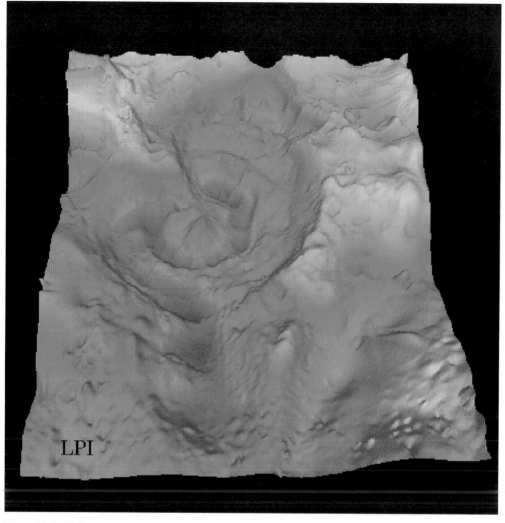

Plate 10.3. False-color representation of a gravity anomaly map of the area of the Chicxulub Crater. North is up. The crater is shown as the nearly circular, low feature in the center. The Chicxulub Crater has no central topographic feature. The other features are regional gravity anomalies and some of these (e.g., the fan shape to the north) interfere with the circular pattern. (Image by V. L. Sharpton. Courtesy: Lunar and Planetary Laboratory.)

Plate 10.4. Image of a 0.32 mm shocked grain from a drill hole in the Chicxulub Crater taken in cross-polarized light. The grain shows evidence of shock metamorphism. (Courtesy of Alan Hildebrand, University of Calgary.)

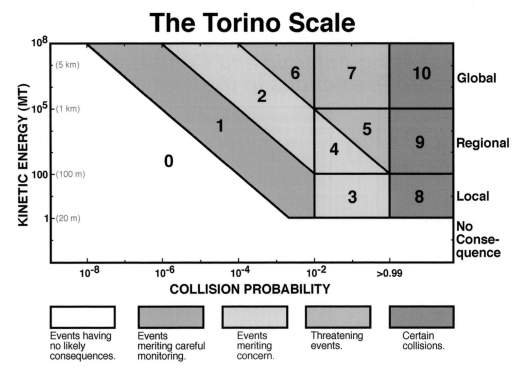

The Torino Scale

KINETIC ENERGY (MT)

10^8
(5 km)
10^5 (1 km)
100 (100 m)
1 (20 m)

6
2
7
10   Global

1
4
5
9   Regional

0
3
8   Local

No
Conse-
quence

10^{-8} 10^{-6} 10^{-4} 10^{-2} >0.99

COLLISION PROBABILITY

Events having
no likely
consequences.

Events
meriting careful
monitoring.

Events
meriting
concern.

Threatening
events.

Certain
collisions.

Plate 10.5. The Torino Scale Diagram. (Copyright © 1999 Richard P. Binzel, Massachusetts Institute of Technology.)

THE TORINO SCALE

Assessing Asteroid and Comet Impact Hazard Predictions in the 21st Century

Events Having No Likely Consequences	**0**	The likelihood of a collision is zero, or well below the chance that a random object of the same size will strike the Earth within the next few decades. This designation also applies to any small object that, in the event of a collision, is unlikely to reach the Earth's surface intact.
Events Meriting Careful Monitoring	**1**	The chance of collision is extremely unlikely, about the same as a random object of the same size striking the Earth within the next few decades.
	2	A somewhat close, but not unusual encounter. Collision is very unlikely.
Events Meriting Concern	**3**	A close encounter, with 1% or greater chance of a collision capable of causing localized destruction.
	4	A close encounter, with 1% or greater chance of a collision capable of causing regional devastation.
	5	A close encounter, with a significant threat of a collision capable of causing regional devastation.
Threatening Events	**6**	A close encounter, with a significant threat of a collision capable of causing a global catastrophe.
	7	A close encounter, with an extremely significant threat of a collision capable of causing a global catastrophe.
	8	A collision capable of causing localized destruction. Such events occur somewhere on Earth between once per 50 years and once per 1000 years.
Certain Collisions	**9**	A collision capable of causing regional devastation. Such events occur between once per 1000 years and once per 100,000 years.
	10	A collision capable of causing a global climatic catastrophe. Such events occur once per 100,000 years, or less often.

Plate 10.6. Explanation of the Torino Scale. (Copyright © 1999 Richard P. Binzel, Massachusetts Institute of Technology.)

Plate 12.1. Comet Halley over the Thames in 1759 as painted by Samuel Scott. (Smithsonian Institution Archives.)

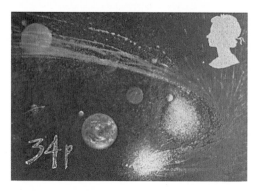

Plate 12.2. Halley postage stamps.

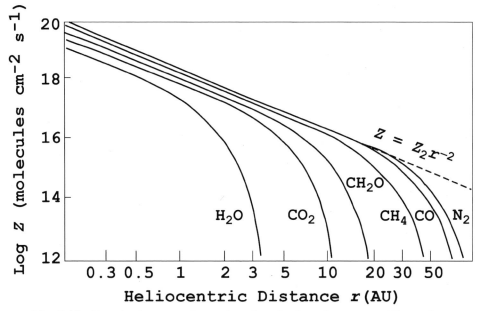

Fig. 7.13. Vaporization rates for various ices in the solar system. The surface temperature of the nucleus is calculated for an albedo of 0.1 and a rapidly rotating nucleus. (After A. H. Delsemme, University of Toledo.)

have been attempts to determine the distribution for sizes since 1982. The estimates come from a variety of methods and some of these, such as the conversion from an absolute magnitude assuming an albedo, can be uncertain. Still, the distribution of sizes is important and should be attempted. Results from different investigator groups (e.g., Fernández *et al.* 1999; Lowry *et al.* 2003) are often discordant. Zahnle *et al.* (1998) have discussed the distribution function in the context of their investigation of cratering rates on the Galilean satellites and adopted the distribution in (7.4) and (7.5) "for specificity . . . ." Given the values in the literature, the exponent in the distributions given by (7.4) and (7.5) should be considered uncertain by ±0.5. See Tables 7.1 and 7.2 for lists of sample cometary nuclei. Finally, note that many nuclei appear to be elongated and typical ratios are ∼2 : 1 : 1.

7.3.2 Surface layers and sublimation

The sublimation of ices to produce the gases and dust that form the coma, hydrogen cloud, and tails is the critical process that makes an object a comet. The principal ice in comets is taken to be water ice. The evidence for this conclusion is solid. The major gases observed in comets are water (H_2O) and its dissociation products, H and OH. Another piece of evidence is the vaporization rate for various snows as shown in Fig. 7.13. Water ice is the only common ice with a dramatic increase in vaporization rate around 3 AU as required by the observed properties of comets.

To understand the physical situation, consider the energy balance for the incident solar radiation. The basic situation is given by:

$$\{\text{Energy received from the sun}\} = \{\text{Radiation back to space}\}$$
$$+ \{\text{Vaporization of ices}\}\{\text{Heat wave into the interior}\}. \qquad (7.6)$$

To begin the discussion, consider a simple situation where the heat wave into the interior is neglected. The discussion follows *Introduction to Comets*, 1st Edn. Then, the terms in (7.6) are given by:

$$F_0(1 - A_0)\frac{\cos\theta}{r^2} = (1 - A_1)\sigma T^4 + Z(T)L(T). \qquad (7.7)$$

Here, F_0 is the solar constant or solar flux at earth $= 2.0$ cal cm^{-2} min^{-1}; A_0 is the albedo at visual wavelengths where most of the sun's radiant energy is absorbed; θ is the solar zenith angle on the nucleus ($\theta = 0°$ at the subsolar point); and r is the heliocentric distance. The radiation back to space is $(1 - A_1)\sigma T^4$ for a gray body; for a black body $A_1 = 0$, and the usual black-body equation applies; A_1 is the albedo in the infrared region (10–30 µm) where the thermal radiation takes place; σ is the Stefan–Boltsmann constant; and T is the temperature. $Z(T)$ is the vaporization rate here in moles cm^{-2} s^{-1}; and $L(T)$ is the latent heat of vaporization in calories per mole.

Equation (7.7) can be integrated over the surface of the nucleus to produce

$$F_0(1 - A_0)\frac{S}{r^2} = 4S(1 - A_1)\sigma T^4 + 4SZ(T)L(T). \qquad (7.8)$$

Here, $S = \pi r_c^2 =$ the cross-sectional area of the nucleus and with $Q = 4SZ =$ the total vaporization rate, we have

$$F_0(1 - A_0)\frac{S}{r^2} = 4S(1 - A_1)\sigma T^4 + QL \qquad (7.9)$$

The latent heat of vaporization for water is 11 700 calories per mole at 150 K and 11 220 calories per mole at 250 K. Figure 7.14 shows the temperature for the subsolar point on a nucleus composed of water snows as a function of heliocentric distance. The temperature range of 150 K to 250 K covers the region of interest and a value of 11 500 calories per mole is appropriate for comets.

Equation (7.9) shows the energy regimes for a comet approaching the sun. Far from the sun, the vaporization rate Q is very low and the energy received from the sun is partially re-radiated (thermal radiation) and some goes to heat the nucleus. As the comet approaches the sun, vaporization is still negligible and the temperature increases to balance increased input from the sun. Eventually, the temperature

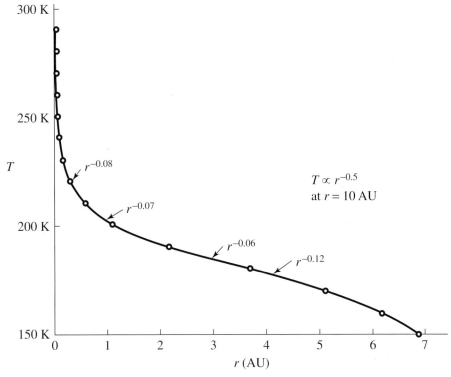

Fig. 7.14. Temperatures at the sub-solar point on a nucleus composed of water snow as a function of heliocentric distance. The local rate of variation is indicated along the curve; note that $T \propto r^{-1/2}$, the normal variation for non-sublimating materials, is reached only for $r = 10$ AU. (Data from Delsemme 1966.)

increases sufficiently to produce vaporization. Then, the solar energy input produces both vaporization and thermal emission. When the comet moves still closer to the sun, essentially all of the solar energy goes into vaporization and Q varies as r^{-2}. This situation occurs by 1 AU and vaporization of water ice becomes dominant for most comets around 3 AU. This is the main driver of cometary activity. These regimes are shown schematically in Fig. 7.15. Observations of comet Halley at the last apparition (Wyckoff *et al.* 1985; Meech *et al.* 1986) showed the onset of sublimation and the development of a coma at 6 AU. This is compatible with the more realistic models that are described below. The earlier activity is produced by sublimation from deeper layers triggered by the inward flow of heat. Some activity can occur at even larger distances due to sublimation of carbon monoxide ice. The following discussion outlines the further development of our understanding of cometary activity.

Figure 7.15 also helps introduce the complications that must be considered in the simple model. First, the sublimation of ices produces a porous dust mantle. This

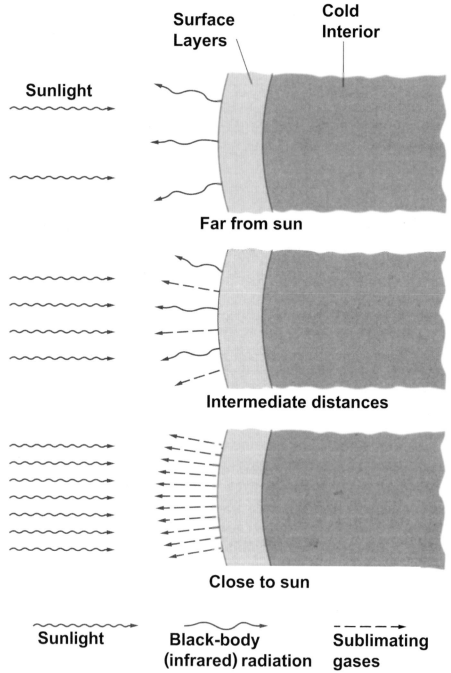

Fig. 7.15. Energy-balance regimes at different distances from the sun for the simple case of no heat wave from the surface layers into the interior.

mantle insulates the ices beneath the surface and (perhaps) regulates the rate of sublimation. The sublimation is believed to take place at a few centimeters below the surface and the gases percolate through the porous dust layer to escape. Energy for the sublimation is transported down to the ices. This scenario is compatible with the surface temperature of around 320 K and a temperature near 215 K for the sublimating ices.

Second, the sublimation does not take place uniformly over the surface. From the example of comet Halley, it is clear that sublimation takes place only in specific areas. The observation that the jets in comet Halley are active only on the sunward side implies a thin dust crust in the active areas. The concentration of activity into specific areas provides a natural explanation for surface features and can also contribute to the development of an irregular nuclear shape. The fraction of the surface that is actively sublimating probably ranges from 100% for a new comet to about 10% for an older comet such as comet Halley to close to zero for an extinct comet.

The general turn-on of cometary activity around 3 AU (but see the discussion above and below) is consistent with the thermodynamic properties of water and the sublimation processes just described. Physically, the greatly increased vaporization required by (7.9) as the comet nears 1 AU is explained by the properties of water. The equilibrium pressure P_E of a vaporized gas in equilibrium with vaporizing snows is given by

$$P_E = N_E kT, \qquad (7.10)$$

where N_E is the gas density at equilibrium. The vapor in equilibrium is said to be saturated. The equilibrium state is maintained by a balance between the vaporizing flux (Z^+) and the condensing flux (Z^-). All molecules that impact the snow surface condense according to the kinetic model of condensation. This means that Z^- can be estimated from the density in the vapor, the molecule's mean speed, and geometrical factors. This quantity is the number of collisions of gas molecules with the surface or

$$Z^-(\text{equilibrium}) = (1/4)N_E\langle V \rangle, \qquad (7.11)$$

where the mean speed for a Maxwellian distribution is

$$\langle V \rangle = (8kT/\pi m)^{1/2}. \qquad (7.12)$$

The vaporizing flux does not change if the saturated vapor is replaced by a vacuum. Hence, we can rewrite (7.10) and (7.12) as

$$Z = \frac{P_E}{(2\pi mkT)^{1/2}}. \qquad (7.13)$$

The vapor pressure for ice is listed in standard chemical tables and shows a steep variation with temperature. Near 200 K, a 20 K or 10% increase in temperature increases P_E (and roughly Z) by a factor of 20. Thus, a minor change in temperature produces the balance implied by the energy equation.

The variation in vapor pressure is described by Clapeyron's equation when the phase change takes place at constant temperature and pressure, i.e.,

$$\frac{dP_E}{dT} = \frac{L}{T(V_g - V_s)}. \tag{7.14}$$

Here, L is the latent heat of vaporization and V_g and V_s are the volumes of the gas and the solid, respectively. In the comet case, $V_g \gg V_s$, and V_s can be ignored. If V_g is written in terms of the ideal gas law, $V_g = RT/P_E$, we find

$$d \ln P_E = -\frac{L}{R} d(1/T). \tag{7.15}$$

Because L is approximately constant, this equation integrates to

$$P_E \propto e^{-L/RT} \propto e^{-5740/T} \text{ (for } H_2O, T \approx 200\,K). \tag{7.16}$$

Equation (7.16) exhibits the extreme temperature sensitivity of P_E, and by equation (7.13), the vaporization rate to temperature.

The preceding discussion presents the traditional view and some physics essential to understanding the production of gas and dust from comet nuclei. But, complications not included in the traditional view are important and must be considered for an understanding of the structure of comet nuclei.

In the general time period during and immediately after the intensive activities associated with Halley's Comet in 1985–1986 and before the activities associated with comets Hyakutake in 1996 and comet Hale–Bopp in 1997, our view of the comet nucleus and the possible complications began to change. The post-Halley view was summarized in *Physics and Chemistry of Comets*, W. F. Huebner, ed. (1990) and in *Rendezvous in Space: The Science of Comets*, by the present authors (1992). An excellent review of the current view (containing extensive references) has been given by Prialnik (2002). A fundamental question is the form of water ice, the dominant volatile constituent in comets, in the interior of the nuclei. The so-called minor constituents such as CO, CO_2, C_2, C_3, HCN, etc. need to be stored in the ice either as clathrate hydrates or as trapped ice. While clathrate hydrates were believed to be important in comets, they are generally formed under high-pressure conditions that do not apply to comets (Keller 1990; Lewis 1997). On the other hand, Mekler and Podolak (1994) have shown that amorphous ice, stable over the age of the solar system, requires temperatures of 77 K or less. They examined models of the solar nebula and found this condition is met beyond 7 AU. Thus, cometary ice in the interior is probably amorphous because comets formed under low-pressure and low-temperature conditions. Ices formed by condensation on a surface at low temperatures do not have energy available to change into the crystalline forms that minimize energy. These transitions (first to cubic and then to hexagonal ice) can only occur at higher temperatures as outlined below. Further, the presence of amorphous ice probably provides the best explanation for cometary outbursts far from the sun as we will discuss below (section 7.3.3).

The dominance of amorphous ice in the initial composition of comets means that models of the interior will need to keep track of water in the form of amorphous ice, crystalline ice and water vapor. The amorphous to crystalline ice transition provides a significant sub-surface source of energy. Prialnik (2002) has noted that there appear to be three energy sources for nuclei. The first is solar radiation as described above. This source deposits energy at the surface or in the near-surface layers. It affects the interior by a wave of heat moving inward. The second is a radioactive source from short-lived isotopes, primarily ^{26}Al (Wallis 1980; Prialnik and Podolak 1999). This source is important mostly when the nucleus is far from the sun. An increase in core temperature could lead to diffusion of volatiles. The third is the energy release in the transition from amorphous ice to crystalline ice. This occurs in two transitions; amorphous to cubic ice at about 137 K and from cubic to hexagonal ice at about 160 K. Currently, the calculations do not distinguish between the two forms of crystalline ice and treat the system as one crystallization process when the temperature exceeds 137 K. The trapped minor constituents are partially released during the crystallization process ($T \approx 120$ to 140 K) and the remainder is released when the crystalline ice sublimates. Thus, gas release takes place in the interior of the nucleus as well as in the surface layers. Some of the complexity is illustrated schematically in Fig. 7.16.

Our ideas about the nature of the comet's bulk structure had also evolved. While the fragile nature of the nucleus (which implied a porous structure) was generally understood, perhaps the high value of the porosity and its importance in understanding cometary activity was not. The low bulk density of nuclei (e.g., Rickman 1989) and the appearance of gases that must come from deeper layers support this view. Of course, most models treat the nucleus as spherical, which is certainly not the case.

The gases in the interior flow to the surface because of pressure. Generally, the path to the surface is tortuous. Some idea of the possible complexity of the porous structure is illustrated in Fig. 7.17. The parameters necessary for model calculation are the porosity (see below), the pore size, the permeability (the measure of fluid flow through rocks and porous material), the surface-to-volume ratio, and the appropriate value for the thermal conductivity. Also, note the recondensation of liberated gases

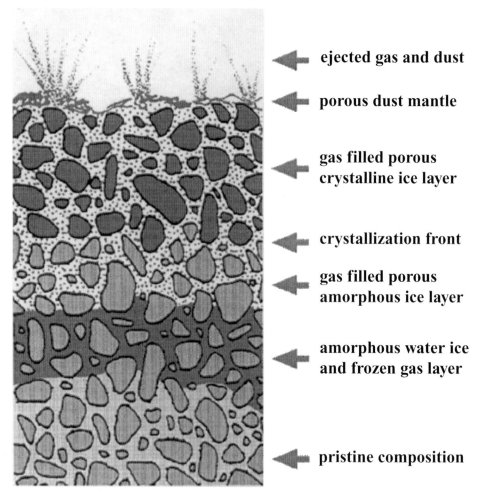

ejected gas and dust

porous dust mantle

gas filled porous
crystalline ice layer

crystallization front

gas filled porous
amorphous ice layer

amorphous water ice
and frozen gas layer

pristine composition

Fig. 7.16. Schematic layered structure of a cometary nucleus from the porous dust mantle down to the pristine composition. The vertical scales are arbitrary. (Courtesy of D. Prialnik, Department of Geophysics and Planetary Science, Tel Aviv University, Israel; From *Earth, Moon, and Planets*, Vol. 77, 1997– 1999, pp. 223–30, Modeling gas and dust release from comet Hale–Bopp, by Prialnik, Figure 1. Copyright © 1999, with kind permission of Kluwer Academic Publishers.)

can take place during the outward flow and almost surely takes place for inward flow to cooler areas.

Models must also account for dust particles entrained in the gas flow. The critical size is determined by a balance between the drag and gravitational forces. Particles smaller than the critical size are carried in the flow, but particles larger than the critical size are left behind. In addition, particles larger than the local pore size cannot be carried in the flow. The larger particles trapped in the near-surface layers can

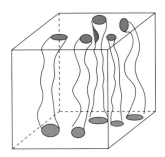 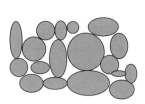 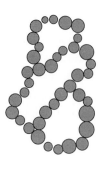

Fig. 7.17. Sample models for the porous structure of comet nuclei. Left: tortuous capillaries that do not cross; middle: rubble pile; right: porous aggregate of grains. (Courtesy of D. Prialnik, Department of Geophysics and Planetary Science, Tel Aviv University, Israel; From *Earth, Moon and Planets*, Vol. 89, 2002, pp. 27–52, Modeling the comet nucleus interior: Applications to comet C/1995 O1 Hale–Bopp, by Prialnik, Figure 3. Copyright © 2002, with the kind permission of Kluwer Academic Publishers.)

form a dust mantle. Such a mantle changes the flow of gas and the energy balance. Determining the thermal conductivity in a reasonably realistic model is complex.

The list of physical processes and parameters to be specified, e.g., porosity and the size distribution of the dust particles, leads to an interesting philosophy of model building. The models are not predictive in the sense of using a set of initial parameters to calculate the behavior of a comet. Rather, the specific parameters that allow the behavior to be reproduced are determined and, thus, properties of comet nuclei otherwise inaccessible to observation are derived. Some conclusions for comets Halley and Hale–Bopp are described below.

For comet Halley (Mekler *et al.* 1990; Prialnik 1992), models were calculated for an initial interior temperature of 30 K, a pore radius of 10 μm, and a composition (by mass) of 75% amorphous ice (including 10% trapped CO) and 25% dust ($\rho \approx 2.65$ g cm^{-3}). The bulk density was 0.4 g cm^{-3}. The porosity is defined as the fraction of the volume occupied by the pores. These parameters yield an initial porosity of 0.63. The principal conclusions were that: (1) temperatures can be maintained at approximately 170 K down to depths of tens or hundreds of meters while crystallization is going on; (2) the crystallization is largely determined by the gas flux contribution to the heat flow; (3) the heat of crystallization contributes about a third of the energy input; and (4) the ratio of CO to H_2O is much higher in the ejected material than in the original composition.

An additional finding was that the thermal evolution involved cycles or spurts. The crystallization of the lower amorphous ice layers takes place when the overlying crystalline ice layer is thinned due to sublimation. This produces a thicker crystalline layer which produces a period of greatly reduced activity. The cyclic nature of the process can be enhanced by the fact that real comets are not spherically symmetric and have surface irregularities. Thus, thinner or thicker local layers are produced naturally. See the discussion of comet Halley's outburst (below).

For comet Hale–Bopp, the models indicate that the nucleus probably did not form a dust mantle. The dust particles must be smaller than the pore size (or an equivalent distribution function) and the escaping dust particles produced the exceptional dust tail (section 4.2.4.2 and Plate 4.1). As for comet Halley, the abundances in the coma are not representative of the interior.

An interesting observation concerns the onset of activity for comets with an outer amorphous ice layer (Prialnik 1997). Such comets should exhibit activity at heliocentric distances of 6.5 ± 0.5 AU. Crystalline ice is a much better medium for heat conduction than amorphous ice. At low temperatures, when the heating and the growth rate of the crystalline layer are slow, the heat generated flows to the surface. As the temperature increases due to solar heating, the crystallization rate increases and a runaway condition can exist. At higher temperatures, ice is sublimated from the pore walls. This is a heat sink and the process is brought under control.

A model of comet Hale–Bopp was developed by Enzian (1999) with the principal purpose of understanding the CO production rates. The results are shown in Fig. 7.18. The assumptions used in the model are: an albedo of 0.04; a radius of 35 km; amorphous ice, 40% by mass; carbon monoxide trapped in the amorphous water ice, 5% by mass; carbon monoxide ice, 5% by mass; and dust, 50% by mass. As shown in Fig. 7.18, the agreement is quite reasonable. Activity starts around 7 AU driven by crystallization of amorphous ice; no outbursts were found. The gradual increase in activity at large distances is due to sublimation of the carbon monoxide ice.

7.3.3 Outbursts

Occasionally and suddenly, the total brightness of a comet can increase by 2 to 5 magnitudes or an increase in luminosity by a factor of 6 to 100. And, larger outbursts have been recorded. Typically, outbursts last 3 to 4 weeks. Solid physical understanding may be yet to come, but they could be driven by crystallization of amorphous ice as described below. Outbursts occur throughout the inner solar system with no believable dependence on heliocentric distance. There is evidence for a preference after perihelion.

Many comets have outbursts and one of the most famous is comet 29P/ Schwassmann–Wachmann 1 as shown in Fig. 7.19. This comet is in a nearly circular orbit with a heliocentric distance of 6 AU. Typically, a shell of material (not necessarily symmetrical) expands at speeds of 100–500 m s^{-1} and produces a planet-like disk. Then, the disk fades away. The observational evidence is consistent

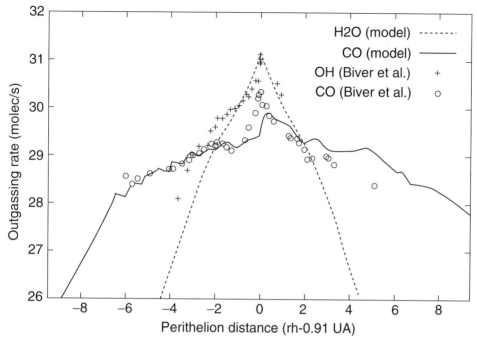

Fig. 7.18. Model calculations of the outgassing rates for H_2O and CO and a comparison with the observations. (Courtesy of A. Enzian, CNES, France; see Enzian 1999.)

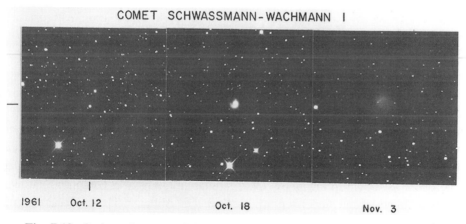

Fig. 7.19. Outburst in comet Schwassmann–Wachmann 1 in 1961. The images have been matched in scale, orientation, and density. For the October 12 image, the lines in the margin indicate the location of the star-like comet image. (Courtesy of E. Roemer, University of Arizona; official U.S. Navy photograph.)

Fig. 7.20. Outburst in comet Halley as observed by O. Hainaut and A. Smette on February 12, 1991. The central part of the dust cloud is over 300 000 km across. (European Southern Observatory.)

with scattering from dust particles with diameters of about 0.5 μm. An observational study (Jewitt 1990) found that the comet is continually active. In addition, Senay and Jewitt (1994) detected carbon monoxide (CO) from the comet. The amount of CO production is sufficient to produce the observed coma. But, an energy source is needed.

Comet Halley displayed an extraordinary outburst on February 12, 1991, when it was about 300 times brighter than expected. It had an extended, nebular appearance as shown in Fig. 7.20. The "nebular" material was composed entirely of dust. The eruption probably took place in December 1990. The heliocentric distance of the comet was 14.3 AU, a fact which contributes to the puzzle of outbursts.

Outbursts are quite common and sometimes are associated with splitting of the comet. Comet 73P/Schwassmann–Wachmann 3 (Fig. 7.21) had an outburst and a major brightness increase between November 7 and November 18, 2000. The December 3 image showed the comet split into three pieces. Also, C/1999 S4

(a)

(b)

Fig. 7.21. Outburst in comet Schwassmann–Wachmann 3. Upper: November 7, 2000; middle: November 18, 2000; lower: December 3, 2000. The last image shows that the comet has split into at least three pieces. (Courtesy of T. Kojima, Japan.)

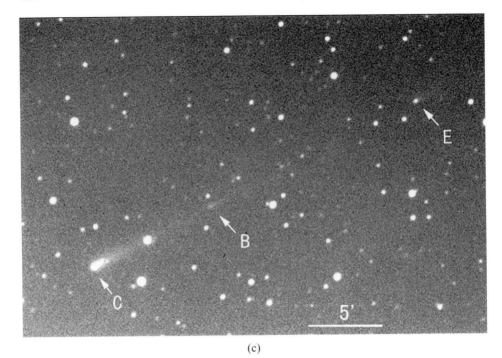

(c)

Fig. 7.21. (*Cont.*)

(LINEAR) showed outburst-type brightenings in June and July 2000 (Fig. 7.22), indicating enhanced dust production. These events were apparently the beginning signs of the nucleus splitting into cometesimals in late July 2000, as discussed below.

The kind of physical mechanism needed is something to produce large amounts of dust, perhaps by disrupting an area of the crust layer. Hughes (1975, 1991) has described many possibilities. A straightforward possibility that must occur is an impact by an interplanetary boulder. This would produce a crater with ejection of dust and snow and expose fresh ices. The areas would eventually return to normal. The dust cloud, however, would probably not be symmetrical.

The outburst problem has produced some novel suggestions. First, the nucleus at large distances from the sun could be charged by the solar wind. The solar wind is electrically neutral and the day side of the bare nucleus receives an equal number of protons and electrons. The night side of the nucleus preferentially receives electrons because the random speeds of the solar-wind electrons are much higher than those of protons. A charge builds up until an equilibrium is established by the net charge attracting protons to the night side. The charge can cause dust to levitate above the surface. A sudden change in solar-wind conditions could cause the dust to be blown

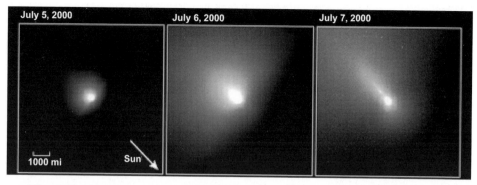

Fig. 7.22. Outburst in Comet LINEAR (C/1999 S4) on July 5, 6, and 7, 2000. (Courtesy of NASA and H. Weaver, Johns Hopkins University; see also Weaver *et al.* 2001.)

off (Flammer *et al.* 1986). This mechanism cannot work when there is significant coma and the solar wind does not strike the surface. Thus, this mechanism could work for heliocentric distances larger than 6 AU and could apply to the outburst in Halley's Comet.

Second, there are relatively simple possibilities. Sub-surface gas pockets could burst through the mantle. Pockets of reactive gases could come into contact and produce a high-pressure volume. Or, the basic non-uniform nature of the cometary surface region could be responsible by exposing layers of fresh ices to sunlight.

Crystallization of amorphous ice has been recognized for years as a possible energy source for outbursts. See Prialnik (2002) for references. The basic process has been described earlier in terms of energy sources for sublimation. Crystallization of amorphous ice is a plausible source if it can be shown to operate at large heliocentric distances. Prialnik and Bar-Nun (1992) have studied the outburst of comet Halley described above. The model attributes the outburst to crystallization of amorphous ice at depths of a few tens of meters. The process involves a pressure build-up, a situation that could take place in regions of small pore size or low permeability. The pressure could exceed the tensile strength of ice. Thus, the gas could crack open wider channels and flow to the surface. The flow should subside with a time scale of a few months. This is compatible with the outburst in comet Halley.

The model just described could well be the dominant one for cometary outbursts. Still, we should recognize the uncertainty in the situation and not be surprised if other mechanisms are subsequently found to operate. The splitting process (if understood) may provide a class of outburst events that do not require additional mechanisms.

7.3.4 Rotation

Rotation rates of nuclei might supply valuable information on the origin of comets, the dynamical history of the nuclear bodies, and the internal structure. The history of the subject is somewhat checkered. Until about 1980, the outgassing of material was considered to be essentially isotropic. The realization that the outgassing took place on the sunward side and from a small number of active areas opened new opportunities.

Larson and Sekanina in the early 1980s extensively studied comet Halley using images from the 1910 apparition. Sekanina and Larson (1984) noted the different shapes of dust ejecta from a small active region on the equator of a nucleus spinning at two greatly different periods. The shapes of ejecta for nuclei with rotation periods of 4.8 hours and 48 hours are quite different, and similar to features seen in comets Donati (C/1858 L1) and Halley in 1910, respectively. Figure 7.23 shows a relatively simple situation of high latitude radial emission forming a conical surface and the effects of radiation pressure.

However, the real situation is usually quite complex. The spin axis is usually not perpendicular to the comet's orbital plane. The size of the active region may be finite. The spin axis may not lie in the plane of the sky as seen from earth and foreshortening can be important. And, multiple active regions can be a major complication. See Jewitt (1997).

Some measure of a nuclear rotation period might be obtained simply from a photometric light curve. This could supply qualitative information. But, as becomes apparent from the studies of the rotation of Halley's nucleus, the rotation can be quite complex.

The spacecraft sent to Halley's Comet in 1986 and the direct imaging of the nucleus by *VEGA-1* on March 6, *VEGA-2* on March 9, and *Giotto* on March 14 provided a superb opportunity to study the rotation of a comet nucleus. But, as things turned out, the solution was not as straightforward as expected. Part of the confusion was that some spacecraft data pointed to a period of 2.2 days while the ground-based data indicated a period of 7.4 days. A major problem was that the orientation of the long axis of the nucleus at the time of the *VEGA-1* observations was reversed from the one originally proposed. Considerable effort went into sorting out this problem and the solution was published by Belton *et al.* (1991) and Samarasinha and A'Hearn (1991) in back-to-back papers in *Icarus*.

The coordinate system and the quantities used to describe the model for Halley's Comet are shown in Fig. 7.24. The dashed line indicates the long axis of the nucleus; **M** is the total angular momentum vector; and **S** is the instantaneous spin vector. The nucleus is taken to be a prolate spheroid with dimensions $17 \times 8.5 \times 8.5$ km. This model has the ratio of maximum to minimum moments of inertia of 2.28.

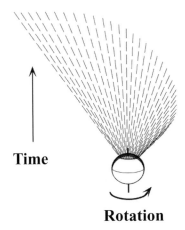

Time

Rotation

Sun

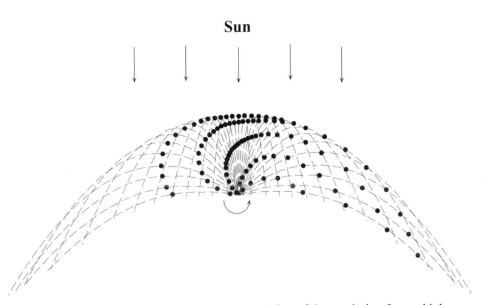

Fig. 7.23. Schematic figure showing the evolution of dust emission from a high-latitude source (above) and the distortion of the trajectories by solar radiation pressure (below). (Courtesy of S. M. Larson, Lunar and Planetary Laboratory, University of Arizona; to appear in the *Proceedings of the Nanjing Conference on Comets*, M. A'Hearn, ed.)

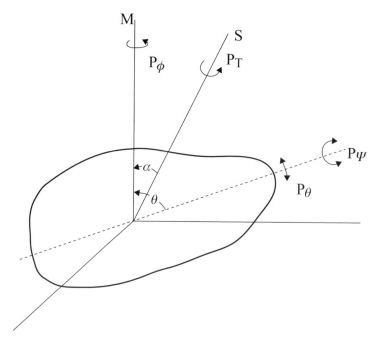

Fig. 7.24. The coordinate system used to describe the rotation of comet Halley's nucleus; see text for discussion. (Courtesy of M. J. S. Belton, Belton Space Initiatives; Reprinted from *Icarus*, Vol. 93, by Belton, Julian, Anderson and Mueller, The spin state and homogeneity of Comet Halley's nucleus, pp. 183–93. Copyright 1991, with permission from Elsevier.)

An acceptable model needed to satisfy the imaging data from the encounters, the photometric light curve, and the properties of the jets (recall the simple examples given above), CN shells, and C_2 production rates. The model uses five, localized active areas on the surface and they are active only when illuminated by the sun. The jets are assumed to provide no net torque and no external torques are assumed. Thus, **M** is constant in space.

The long axis of the nucleus makes an angle of 66° to **M**, the total angular momentum vector, and rotates around **M** with a period, P_ϕ, of 3.69 days. The spin around the long axis has a period, P_ψ, of 7.1 days. There is no motion in the θ coordinate. The total spin vector **S**, is inclined to **M** by 21.4° and freely precesses around **M** with a period of 3.69 days. This complex situation is illustrated in Fig. 7.25.

Comparison of this model with the extensive observations produces good, but not perfect agreement. Still, in view of the complexity of the problem and the wealth of data to be explained, it is excellent. The dynamical properties of the nucleus and its dimensions are consistent with a constant internal density. Note that the nucleus

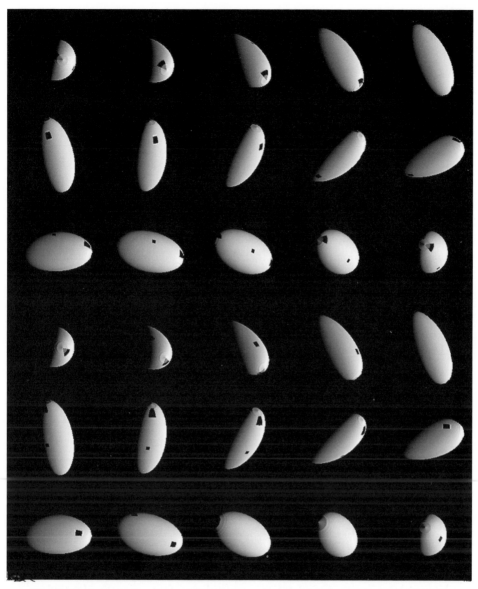

Fig. 7.25. Schematic of one full rotation sequence for the nucleus of Halley's Comet. The sequence should be read starting at top left and moving to the right. The interval between images is 0.25 days and the 30 images span 7.25 days. The sequence repeats after *approximately* 7.25 days. The five active areas are marked as low albedo features. See Belton *et al*. 1991. (Courtesy of M. J. S. Belton, Belton Space Initiatives.)

of comet Halley is not in the lowest rotational energy state for a given angular momentum (which is rotation only around the short axis; Belton 1991). See below.

The model adopted has the energy of its spin state well above the minimum allowed for its angular momentum value. The relaxation time due to frictional dissipation in the comet's interior is believed to be relatively short. Peale and Lissauer (1989) have suggested

$$\tau = 10^6 Q \text{ (years)} \tag{7.17}$$

for the dissipation time, where Q is a dimensionless measure of energy dissipation per rotation cycle. Peale and Lissauer consider Q to be in the range 1 to 10^2. Thus, the high energy state is probably not primordial, but it persists for times much longer than the periods of many comets. Possible causes of the high energy state are: torques due to jets at succeeding perihelion passages; numerous minor splitting of the nucleus; a major splitting of the nucleus; or a combination of all of these possibilities.

The apparently slow rate of damping toward the lowest rotational energy state and the possibility that comet nuclei may generally be in an excited state offers the possibility of spin-up. Samarasinha and Belton (1995) find that spin-up can happen for a Halley-like nucleus if the active areas are the same for multiple orbits. Nuclear spin-up could be a major cause of splittings.

Let us now revisit the question of the determination of rotational properties from coma structure or brightness variations. Sekanina (1981) and Whipple (1982) have reported rotational periods for approximately 60 comets. In view of the complications seen for Halley's Comet, these should be considered indicative rather than specific. With this caveat, the average rotation period for comets appears to be longer than for asteroids.

The average rotation periods were used by Whipple to constrain the density and/or the tensile strength as follows. For a rotating sphere, a particle on the equator will have orbital speed when

$$P_{crit} = 3.30/\rho^{1/2} \text{ (hours)}. \tag{7.18}$$

For $\rho = 1.3$ g cm^{-3}, $P = 5.5$ hours and for $\rho = 0.65$ g cm^{-3}, $P = 7.8$ hours. Some reported periods would violate this constraint implying internal strength or a calculation that is too simple. Jewitt and Meech (1988) extended this idea to prolate spheroids. For this case, the critical period is larger, i.e., for the same density, the rotation must be slower for the nucleus to be stable.

The detailed studies by Larson and Sekanina for Halley's Comet fare as follows. Their rotation period was about 52 hours; note that Sekanina and Larson (1986) warned that "one must still exercise caution" and specifically noted several discordant determinations. Their value of 52 hours is not compatible with the details of the rotation described above. A success of their models was the now verified hypothesis that the dust emission came from jets active only on the sunlight side of the nucleus. See Larson (2002) for a review of nucleus spin states as determined from ground-based observations.

7.3.5 Interiors

The sub-surface layers of comets have not been probed directly and inferences must be drawn primarily from the gas and dust emitted through the surface layers. The composition of the interior must be reflected in the composition of the coma gas and dust described in Chapter 5. Of course, the main constituents of most comets are believed to be water ice with impurities and dust. The picture built up is that the nucleus is a low-strength structure with high porosity and a density perhaps as low as 0.3 g cm^{-3}. How do these properties vary from comet to comet and what does the evolution of properties indicate about internal properties of nuclei?

A'Hearn *et al.* (1995) have carried out a monumental study of 85 comets from 1976 to 1992 based on narrow-band photometry. The filters chosen covered CN, C_2, C_3, OH, and NH as well as continuum regions at several wavelengths between 3290 and 5240 Å. The overall approach to the investigation stressed standard observing approaches and homogeneous data reduction. The reader is invited to consult the original paper for details. The main conclusions for the interiors of nuclei are summarized here. Caveats are that the results may not apply to all comets and that differences between the interior composition and that of the coma are not too severe.

The compositions of most comets as indicated by the species sampled are surprisingly uniform. The same result is also apparent from the ultraviolet spectra of comets. In addition, without some unusual event such as a splitting of the nucleus, an individual comet's production of gas and dust as a function of position in its orbit is essentially the same from one orbit to the next. This observation implies that the average relative abundance of gas and dust is uniform with depth. Thus, the interior is not differentiated, either as a result of formation conditions or as a result of subsequent evolution.

A'Hearn *et al.* divided their sample into old and new comets in the dynamical sense based on the original value of the reciprocal major axis $(1/a_0)$; they chose a value of $(1/a_0) \leq 50 \times 10^{-6}$ AU^{-1} for new comets and also introduced several other groups. A major finding was that a comet's composition did not depend on dynamical age.

Exceptions to compositional similarity were found. A well-defined class of comets is substantially depleted in C_2 and C_3. Almost all members are part of the Jupiter family of comets. But, not all Jupiter-family comets show the depletion.

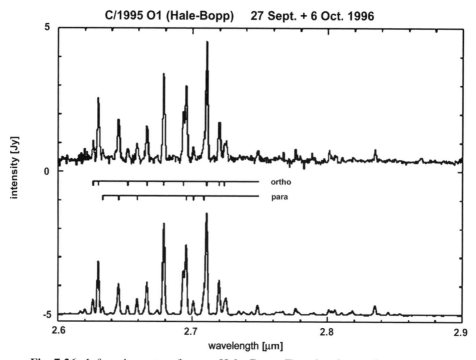

Fig. 7.26. Infrared spectra of comet Hale–Bopp. Top: the observed spectrum, an average of observations taken on September 27 and October 6, 1996. The ortho and para lines of water are marked. Bottom: the synthetic spectrum for $T_{rot} = 28.5$ K and an OPR $= 2.45$. (Courtesy of J. Crovisier, Observatoire de Paris-Meudon; see Crovisier *et al.* 1997.)

The formation temperature for cometary water and possible interior temperature obtained for comet Halley was described in section 7.1. In addition, Crovisier *et al.* (1997) measured the ortho to para ratio (OPR) in the infrared spectrum of comet Hale–Bopp (Fig. 7.26). They determined a temperature of 28.5 K and concluded that the comet was formed at temperatures of ~25 K. If this measurement were carried out on a large number of comets, a valuable constraint on the physical conditions for formation of comets would be established.

Masses of comet nuclei would be best determined from their gravitational perturbations on a passing spacecraft. No favorable circumstances have occurred so far. Not only must the spacecraft be carefully tracked, but the comet must be sufficiently distant so that the spacecraft is not perturbed by the outflow of gas and dust.

Indirect methods must be used. An older "method" was simply to assign a density of 1 g cm^{-3} and calculate the mass as roughly $4r^3$. This crude estimate is useful; it would be improved if the density were better known.

An alternate approach is to attempt to derive the mass from an analysis of non-gravitational forces as discussed in Chapter 2. While the modeling of the nongravitational effects can be used for orbit calculations quite nicely, it is not clear that the nuclear mass can be reliably extracted. Estimates for comets P/Halley and P/Kopff, for example, often disagree. The best estimates for Halley combined with known dimensions seem to cluster around 0.3 g cm^{-3}, now considered to be a plausible value. Bear in mind that the extremes range from 0.1 to 0.9 g cm^{-3}. See Rickman (1989) for a review and extensive references. At the lower end, these densities could imply rather fragile bodies. Perhaps the mass can be simply estimated as r^3!

An approach to estimating the nucleus density has been used for the case of P/Tempel 2 (Jewitt and Meech 1988; Jewitt and Luu 1989). Based on the shape and rotation period, a critical density was calculated that would put the comet just in a state of internal tensile stress and, thus, unstable to centripital disruption. Recall the approach by Whipple shown in (7.18). The basic idea follows from the belief that comets have a very low tensile strength. In other words, almost any stress should pull them apart; see just below. The rotating comets would impart centripital forces that could be balanced by self-gravity. Jewitt and Meech (1988) carried out the calculation for the apex of a prolate spheroid rotating around a minor axis. For P/Tempel 2, they obtained a critical density of 0.3 g cm^{-3}. Since uncompacted snow has a density of 0.1 g cm^{-3}, they concluded that the nuclear snows must be compacted. If the rotations of most comets are as complicated as comet Halley's, the details may need refinement.

The tensile strength of comets should contain clues about the nature of the interiors. Perhaps the most significant fact is the fairly frequent splitting of comets, often for no obvious cause. Comet Shoemaker–Levy 9 was the exception: see section 10.3.2.

Comet Biela in the nineteenth century provided an example of comet splitting and possible complexities. The comet was seen in 1832 and had a period of about 6.6 years. The 1839 apparition was not favorable. At perihelion, the comet was on the opposite side of the sun and was not seen. At the 1846 apparition, Biela's Comet returned as two comets, close together on the sky, each with a tail (Fig. 1.6). The brightness of the two components varied with a different component being brighter at different times, and sometimes a luminous bridge between the two components was reported. Similar behavior was seen in the 1852 apparition except that the two comets were about 2 million kilometers apart. On the next pass in 1859, the observing conditions were poor. But, the observing conditions in 1866 were excellent and the comet was not observed. Comet Biela was never seen again.

Comet West in March of 1976 split into four nuclei as shown in Fig. 7.27. The splitting into a small number of discrete bodies may indicate that the building blocks – the cometesimals – may not be tightly bound together. Thermal stresses or

Fig. 7.27. Splitting of the nucleus of comet West (1976 VI). The images (left to right) were taken on March 8, 12, 14, 18, and 24, 1976. (New Mexico State University Observatory.)

rotation may cause them to come apart. On the other hand, some comet nuclei such as Halley's have survived numerous inner solar system passes. Generally, if there is a spectrum of strengths, the weaker nuclei can be expected to split soon after beginning inner solar system passes while the stronger nuclei can survive many passes.

Comet LINEAR (1999 S4) split into several small pieces or cometesimals during its passage around the sun in late July 2000. Figure 7.28 shows a *Hubble Space Telescope* (*HST*) image of the event. This event, recorded by *HST* and the Very Large Telescope (VLT), provided a superb opportunity to study the disintegration of a cometary nucleus. As shown in Fig. 7.22, the disintegration sequence began on July 5, 2000 with an outburst. At first, the comet appeared to disappear, but high resolution images (Fig. 7.28) by *HST* (August 5) and VLT (August 7) revealed 15 fragments or cometesimals (Weaver *et al.* 2001). By August 14, 2000, the fragments were no longer visible. Photometric analysis of the fragments yielded diameters of about 100 meters. Objects much smaller than this value would be at or below the limit of detectability for *HST* and VLT. If as seems likely, the fragments are cometesimals, their size is somewhat smaller than the typically quoted value of 500 meters.

The splitting examples given above show no straightforward correlation with any parameter related to the orbit such as perihelion distance, time before or after perihelion, inclination, or distance above or below the ecliptic plane. These more-or-less random or spontaneous splittings occur for about 10% of dynamically new comets during their first perihelion passage; for about 4% of long-period comets making a subsequent return; and for about 1% of short-period comets (Weissman 1980; Sekanina 1982).

Another class of splittings occurs when a comet passes close enough to a planet or the sun to be within the Roche limit (see section 2.7) for tidal disruption. These events can be utilized to compare the induced tidal stresses to strengths of materials. This has been carried out for the sungrazing comets 1882 II (C/1882 R1) and 1965 VIII (C/1965 S1) and for 16P/Brooks 2, which passed close to Jupiter in 1886, and comet Shoemaker–Levy 9 (D/1993 F2), which passed close to Jupiter in July 1992.

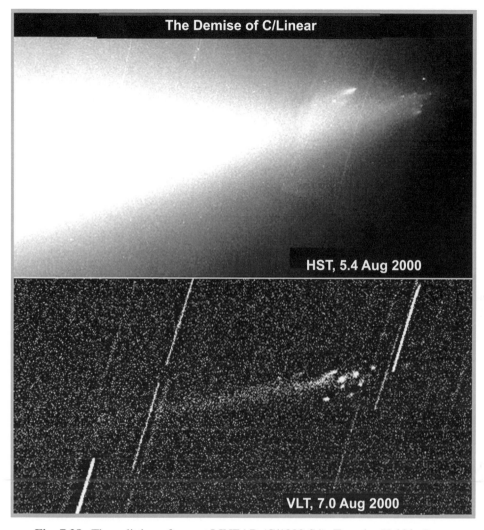

Fig. 7.28. The splitting of comet LINEAR (C/1999 S4). Top: the *Hubble Space Telescope (HST)* view on August 5, 2000 shows the bright dust tail (extending to the left) that was primarily the product of the destruction of the nucleus around July 22, 2000. The large remnants are seen near the tip of the tail. Bottom: an image taken approximately 1.6 days after the *HST* image by the Very Large Telescope (VLT) of the European Southern Observatory. Image processing was used to suppress the light from the tail and enhance the visibility of the fragments. They are clearly seen. The streaks are star trails and the image subtends a region 103 000 km by 58 000 km at the comet. (H. Weaver, Johns Hopkins University; C. Delahodde, O. Hainaut, R. Hook, European Southern Observatory; Z. Levay, Space Telescope Science Institute; and the *HST*/VLT observing team; NASA/ESA, ESO.)

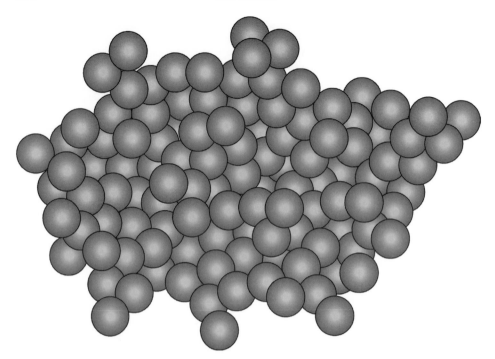

Fig. 7.29. A simple comet nucleus made by assembling uniform cometesimals at random locations of an original seed particle and then the resulting structure. The cometesimals are assumed to stick perfectly and the aggregate consists of approximately 10^3 particles. (Courtesy of K. Meech, University of Hawaii; after Jewitt and Meech 1988.)

The cometary tensile strengths range from 10^2 to 10^4 N m^{-2} (the units of tensile strength are force per unit area); see Rahe *et al.* (1994) and Prialnik (2002). By comparison, solid water ice has a value of $\sim 2 \times 10^6$ N m^{-2} and rocky materials have values $\sim 4 \times 10^6$ N m^{-2}. Steel comes in at $\sim 4 \times 10^8$ N m^{-2}. The specific values for comets should not be taken too seriously. But, by any standards these are low material strengths. Note that these values probably represent the strength holding the individual cometesimals together. The internal strength of the cometesimals may be much higher.

Evidence available indicates: a "fairly" uniform interior consisting of volatile ices (usually mostly H_2O ice) and dust, a rotating nucleus, an elongated nucleus, and a weak or fragile structure. Jewett and Meech (1988) have carried out simulation of the shapes of nuclei by assembling small cometesimals onto an initial seed particle from random directions. The particles are assumed to stick perfectly and nuclei with 10^3 particles were grown. A sample is shown in Fig. 7.29. The results from this simple approach resemble the observationally determined shapes of comets and strengthen the belief that the nucleus is composed of building blocks, the

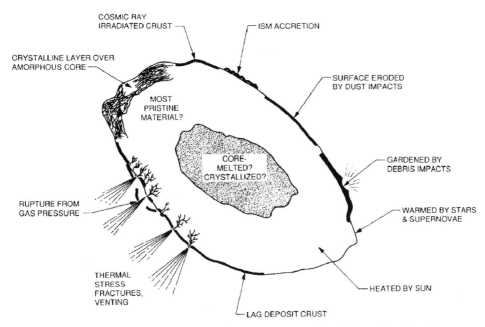

Fig. 7.30. A summary schematic of a cometary nucleus illustrating the physical processes at work. (Courtesy of P. R. Weissman, NASA–Jet Propulsion Laboratory.)

cometesimals. Remember, however, that collisions and anisotropic mass loss could also be important in determining the shape.

Detailed knowledge of the interior of comets must wait until results are available for a direct mission such as *Deep Impact*, described in Chapter 11. In the meantime, Fig. 7.30 shows a nucleus schematic that emphasizes processes operating on the nucleus rather than specific details which must be regarded as largely unknown. Many scientists feel that the basic building blocks are the cometesimals roughly 0.5 km in diameter or perhaps smaller.

7.4 The fate of comet nuclei

Some comet nuclei are ejected from the solar system by gravitational perturbations. Others are still in the Oort cloud or the Kuiper belt. But, any comet that makes repeated passes through the inner solar system is doomed.

Exposure to the sun's radiant energy produces sublimation. The gas and dust liberated from the nucleus is lost to the comet. The mass loss per inner solar system pass has been estimated at 0.01 to 0.1% of the total mass. Simplistically, nuclei can be expected to last for 1000 to 10 000 passes. In terms of size, the shrinkage per pass can be estimated by differentiating the expression for the mass of a spherical object and dividing by the original expression to obtain:

$$\frac{\mathrm{d}M}{M} = 3\frac{\mathrm{d}r}{r}. \tag{7.19}$$

For $\mathrm{d}M/M$ of 0.0003, $\mathrm{d}r/r$ is 0.0001. A 10 km radius nucleus would shrink by 1 meter per pass if the loss were uniformly distributed. If the mass loss were concentrated in 10% of the total area, the loss could amount to about 10 meters in the active areas. The numbers themselves should not be taken too seriously, but they illustrate the fact that the mass loss per pass is rather superficial.

Still, the loss is inexorable and ultimately all the volatiles are lost. The result is a body of refractory materials or dust. If the structure is strong enough, the result is an extinct comet. While our knowledge of extinct comets is necessarily incomplete, there is no doubt that a substantial population exists and that some are among the objects considered to be asteroids; see Chapter 8.

If the nuclear structure is weak, the nucleus can disintegrate. The final remnant consists of the larger particles that experience essentially no radiation pressure. Gravitational perturbations distribute the particles along the orbit and change the orbit. These particles are the meteoroids that produce meteor showers in the earth's atmosphere; see Chapter 10.

7.5 Summary

Nuclei have gone from an unseen abstraction that provides material for comas and tails to specific objects of considerable complexity. Surface details have been determined by space missions to comets Halley and Borrelly. These nuclei are dark, not close to spherical, and emit material in jets that are active on the sunward side. Sizes have been determined by several methods and the range is large. Cometary activity is basically produced by water–ice sublimation. Most comets have a dust crust on at least a portion of the surface.

The deep interior is being probed by attempts to understand the heliocentric variation of cometary activity, outbursts, rotation, and the disintegration of the nucleus. Recent examples are the models of comet Halley's complex rotation and the breakup of comet LINEAR (S/1999 S4). The interior is chemically homogenous, cold, and porous. Phase transitions of ice are an energy source. Nuclei are probably assemblages of cometesimals.

The complex processes operating on comet nuclei are being modeled and the results compared with observations. Of course, comets in the inner solar system put on an impressive show, but their fate is destruction.

8

Asteroids

8.1 History

On January 1, 1801, the asteroid Ceres was discovered by G. Piazzi at Palermo, Sicily. Initially, the discovery appeared to confirm the regular spacing of the planets as represented by the Titus–Bode Law by finding the "missing planet" in the gap between Mars and Jupiter. The apparent confirmation was short-lived because of the discovery of Pallas in 1802, Juno in 1804, and Vesta in 1807. Clearly, there was not a single large planet in the gap. The absence of significant disks observed for these objects in the early telescopes implied a "small" size. This prompted W. Herschel to use the term asteroid because of their stellar appearance. Minor planet is an alternate term for asteroid. We prefer asteroid and feel that minor planet is better used to describe smaller versions of the major planets, i.e., bodies that are approximately spherical and have a planet-like internal structure. Both terms are widely used. For the history of asteroid studies, see Cunningham (1988) and Peebles (2000). For a new compendium on asteroids, see *Asteroids III* (Bottke *et al.* 2003).

Table 8.1 lists some properties of these early discoveries. These are among the larger asteroids known. The searches continued through the nineteenth century and the development of photography roughly around the turn of the twentieth century greatly increased the rate of discovery. We now know that binary asteroids are fairly common (Burns 2002). Searches are currently carried out over large areas of the sky with CCD detectors. The rate of discovery is high, particularly for Near-Earth Asteroids (NEAs). Comet orbits can be either prograde or retrograde. Most asteroids orbits are prograde and Fig. 8.1 shows the positions projected into the plane of the ecliptic of the major planets and approximately 1500 asteroids for January 1, 2002. Note that the numbered asteroids are those with known orbits that allow this type of presentation; thousands more are known. The figure shows that

Table 8.1 *First discovered asteroids*

Asteroid	Number	Diameter (km)	Semi-major axis (AU)	Eccentricity of orbit
Ceres	1	913	2.77	0.097
Pallas	2	523	2.77	0.180
Juno	3	244	2.67	0.218
Vesta	4	501	2.36	0.097

Source: Data from *Asteroids II* (Binzel *et al.* 1989.)

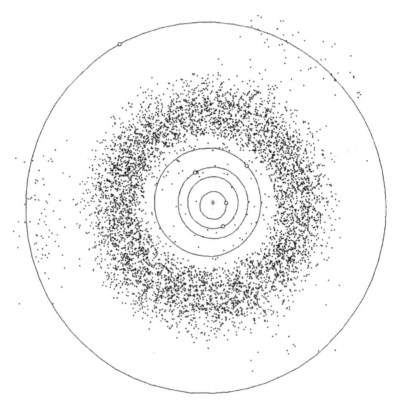

Fig. 8.1. Positions of asteroids projected into the plane of the ecliptic showing the main belt and the Trojans on January 1, 2002. The plot shows approximately 1500 asteroids, positions from the Minor Planet Center. The orbits of the planets are plotted with the main belt asteroids falling between the orbits of Mars and Jupiter.

most asteroids are concentrated into the *main belt* between Mars and Jupiter. The figure also shows the relatively small number of asteroids that cross or approach the earth's orbit and the *Trojan asteroids* ahead of and behind Jupiter in the L_4 and L_5 Lagrangian points in the sun–Jupiter system. Also, there are: the *Apollo*

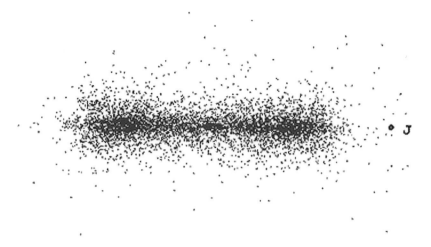

Fig. 8.2. Positions of asteroids projected on a plane perpendicular to the ecliptic on January 1, 2002, showing the concentration to the ecliptic plane. The plot shows approximately 1500 asteroids. The positions are from the Minor Planet Center. The sun is at the center and Jupiter is marked (J).

asteroids with semi-major axes $a \geq 1.0$ AU and perihelion distances $q \leq 1.017$ AU; the A*mor asteroids* have $a > 1.0$ AU and $1.017 < q \leq 1.30$ AU; and the *Aten asteroids* have $a < 1.0$ AU and aphelion distances $Q > 0.983$ AU. The significance of the heliocentric distances of 0.983 AU and 1.017 AU is that they are the perihelion and aphelion distances of the earth. The C*entaur asteroids* have eccentric orbits in the outer solar system, generally between Saturn and Neptune; note that some if not all Centaurs might be cometary bodies. Figure 8.2 shows the positions projected onto a plane perpendicular to the ecliptic plane, showing that asteroids are generally low inclination objects.

Figure 8.1 does not show the structure in the semi-major axes of the asteroids that provided early clues to the dynamical processes at work in the asteroid belt. Figure 8.3 shows the distribution of semi-major axes for the numbered asteroids. The distribution is not smooth, but shows major gaps. Notice, for example, the deep minimum where the asteroid orbital period would be $1/3$ the orbital period of Jupiter at 2.5 AU. These gaps were discovered in the 1860s by D. Kirkwood. Resonances with Jupiter could produce removal of the asteroids, but understanding of the process was a recent event (see section 2.6 and below).

A major development in the study of asteroids has been their exploration by spacecraft missions. These missions have collected considerable data, including detailed imaging of their surfaces. Figure 8.4 shows a gallery of images of asteroids obtained from nearby spacecraft. The asteroid 951 Gaspara was the first to be

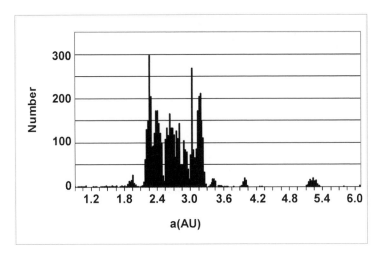

Fig. 8.3. The distribution of semi-major axes for 1500 numbered asteroids showing the Kirkwood gaps, e.g., the 1/3 resonance at 2.5 AU. Data from the Minor Planet Center.

visited by a spacecraft, *Galileo*. Gaspara is approximately $19 \times 12 \times 11$ km and is heavily cratered. *Galileo* also visited 243 Ida (Belton *et al.* 1996) and the images revealed a tiny satellite asteroid, Dactyl. Ida is approximately $56 \times 24 \times 21$ km; it is heavily cratered, has large boulders, and has a layer of fine material. Dactyl, the first confirmed asteroidal satellite, is about 1.5 km across (Chapman *et al.* 1995) and orbits Ida at a distance of approximately 85 km (Belton *et al.* 1995). Both 243 Ida and Dactyl are S-type asteroids and are probably members of the Koronis asteroid family. Dactyl does not resemble an angular rock fragment. Dactyl has minor but distinct spectral differences from Ida that are likely due to differences in space weathering (Clark *et al.* 2001); see section 8.3.3. The *NEAR* (*Near Earth Asteroid Rendezvous*) – *Shoemaker* spacecraft studied 253 Mathilda during a flyby, which came within 1200 km of the asteroid, and collected over 500 images (Veverka *et al.* 1997). Mathilda is approximately 52 km in diameter and has a very low albedo of 0.03. The low density of 1.3 g cm^{-3} (from gravitational perturbations revealed by doppler tracking; Yeomans *et al.* 1997) implies that it is porous. It may be a large rubble pile. The surface features are dominated by very large craters, which range in size from the limit of resolution (<0.5 km) up to 30 km. Gaspara, Ida, and Mathilde are main belt asteroids and have asteroid type, explained below, of S, S, and C, respectively.

The *NEAR–Shoemaker* spacecraft studied the near-earth asteroid 433 Eros. It orbited the asteroid for a year, mapped the surface at high resolution (Veverka *et al.* 2000), and on February 12, 2001 was the first spacecraft to land on an asteroid

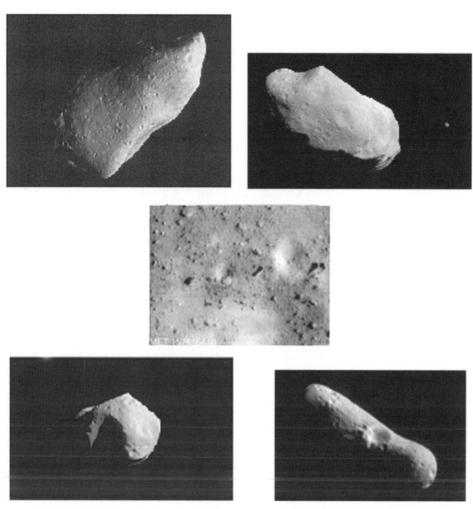

Fig. 8.4. Gallery of asteroid images from spacecraft. Upper left, 951 Gaspara (19 by 12 by 11 km) imaged from *Galileo*. Upper right, 243 Ida (35 km) and its tiny satellite Dactyl (1.5 km) at the right edge of the frame, as seen from *Galileo* as it passed at about 10 000 km. Lower left, 253 Mathilde (59 by 47 km) from *NEAR–Shoemaker* as it passed at a distance of 2400 km. The large crater is 10 km deep. Lower right, 433 Eros (33 by 13 by 13 km) from *NEAR–Shoemaker*. Center, view of 433 Eros taken from the *NEAR–Shoemaker* spacecraft as it descended to the surface. (NASA.)

(Veverka *et al.* 2001a). Eros is peanut-shaped with a maximum diameter of about 34 km and smaller dimensions of about 13 km × 13 km (Zuber *et al.* 2000). Its density is 2.7 g cm^{-3} (Yeomans *et al.* 2000). The surface is heavily cratered and shows pits or grooves (Veverka *et al.* 2001b; Cheng *et al.* 2001). The surface also shows ejecta blocks, a relative lack of small craters, and a smooth appearance apparently produced by fine material infilling some surface features. Eros is an S-type asteroid (see below).

The *Deep Space 1* mission that has produced important new results for comets (sections 6.3.2.3 and 7.2) via the flyby of comet 19P/Borrelly has also provided the first direct detection of an asteroidal magnetic field (Richter *et al.* 2001) at the asteroid Braille on July 29, 1999.

Several asteroids (e.g., Toutatis) have been imaged by large radars such as those at JPL's Goldstone facility and at Arecibo. This capability can be valuable for studying asteroids that come sufficiently close to earth (e.g., Ostro *et al.* 1999).

8.2 Why study asteroids?

The study of asteroids is currently in a renaissance, stimulated by space missions as well as advances in meteoritics, observations, and theory (Binzel 2000). But, why should people interested in comets study asteroids? The general answer is that, although the study of asteroids and comets evolved largely as distinct disciplines, they have much in common and the demarcation between the two is not always clear. For an introduction to current asteroid studies, see Chapman (1999), Asphaug (2000), and Binzel (2001).

On most formation scenarios, the building blocks of planet formation can be expected to be a continuum. Rocky planetesimals should dominate the inner solar system (the region of Mercury, Venus, Earth, and Mars) and icy cometesimals should dominate the outer solar system (the region of Jupiter, Saturn, Uranus, Neptune, Pluto (an icy planet), and beyond). (A class of icy objects beyond Neptune, of which Pluto is a member, will be discussed in Chapter 9.) The transition with heliocentric distance between icy asteroids and icy bodies that are called comets is not likely to be sudden. In the following sub-sections, we present specific examples illustrating the lack of clear demarcation.

8.2.1 Orbital properties

A plot of semi-major axis vs. eccentricity, Fig. 8.5, for comets and asteroids is instructive. Note that asteroids occupy the low eccentricity/low semi-major axis part of the diagram, while comets generally occupy the high eccentricity/high semi-major axis part of the diagram. The so-called "boundary" between asteroids and comets is only approximate and there are major exceptions.

Wetherill (1991) has presented arguments showing that comets can dynamically evolve into asteroidal orbits. This process takes much longer than the time for comets to expend their volatiles and become dormant. Thus, some

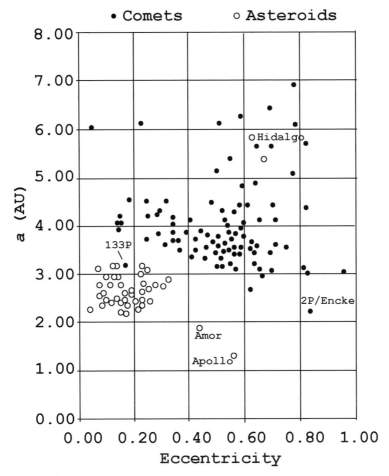

Fig. 8.5. A plot of semi-major axes vs. eccentricity for asteroids and comets. Note that the demarcation between asteroids and comets is not always clear.

asteroid-appearing objects can have a cometary origin. A significant fraction of objects in earth-crossing orbits, specifically the Apollo objects, may be old, non-active comets.

8.2.2 2060 Chiron and 4015 Wilson–Harrington

The discovery of 2060 Chiron by C. Kowal in 1977 (see Kowal 1996) helped bring the "distinctions" between asteroids and comets into focus. It was initially considered an asteroid and had a low albedo like other outer solar system objects. Chiron's orbital period is 51 years, the semi-major axis is 13.7 AU, and the eccentricity is 0.38. Thus, the orbit is mostly between Saturn and Uranus, fairly eccentric, and unstable.

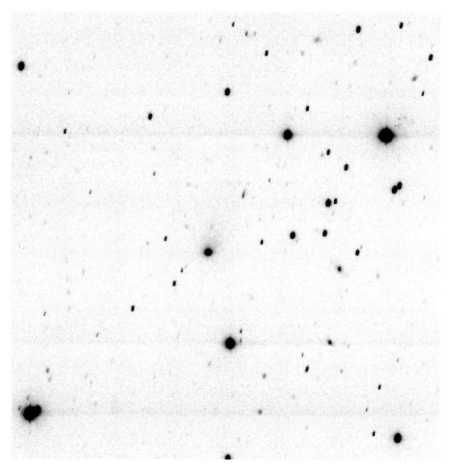

Fig. 8.6. Image of 2060 Chiron taken on January 6, 1992, showing an asymmetric distribution of dust in the coma. (Courtesy of Karen Meech, University of Hawaii.)

Chiron's orbit brought it closer to the sun in the 1980s and this period of time corresponded to the development and wide-spread use of CCDs in astronomy. Chiron displayed brightness variations in 1987 and 1988. In 1989, Meech and Belton (1990) detected a coma. A sample image is shown in Fig. 8.6. In addition, CN band emission was detected in 1990 (Bus *et al.* 1991).

Extensive observational evidence firmly establishes Chiron as a comet. Chiron's estimated diameter is about 200 km. If thrown into the inner solar system, it would become a spectacular comet. Chiron is the first discovered member of the Centaurs, objects that travel in eccentric orbits in the outer solar system. These are cataloged as asteroids, but most would probably become comets if sent into the inner solar system.

Any doubt about the cometary nature of 2060 Chiron should have been removed with the modeling study by Prialnik *et al.* (1995). They show that the activity pattern is consistent with gas release in a porous matrix composed of dust and amorphous water ice containing a small amount of trapped CO. The crystallization of the amorphous ice releases the occluded (trapped) CO molecules. The crystallization occurs in spurts and is triggered by thermal pulses reaching the CO source area some thousand or more meters below the surface. Finally, Prialnik *et al.* (1995) show that the current activity could be maintained for a very long period of time.

Asteroid 4015 is a marvelous example of a comet nucleus posing as an asteroid. The asteroid discovery took place in 1979 and the object was first designated 1979 VA (see Chapter 2 for an explanation of asteroid nomenclature). When sufficient astrometric observations determined a good orbit (and the designation 4015), the object's past position could be calculated. T. Bowell found the object on Palomar Sky Survey images taken in 1949. But the object displayed a plasma tail (Fernández *et al.* 1997). B. Marsden noted that the earlier object was comet P/Wilson–Harrington. The object is now known as 4015 Wilson–Harrington.

The Palomar Sky Survey image on November 19, 1949 is shown in Fig. 8.7. Palomar Sky Survey images taken later in November 1949 show an asteroidal-appearing object. Thus, the November 19 image seems to have recorded the comet's last activity, at least for a number of years.

Additional observations indicate a diameter of about 3 km and a low albedo. The spectrum is similar to spectra of C-type asteroids (see below).

The story of 4015 Wilson–Harrington shows the value of following up on discoveries that may, at first sight, seem routine. Astronomers have had the extraordinary good fortune to observe a comet becoming an asteroid. Together with Chiron and the asteroid 3200 Phaeton (the source of the Geminid meteor shower; Chapter 10), the asteroid–comet distinction can be regarded as securely blurred.

8.2.3 Definitions revisited

Before continuing the discussion, a brief review of definition relating to comets and asteroids is in order. This discussion follows a similar one by L.-A. McFadden (1993). In terms of place of formation, comets are (icy/snowy) bodies formed in the outer solar system while asteroids are (rocky) bodies formed between Mars and Jupiter. Of course, the boundary between these cases is probably not sharp.

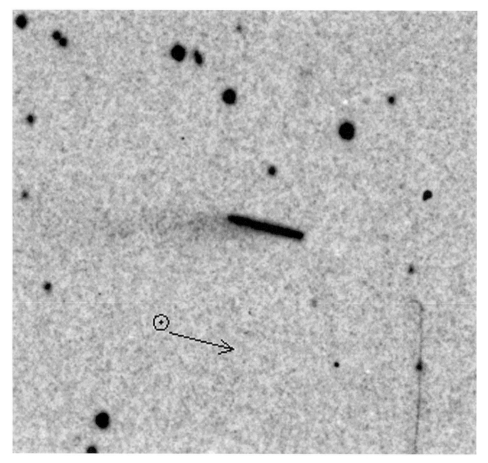

Fig. 8.7. Palomar Sky Survey blue image of comet Wilson–Harrington taken on November 19, 1949, showing the plasma tail. The sunward direction is marked. (Courtesy of Yan Fernández, University of Hawaii.)

Observationally, the situation is much more difficult. An object observed with a coma and/or a tail is called a comet. Such an object should be called an *active comet*. If a cometary body still has (sub-surface) volatiles, but is not currently active, it should be called a *dormant comet*. Activity may occur in the future. A cometary body that has lost all volatiles should be called an *extinct comet*.

Dormant and extinct comets are obviously the ones that can be confused with asteroids. The development of an observational diagnostic that permits a reliable discrimination would be very valuable. In any event, the small bodies of the solar system – asteroids, comets, and satellites – provide opportunities for studying specific phenomena and inter-relationships.

8.3 **Properties of asteroids**

Asteroids are studied in a variety of ways. Astrometry and orbit determination provide clues on formation, evolution, and dynamical processes. Photometry and spectrophotometry supply sizes and information on surface properties, including mineralogy. Meteorites bring samples of high-density asteroids to the earth's surface, where laboratory studies can be carried out. Experimental and theoretical studies examine the collisional processes that are so important in understanding asteroids. Finally, spacecraft missions to asteroids, with detailed surface mapping, density determinations, and compositional results, are providing a major stimulus to the field.

8.3.1 Orbits and dynamics

Basic data on orbital properties were presented in sections 8.1 and 8.2.1, and the dynamics is discussed in section 2.6. The *Kirkwood gaps*, shown in Fig. 8.3, were understood long ago in terms of commensurabilities with Jupiter's orbital period. Asteroids in the famous 3 : 1 orbital period resonance with Jupiter have similar, repeating geometrical circumstances during the times of closest approach. Thus, the orbital perturbations by Jupiter are in the same direction and accumulate; the result is that the asteroid is removed from the orbit with the commensurability. The same argument applies to the other major Kirkwood gaps corresponding to the 5 : 2, 3 : 7, and 3 : 5 commensurabilities. If there is no commensurability, Jupiter's orbital perturbations are in random directions and do not accumulate. Thus, the long-term effect on the asteroid's orbit is small.

While the generic mechanism for removal of asteroids with period commensurabilities with Jupiter was known, the details were not understood until studies by J. Wisdom in the 1980s (e.g., Wisdom 1985). The suspect locations are chaotically unstable. An asteroid at the 3 : 1 commensurability undergoes large and unpredictable changes in eccentricity on short time scales, and it is removed from the gap area. Also see section 2.6.1. Note that currently only the 3 : 1 resonance has been successfully modeled in detail. The large chaotic zones in the asteroid belt also provide a mechanism for delivery of asteroidal material, meteorites, to earth. The chaotic zones in the asteroid belt produced by Jupiter supply insight into the history of the material in the main asteroid belt. The material is not the debris from a large disrupted or exploded planet. The material is from a planet or planets that were unable to form because of Jupiter.

The similarities of orbits of some asteroids lead K. Hirayama (1918) of the Tokyo Observatory to suggest the concept of *asteroid families*, i.e., asteroids with rather similar semi-major axes, eccentricities, and inclinations. This proposal was early

Table 8.2 *Some major asteroid families*

Name	a(AU)	i(°)	Asteroid type
Koronis	2.9	2.0	S
Eos	3.0	10.0	Intermediate, S–C
Themis	3.1	1.5	C

The orbital parameters are typical values for each family.
Data from Kowal (1996).

evidence of the importance of collisions in the asteroid belt. We now understand that families of asteroids can result when the collision process disrupts the parent body, but does not impart enough energy to send the fragments into significantly different orbits. The processes of collisions are discussed below. Some major families are listed in Table 8.2.

Another process for altering asteroid orbits was suggested by I. Yarkovsky. The infrared radiation from the surfaces of asteroids carries away momentum. If the temperature is uniform over the surface, no change in momentum results. But, if this is not the case, the momentum of the asteroid is changed. If the axis of rotation is roughly perpendicular to the plane of the ecliptic, the morning side of the asteroid is cooler than the afternoon side. Thus, momentum is added or subtracted from the orbit depending on the sense of rotation with respect to the orbit. This is analogous to the action of mass loss in comets producing the non-gravitational forces. The picture described is the diurnal Yarkovsky effect. It can also operate as a seasonal effect.

The effect is minute and unimportant for comets. But, over time, the orbits of asteroids can be altered. The effect is not important for objects larger than a few kilometers in radius (because surface area ÷ mass decreases with increasing radius). But, for smaller bodies in the range 10 centimeters to 100 meters, a 0.1 AU change in semi-major axis is produced in tens of millions of years. Such changes move asteroids or fragments in stable orbits into the resonances described above. They are then perturbed into highly eccentric orbits that produce ejection from the solar system or allow them to be sampled at earth as meteorites.

With the mechanisms for removal of asteroids from the main belt just described, a logical question is: why are there any asteroids left? On average they have been removed. Today, we are seeing no more than one part in one thousand of the original mass. Although there are many asteroids, the total mass is quite small, estimated at 2×10^{24} grams or roughly 1/37th the mass of the moon. And, a major fraction of this mass is in Ceres.

8.3.2 Photometry and spectrophotometry

Asteroids' dimensions are determined by the "standard method" outlined in Chapter 7. This method uses simultaneous visual and infrared photometry to determine the size and the albedo. Sizes can also be determined from rare occultations of stars by asteroids and from radar soundings of asteroids that come close enough to earth. Of course, direct imaging by spacecraft is now available and should contribute a steady flow of a small number of well-determined sizes.

The size distribution can be approximately represented by

$$n(a) \propto a^{-3/2} \quad \text{or} \quad N(a) \propto a^{-5/2}, \tag{8.1}$$

where $n(a)$ and $N(a)$ are the differential and integral distributions. These distributions should not be taken seriously for the larger asteroids (Table 8.1) and the distributions are steeper for radii less than about 5 km.

Albedos on the surfaces of asteroids do not appear to exhibit large variations. Hence, we assume that most brightness variations mean that the observed cross-sections are varying. Some vary by as much as a factor of 2. If the shapes are represented by tri-axial ellipsoids, the typical shape is

$$a : b : c = 2 : \sqrt{2} : 1. \tag{8.2}$$

Note that many asteroids are irregular. The densities of asteroids are estimated to be in the range 1.0 to 3.5 g cm^{-3}. Rotation periods show a large range with a typical value of about 9 hours.

The geometric albedos (see Chapter 3) show a very large range from a low of about 0.02 to nearly 0.50. There is a low value peak around 0.04 and a high value peak around 0.15. These are separated by a minimum near 0.07. The idea of different classes of asteroids follows from this relatively crude approach. When the albedos are combined with U–V colors as shown in Fig. 8.8, several classes of asteroids can be distinguished. This result naturally leads to the idea of asteroid taxonomy as developed by D. Tholen and others.

There are several asteroid taxonomies in the literature. Here the system initially developed by D. Tholen (1984) is described. This system and the others (e.g., the Bowell, Barucci, and Tedesco taxonomies) are described by Tholen and Barucci (1989).

The Tholen Taxonomy is based on the Eight-Color Asteroid Survey (ECAS) and albedos. The goal was to provide a classification system, a taxonomy, that places the objects into distinct classes. The process starts with a principal component analysis. Consider a sample of objects for which two observational quantities have been measured, say x and y. If they are correlated, the coordinate system can

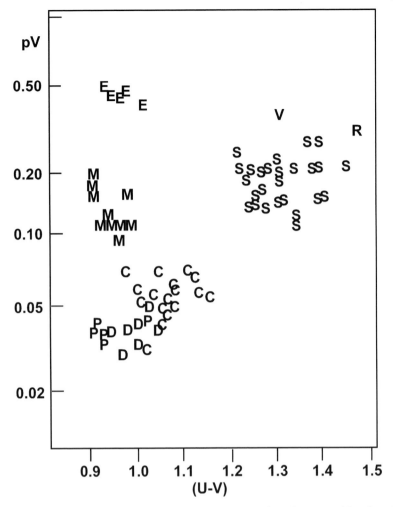

Fig. 8.8. A plot of visual albedos, pV vs. (U–V) colors for asteroids, showing the clustering of points for several classes of asteroids and the importance of the albedo. The points are plotted with their class symbols, S, C, E, M, P, D, V, and R. (Data from *Asteroids II*, ed. R. P. Binzel, T. Gehrles, and M. S. Matthews, Tucson: University of Arizona Press, 1989. The albedos and colors are from Tedesco *et al.*, pp. 1151–61 and the classes are from Tholen, pp. 1139–50.)

be rotated to maximize the variance along the first axis, called the first principal component, and to minimize the variance along the second axis, called the second principal component. This idea is easily generalized to the multiple dimensions of the ECAS data.

However, for the ECAS data, 68% of the variance is contained in the first principal component and 27% in the second principal component. Thus, a two-component

Table 8.3 *Minerals important in reflectance spectroscopy*

Mineral	Formula
Pyroxene (Clinopyroxene)[a,b]	$(Ca, Mg, Fe) SiO_3$
Olivine[b]	$(Mg, Fe)_2 SiO_4$
Serpentine	$Mg_3Si_2O_5(OH)_4$
Asphaltite	mixed, complex hydrocarbons
Graphite	C
Troilite	FeS
Magnetite	Fe_3O_4

[a] Clinopyroxene is presented as an example of the pyroxene group.
[b] Elements in parentheses separated by commas are interchangeable in the mineral structure.

plot contains 95% of the variance. The explanation for this behavior is that there are only two major features in the 0.3 to 1.1 μm wavelength range covered by the ECAS asteroid reflectance spectra – one in the UV and one in the IR.

The basis for the Tholen Taxonomy was a plot of the first and second principal components for the 405 asteroids with the highest quality ECAS data. The plot was done as a minimal tree diagram and the branches of the minimal tree cut to produce the initial classes. The effects of observational uncertainties were introduced and the results iterated until the classifications were stable; see the description of these statistical techniques in Tholen and Barucci (1989).

The albedos of the various classes initially assigned were examined for consistency. The final E, M, and P classes are the same except for albedo. Briefly, this is the origin of the taxonomic types listed in Table 8.4.

Major insights into the minerals found on asteroids can be made from reflectance spectroscopy in the visual and near-infrared spectral regions. Well-known absorption bands are diagnostic of the presence of some of the rock-forming minerals. This approach can be applied to the asteroids that are bright enough and, with advances in instrumentation, this number is large. Of course, there are still issues with the interpretation.

Pieters and McFadden (1994) list the common components of asteroid surfaces that can be remotely identified, listed in descending order of certainty: pyroxene, olivine, phyllosilicates (including serpentine), organic material (including asphaltite), and opaques (including metallic iron, graphite, troilite, and magnetite). Some of the minerals are identified in Table 8.3.

Figure 8.9 shows sample reflectance spectra for these minerals. The opaques, including metals, have no diagnostic bands and generally darken the spectrum.

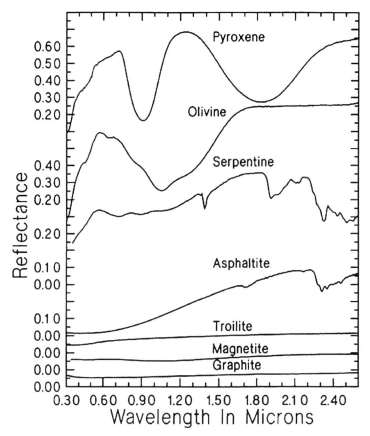

Fig. 8.9. Sample reflectance spectra for mineral components of asteroids and meteorites. (Data from RELAB, Brown University. Courtesy of Lucy McFadden and
Carlé Pieters.)

Also as seen in Fig. 8.9, the silicate minerals, e.g., pyroxene(s) and olivine, show
strong absorption bands. One might imagine that the interpretation of the spectra
from a surface of pure olivine would be straightforward. In nature, the surfaces
contain mixtures of major and minor minerals, and the interpretation of spectra can
be complex.

A linear superposition of spectra could be expected if the major mineral species
were separated on the asteroid. Usually, however, this is not the case and the reflectance spectra originate from areas of commingled mineralogy and constituents
of widely varying size range down to powders. Despite the complications and the
questions of uniqueness, minerals can be identified on asteroids. However, Pieters
and McFadden have noted that:

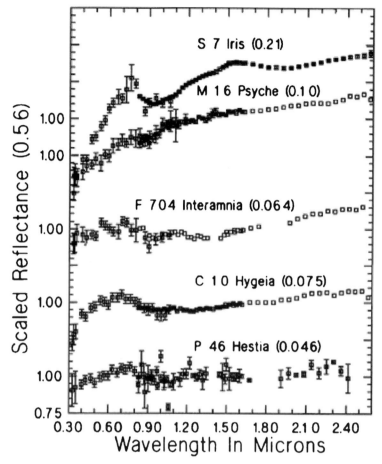

Fig. 8.10. Representative reflectance spectra of several asteroid classes scaled to unity at 0.56 μm. These are composite spectra derived from collections by Chapman and Gaffey (1979), Zellner *et al.* (1985), and Bell *et al.* (1988). Taxonomy is derived from Tholen (1989). Albedos (in the parentheses) are from *IRAS* measurements (Tedesco 1989). See Pieters and McFadden (1994). (Courtesy of Lucy McFadden and Carlé Pieters. Data from the Planetary Data System, Small Bodies Node, University of Maryland and NASA.)

A certain amount of humility, however, must be superimposed on all interpretive approaches . . .

Sample asteroid spectra are shown in Fig. 8.10.

The very dark and almost featureless spectra of some asteroids implies the presence of opaque materials such as iron, graphite, troilite, and magnetite. The asteroid spectra showing absorption bands resembling the olivine and pyroxene reflectance spectra (Fig. 8.9) imply the presence of these minerals. The Type S asteroids are

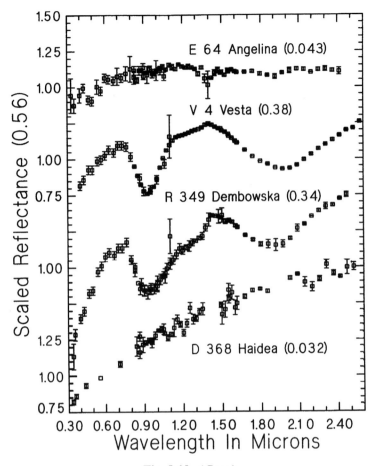

Fig. 8.10. (*Cont.*)

found near the inner edge of the main belt, generally within about 2.5 AU. They have signatures of olivine and pyroxene and a reddish spectrum possibly due to absorption by iron and nickel. They are thought to be the source of ordinary chondrites and stony-irons. In fact, the main asteroid types show a fairly distinct segregation with heliocentric distance as shown in Fig. 8.11. This figure uses the terminology for the main classes of asteroids that is summarized in Table 8.4. The middle belt, roughly between 2.5 and 3.5 AU, is dominated by C-type asteroids (and the less abundant, dark B, G, and F types) and is the likely region for the parent bodies of carbonaceus chondrites. Beyond 3.5 AU are the dark P and D (most Trojans are D type) asteroids. The surfaces must have large quantities of opaque minerals and organic matter. The water present must be ice. These facts are consistent with a formation scenario where the asteroids closest to the sun had silicate melting, those at an intermediate distance had ice melting, and those at the largest distances never

Table 8.4 *Summary of the main types of asteroids*

Type	Peak (AU)	Albedo	Reflectance spectra	Minerals	Example	Meteorite analogue
E	1.9	0.25–0.60	Flat or slightly reddish over 0.3–1.1 µm	Forsterite, (olivine group)	3103 Eger	Aubrites
R	1.9	~0.40	Similar to S, stronger absorptions (olivine)	Pyroxene, Olivine	349 Dembowska	??
S	2.2	0.10–0.22	Reddish shortward of 0.7 µm, weak/moderate absorption near 1 µm & 2 µm	Olivine, Pyroxene	433 Eros	Stony-irons or ordinary chondrites
M	2.7	0.10–0.18	Similar to E, moderate albedo	Metals	16 Psyche	Irons, enstatite chondrites
F	2.7	0.03–0.06	Flat, essentially featureless	Carbon	762 Pulcova	??
C	3.1	0.03-0.07	Similar to F, but with UV absorption below 0.4 µm	Fe silicates, some with OH	10 Hygiea	Carbonaceous chondrites (CM)
P	4.0	0.02–0.06	Similar to E with low albedo	Carbon	87 Sylvia	None
D	>5.0	0.02–0.05	Essentially featureless, redder than P, especially longward of 0.6 µm	Carbon	1143 Odysseus	Kerogens[?], a constituent of carbonaceous chondrites

There are many more minor types including V for Vesta. *Sources:* Tholen and Barucci (1989); Pieters and McFadden (1994); Lewis (1997); Chapman (1999); and Binzel (2001). Peak refers to the maximum in the semi-major axis distribution. The minerals listed are intended to be representative.

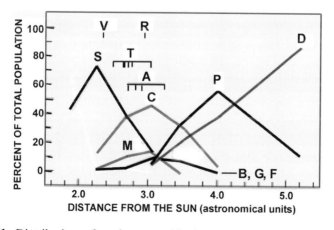

Fig. 8.11. Distribution of major asteroid classes with heliocentric distance. (Courtesy of Jeffrey F. Bell, Hawaii Institute of Geophysics and Planetology.)

had ice melting. Put another way, there is large diversity among asteroids from dry, siliceous bodies close to the sun to carbonaceous, somewhat water-laden (and perhaps icy) bodies far from the sun. The meteorite information in Table 8.4 is discussed in the next section.

8.3.3 Meteorites

Nature delivers samples of the asteroid belt to earth in the form of meteorites, the bodies that enter the earth's atmosphere and survive to the earth's surface. A *fall* is a meteorite that is found because it was seen or heard to fall. A *find* has no witnesses to the fall, but was identified because of its distinctive appearance (surfaces that are glossy and have flow marks) or chemical composition (e.g., the Fe : Ni value). These distinctions are important in understanding how accurately the sample of meteorites in museums represents the asteroid population. Note that some meteorites come from the moon and Mars. For an introduction to meteorite studies, see McSween (1999) and Zanda and Rotaru (2001).

The basic classes of meteorites are as follows. Irons are composed of almost pure metallic iron and nickel. These are the easiest to identify because pure iron is almost never found on earth. Stony-irons are composed of stony material and metallic iron; they are rare. Stones are composed of rocky material (silicates). They are easily mistaken for terrestrial rocks and they are usually found when witnessed. The most common type of stone is the ordinary chondrite, so-called because of their chondrules, beads roughly 1 mm in radius containing silicates. The carbonaceous chondrites are darker stones containing carbon, complex organic compounds, and usually chemically bound water. Stones without chondrites are the relatively rare achondrites. The percentages of these types in falls is: irons, 5%; stony-irons, 1%; stones (chondrites), 86%; and stones (achondrites), 8%.

A gallery of meteorites is shown in Fig. 8.12. The names of meteorites have been historically taken from the town nearest to the place of discovery. An example is the Acapulcoites that are named after a stone that fell in the area of Acapulco, Mexico. This idea fails for meteorites discovered in Antarctica because of the lack of towns. Antarctic meteorites are given a designation composed of letters and numbers. A major site in Antarctica is the Allan Hills region and the one-hundred thirteenth meteorite discovered in 1978 is designated ALH 78113.

A physically significant distinction among asteroids is based on their presumed origin. While some processing such as *aqueous alteration* (mineral transformation produced by reactions with water) has occurred, the material in *primitive meteorites*

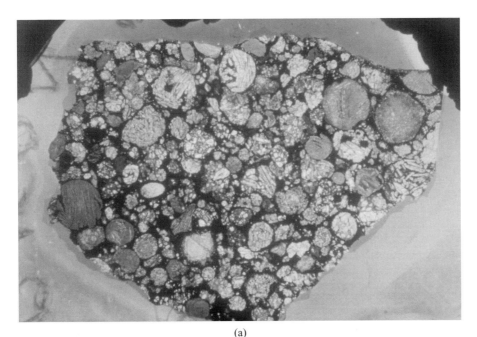

(a)

(b)

Fig. 8.12. Gallery of meteorites. (a) The ordinary chondrite Semarkona (LL3). This is a transmitted light photograph of a polished thin section that is 1 inch across. The section contains many chondrules and the black material is both metal and fine-grained matrix. (b) The Palo Blanco Creek eucrite, a basaltic meteorite. This hand sample is about 10 cm across and shows brecciated texture. (c) The carbonaceous chondrite Allende (CV3). This is a cut surface of a hand sample. The

(c)

(d)

Fig. 8.12. (*Cont.*)
meteorite consists of about 50% dark, fine-grained matrix. Chondrules and CAIs (calcium-, aluminium-rich inclusions) are lighter, millimeter-sized inclusions. (d) The Toluca iron meteorite. The scale cube is 1 cm across. The polished surface shows the Widmanstätten pattern. The inclusions (brown in color) are iron sulfide (troilite). (Courtesy of the Institute of Meteoritics, University of New Mexico.)

has never been subjected to great heat or high pressure. The passage through the earth's atmosphere only heats an outer layer and the interior does not become hot. Most stones are primitive.

The material in differentiated meteorites comes from differentiated parent bodies. The meteorites are fragments. The material was processed and at least heated above its melting temperature. The radioactive isotope ^{26}Al was abundant in the solar nebula early in the solar system's history. Its rapid decay can explain the thermal histories of main belt asteroids. Irons and stony-irons are differentiated meteorites. The irons come from the metallic cores of the parent bodies while the stony-irons generally come from the interface zone between the cores and the mantles of silicate materials. This conclusion implies that there were bodies in the asteroid belt large enough to differentiate; then collisions fractured them, and in many cases, totally disrupted them.

An important type of differentiated stony meteorite is composed of basalt (a type of rock formed from magma) and must come from asteroids that had volcanic activity. The HED meteorites (Howardites, Eucrites, and Diogenites) appear to have a close association with the asteroid Vesta and the discussion returns to this subject below. Meteoriticists have studied samples of meteorites in the laboratory and have developed a detailed classification scheme far beyond the scope of this introductory discussion.

Identification of a meteorite analogue with a class of asteroids (Table 8.4) comes primarily from comparison of the reflectance spectra of meteorites (Fig. 8.13) and asteroids (Fig. 8.10). These spectra show obvious similarities with the spectra of mineral samples.

The asteroid Vesta is not a member of the more common classes of meteorites summarized in Table 8.4 and almost has its own class (V). It is the only large asteroid in the class and approximately 20 smaller members (in the range 5 to 10 km in diameter) are known. The spectrum of Vesta closely matches the spectra of the HED meteorites. These basaltic achondrites came from a body with volcanic activity. But, how were they delivered to earth? First, Binzel and Xu (1993) found that the smaller V asteroids had semi-major axes extending from Vesta to the 3 : 1 resonance in the main belt. As discussed above, the 3 : 1 resonance ejects objects from that zone in a chaotic manner. This zone has been called a "dynamical escape hatch" and it can supply objects to the inner solar system (as well as ejecting objects from the solar system). They also suggested that major impacts on Vesta were responsible for the smaller V asteroids. Thomas *et al.* (1997) obtained *Hubble Space Telescope* images of Vesta that revealed a large impact crater near the south pole (Fig. 8.14). The crater is 460 km in diameter and approximately 13 km deep. Compare the size with Vesta's diameter of 530 km. This major impact excavated a

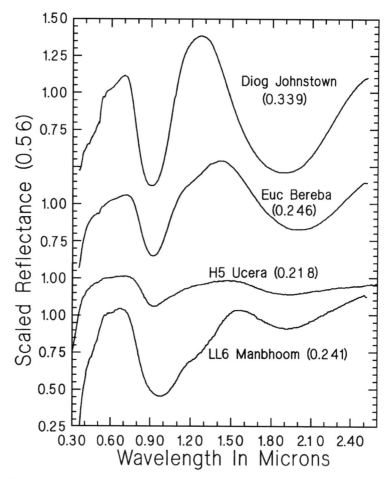

Fig. 8.13. Sample reflectance spectra of meteorites scaled to unity at 0.56 μm; the number in parentheses is the laboratory reflectance at 0.56 μm. (Data from RELAB, Brown University. Courtesy of Lucy McFadden and Carlé Pieters.)

substantial portion of the igneous crust. The exposed layers have colors consistent with mantle minerals such as pyroxene or olivine. The association of the HED meteorites with Vesta is quite convincing.

But, not all of the asteroid-meteorite correspondences are convincing as indicated in Table 8.4. A particularly irritating problem was the correspondence between the most common asteroids, the S type, and the most common meteorite, the ordinary chondrite. The absorption bands in the visible and near-infrared (pyroxene and olivine) fit, but the colors were too red. The idea of space weathering was invoked. The effect of micrometeorites and the solar wind could redden the surface layers. The process releases small particles of iron that produce the reddening.

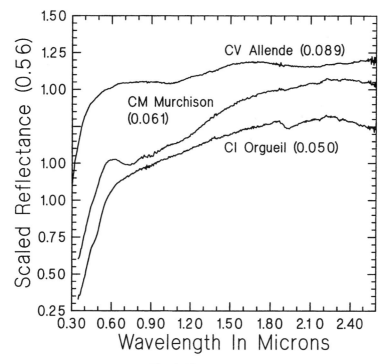

Fig. 8.13. (*Cont.*)

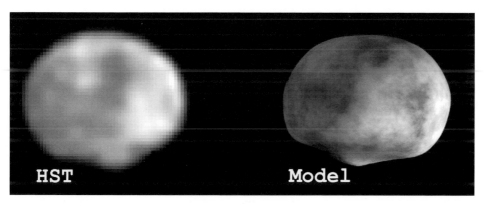

Fig. 8.14. An *HST* image of the asteroid 4 Vesta obtained on September 4, 1997, and a model constructed from the image. Note the sculpted appearance (bottom) that resulted from a large impact crater near the south pole. (P. Thomas, Cornell University; B. Zellner, Georgia Southern University; STScI/ NASA.)

Thus, only the surface layer is reddened and the asteroid below is a "normal"
S type. Confirmation comes from the *NEAR–Shoemaker* spacecraft which orbited
the S-type asteroid 433 Eros. First, images of Eros show craters that have disturbed the outer patina and exposed the lighter layers below. See Clark *et al.*'s
(2001) study of the crater Psyche. Second, the x-ray/γ-ray spectrometer on *NEAR–
Shoemaker* determined elemental abundances for 433 Eros (Trombka *et al.* 2000)
and found that this S-type asteroid has the same basic composition as the ordinary
chondrites (except for depletion in Ca and S). While this approach is promising,
final acceptance should be reserved until understanding of the infrared spectra is
obtained.

While the progress in relating Vesta to the HED meteorites and the S-type asteroids to the ordinary chondrites is impressive, there is still much to do. Note the
lack of correspondence for some types of asteroids in Table 8.4.

Having samples of the asteroid population in laboratories allow ages of the
material to be determined. Radiometric ages can be determined from precise measurements of the parent/daughter ratios for elements subject to radioactive decay.
The common measurements use the decay of uranium or thorium to helium, kryp-
ton to argon, and rubidium to strontium. Almost all values are close to 4.5 billion
years. Shock heating by a collision can lower the apparent ages. The age of 4.5 $\pm$
0.1 billion years is widely regarded as the time since solids condensed from the
solar nebula and began the process of forming larger bodies. In essence, this is the
age of the solar system.

Cosmic ray ages for meteorites provide data for the time between excavation from
the parent asteroid and arrival at earth. The meteorites are bathed in the known cosmic ray flux when they are set adrift in the heliosphere and the cosmic rays penetrate
below the surface. The bombardment by the cosmic rays produces radioactive isotopes and these can be measured. A complication is the material lost by ablation
during its passage through the atmosphere. The measurements can determine the
approximate size of the original meteoroid and the amount lost through ablation.
Cosmic ray ages for most meteorites are in the range 10 to 50 million years. This
is the time it takes for a collision product to come to earth.

8.3.4 Collisions

A full understanding of asteroids (and meteorites) requires an understanding of
collisions and their by-products. The problem is intrinsically messy. Objects of
different sizes, different shapes, different rotations, different material strengths
and different internal structure can collide at different angles. Only the relative
speeds have a limited range because most collisions take place between bodies

in similar orbits. The initial impactors are damaged, fragments can be produced, and these fragments may or may not have enough energy to escape. Despite the obvious complexity, understanding is important because collisional processes are an integral part of asteroids and their evolution.

The basic approach is two-fold. Laboratory experiments can establish a body of empirical data. These data must be scaled from the laboratory size regime to the asteroidal size regime, a scaling of many orders of magnitude. When a numerical theory is developed, the results can then be compared to the observations. For example, a plausible global theory of collisions should reproduce the observed size distribution for asteroids.

Some distinctions in this subject are useful. Impact events can be considered as low velocity or high velocity. The dividing line is the longitudinal sound-wave speed in the target material; for most cases, this value is $\sim$1 km s^{-1}. Low-velocity events are usually relevant to the study of accretion and fragmentation processes. Typical relative impact speeds in the asteroid belt are around 5 km s^{-1} and are therefore high-velocity impacts.

Another distinction, as introduced by Fujiwara *et al.* (1977), involves the range of specific energy of the event (kinetic energy of impact $\div$ target mass). From lowest to highest energy, they recognized: (1) simple cratering; (2) larger cratering with fragments (called spalls) off the target; (3) fracturing of the surface layers leaving a core region; and (4) complete destruction. A physical understanding of the processes that produce the different outcomes as a function of specific energy would be a major step. But, the experimental work alone offers encouragement. Fujiwara *et al.* (1978) measured the shapes of impact-produced fragments. In terms of tri-axial ellipsoids, they found the dimensional ratios of $2 : \sqrt{2} : 1$, in agreement with average asteroidal values.

Scaling the laboratory results to numerical codes that include asteroid sized bodies must consider the flaws or cracks that exist in all natural solids. How the stresses produced by the impact are transferred to the crack and the crack develops under stress are crucial to understanding the fragmentation process. Codes have been developed that model these complex processes. Two regimes have been recognized in the fracture process – a strength regime where material bonds control the fracture and a gravitational regime where gravitational stresses control the fracture.

Agreement between different codes has been found. Asphaug and Melosh (1993) in their investigation of a large impact crater, Stickney, on Phobos, and Melosh and Ryan (1997) found that for large, homogeneous bodies, the energy transferred to the ejecta was insufficient for them to escape. In other words, the asteroid could be shattered but not dispersed. This result implies that some asteroids could be rubble piles that gravitationally re-accumulated after a major impact event. The low density derived for the C asteroid 253 Mathilde supports this result.

Disagreement between different codes emerged, for example, with attempts to model the impact responsible for the fragments from Vesta (see section 8.3.3). Ryan and Melosh (1998) produced the crater and fragments, but they had insufficient energy for large fragments to escape. Asphaug (1997) modeled the event with a 42-km diameter impactor with a speed of 5.4 km s^{-1}. Asphaug's results were the production of the crater and large escaping fragments. The codes were identical except for the algorithm used to initiate the crack growth that ultimately leads to material failure.

The current situation was succinctly summarized by Ryan (2000):

Although some refinements still need to be made for a perfect correlation between theory and observation, impact codes and scaling relations, together with evolution codes, are beginning to provide a better understanding of the evolutionary processes at work in the asteroid belt.

8.4 Summary

This brief overview of asteroids with some attention to their relation to comets reveals objects that offer key insights into the formation of the solar system. The asteroids are not the debris from an exploded planet, but are the remnants of a planet or planets that never formed. Some of the early remnants were large enough to form "minor" planets that differentiated into a core, mantle, and crust structure. Some are relatively unaltered pieces of the solid materials that coagulated to form the solar system.

Collisions between asteroids are important. Original bodies can be destroyed; fragments are produced and some can escape; and some asteroids are re-formed as rubble piles. These processes can repeat. A schematic illustration of these processes for a moderately large asteroid is given in Plate 8.1.

Resonances in the asteroid belt and collisions between asteroids can place asteroids and fragments into high-eccentricity orbits that send the object out of the solar system or to earth as meteorites. Laboratory analysis of meteorites can tell us much and reflectance spectroscopy can establish correspondences between asteroids and meteorites. Some of the meteorites contain chemically bound water and organic matter.

Some relations between asteroids and comets seem clear. Formation scenarios involve a continuum of objects and some apparent asteroids in comet-like orbits are almost surely dormant or extinct comets. Finally, three transition objects are known that have gone from one designation to another. These are the two discussed in this chapter, 2060 Chiron and 4015 Wilson-Harrington, and 7968 Elst-Pizarro. On the whole, the argument for comet scientists knowing something about asteroids, and vice versa, is sound.

9

Origins of comets

Hypotheses concerning the origin of comets must be compatible with two basic facts: (1) the lack of substantial numbers of comets with hyperbolic orbits indicates an origin within the solar system or at least in a system with exactly the same space motion as the solar system; and (2) the composition of comets shows similarities with the composition of interstellar clouds. At the very least, the composition of comets is consistent with condensation from interstellar clouds. Our working hypothesis for the origin of comets is that they are natural products of condensation in the contracting solar nebula. Near the sun this condensation process led to the formation of the terrestrial planets and, at larger distances, the Jovian planets. New comets, thus, may be essentially unchanged products that are representative of conditions in the solar nebula; though, as we will discuss in section 9.2.2, cosmic rays do affect the surface layers of the nuclei of even new comets. Hence, new comets may be a source of information for the early history of the solar system when it was in the process of formation from collapsing interstellar material.

Our understanding of the origin of comets has taken a great leap forward in the last two decades. Today, we recognize three groups of comets: the Jupiter family with $P \leq 20$ years, the Halley family with $20 < P \leq 200$ years, and the long-period comets, with $P > 200$ years. Comet scientists still refer collectively to the Jupiter and Halley families as short-period comets, though there is growing evidence that the two families may originate in somewhat different regions of the solar system. The Jupiter family of comets, like the planetary system, is relatively flat. Jupiter-family comets all move in a prograde sense, and have inclinations with an average value of $11.5°$ and a maximum of about $32°$. While the Halley family is a relatively small group, with roughly a dozen known members, certain statistical properties of the group are clear. For instance, the distribution of inclinations is not isotropic. Instead, the planes of the orbits of comets are inclined to the plane of the ecliptic by less than $70°$, and members of the group move in either a prograde or retrograde sense. Comets moving in the retrograde sense have inclinations listed as between

$90°$ and $180°$. Members of the three groups of comets originate in either the Oort cloud or the Kuiper belt. One objective of this chapter is to present the current state of our understanding of that sentence.

9.1 Evolution of cometary orbits

9.1.1 Long-term evolution of short-period comet orbits

Duncan *et al.* (1988) have carried out an extensive numerical simulation of the evolution of the orbits of comets, taking into account the perturbations of the four major planets, Jupiter, Saturn, Uranus and Neptune. Their objective was to ascertain whether the source of short-period comets is the Oort cloud or a repository of comets beyond Neptune. Through numerical experiments, they found that the calculations involved in modeling the evolution of a comet from a Neptune crossing orbit to an orbit with $q \sim 1.5$ AU proceeds slowly. However, they found that the statistical properties of the resultant orbits are relatively insensitive to the assumed masses of the giant planets. As a result, the authors could increase the rate of evolution of the orbits in their model by increasing the masses of the giant planets in their code by factors up to 40. They concluded that the short-period comets cannot be explained by planetary scattering from a spherical population of comets (i.e., the Oort cloud). They hypothesize a comet belt in the outer solar system. This is a seminal paper. Up until its publication, most comet scientists believed that short-period comets originated from the long-period comets via interactions with Jupiter. It is still possible that a very few short-period comets could have originated from long-period comets interacting with Jupiter.

Levison and Duncan (1994) have developed a new simulation code that is an order of magnitude faster that the Duncan *et al.* (1988) code. This new code is based on a symplectic method (Saha and Tremaine 1993) in which the equation of motion is derived from a Hamiltonian consisting of two parts,

$$H = H_{\text{Kepler}} + H_{\text{interaction}}. \tag{9.1}$$

H_{Kepler} is the normal two-body Keplerian motion and $H_{\text{interaction}}$ contains all the mutual perturbations. They call the method the *Regularized Mixed-Variable Symplectic* (RMVS) method. The code includes effects of all planets except Mercury and Pluto.

This code has been used to integrate the motions of currently known short-period comets both forward and backward in time. Levison and his co-workers find that the median lifetime – the time before half the comets are either ejected or become sungrazers – is 4.5×10^5 years. About 6% of the "lost" comets become sungrazers, and the remaining 92% are ejected from the solar system. A comet switches between the Jupiter family and the Halley family (defined as a switch from $P < 20$ years to $20 < P < 200$ years) as much as a dozen times during its dynamical lifetime, which might suggest that the two families are not dynamically distinct. However, the Jupiter-family comets are characterized by a Tisserand parameter relative to Jupiter of $T > 2$, while the Halley-family members are characterized by $T < 2$ (see section 2.2.2.1 for a definition of the Tisserand parameter). In the simulation over 90% of the comets remained in the same Tisserand family (i.e., $T > 2$ or $T < 2$) throughout their lifetimes. Most of those comets that switched families based on their periods, but remained in the same Tisserand family, tended to retain parameters near $T \sim 2$. A problem with the 4.5×10^5 year dynamical half-life for short-period comets is that many short-period comets will have decayed before the interval is over.

It is interesting to look at the dynamical history of one of the well-known Jupiter-family comets predicted by Levison and Duncan's simulations. For example, 2P/Encke appears to become a sungrazer in about 1.4×10^5 years. They point out that this behavior is not an artifact of the RMVS method, because they have verified the result with a second, independent program. Figure 9.1 shows the temporal behavior of a, q and i for P/Encke.

An interesting result of the simulations is that the flat distribution of inclinations observed for the Jupiter-family comets thickens as it ages. This thickening supports the hypothesis that Jupiter-family comets originate in a belt of comet progenitors somewhere in the outer solar system. This belt, to be discussed in section 9.2.2, is now known as the Kuiper belt, in honor of G. P. Kuiper, a pioneer in studies of the origin of the solar system. In Europe, the belt is often called the Edgeworth–Kuiper belt, to honor the work of Edgeworth.

9.1.2 The Halley-family comets

The dynamical evolution of the orbits of short-period comets is becoming better understood as a result of the work discussed above. There is strong evidence for separate sources for the progenitors of short-period and long-period comets. Levison *et al.* (2001) have carried out dynamic simulations to show that the Halley family of short-period comets may arise in a massive disk-like inner Oort cloud. In the simulations, they treat the Oort cloud comets as massless test particles, the dynamics of which include the gravitational effects of the sun, giant planets, passing stars, and galactic tides. As initial conditions the team assumed the distribution of semi-major axes and cosines of inclinations to be uniformly distributed.

This initial uniform distribution failed to produce the observed inclinations of Halley-family comets. The simulation demonstrated that there is a rough conservation of inclinations of the test particles during capture. This led the Levison team to postulate an idealized two-component Oort cloud, with a boundary between the two

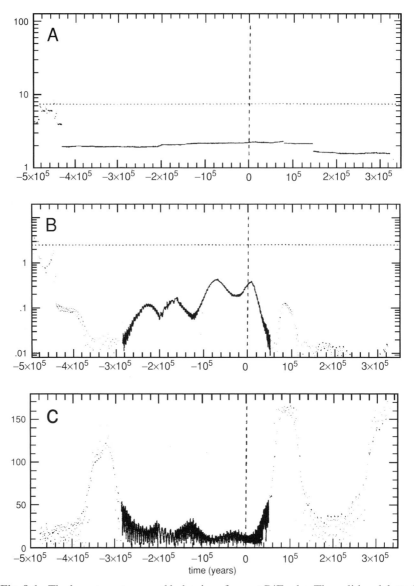

Fig. 9.1. The long-term temporal behavior of comet P/Encke. The solid and dotted curves represent the periods of time "before" and "after" the comet becomes a sungrazer. (A) Semi-major axis; (B) perihelion distance; (C) inclination. (Courtesy of H. Levison: Reprinted from *Icarus*, Vol. 108, by Levison and Duncan, The long-term dynamical behavior of short-period comets, pp. 18–36. Copyright 1994, with permission of Elsevier.)

components at 20 000 AU. Beyond 20 000 AU the cloud is spherically symmetric. However, inside that distance the cloud consists of a massive, flattened disk. The team allowed the median inclination of the disk to be an adjustable parameter, and they ran simulation models for median inclinations between 10° and 50°. They found

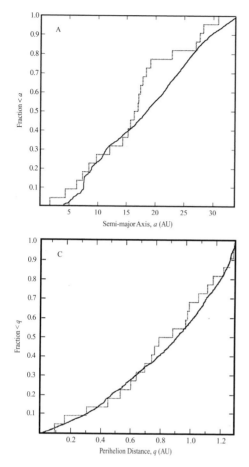

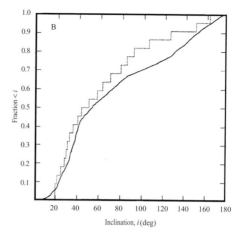

Fig. 9.2. Comparison between orbital element distribution of Halley-type comets (dotted line) and the best-fit model of Levison *et al.* (2001) (solid line). (a) Semi-major axis; (b) inclination; (c) perihelion distance. (Courtesy of H. Levison: From Levison, Dones, and Duncan. 2001. The origin of Halley-type comets: Probing the inner Oort cloud. *Astron. J.* 121: 2253–67. Figure 10, Comparison between orbital elements of Halley-type comets and the best fit model. Reproduced by permission of the AAS.)

best fit to be about 20°, (Fig. 9.2). The model predicts that comets in the inner Oort cloud are much less likely to be captured than comets in the outer cloud. In order to produce the necessary flux of new Halley-family comets, the inner cloud must be more massive than the outer cloud.

To leave this topic on a provocative note, the team found that some Jupiter-family comets could have come from this inner Oort cloud. Clearly, the last word has not been written on this subject. It is certainly suggestive that there are three populations of known comets, and three possible origin sites. It would be naïve to suggest that there are sharp boundaries between the inner and outer Oort clouds.

There is a particularly interesting object in the *Catalogue of Cometary Orbits*. Its period is 50 years, which would make it a Halley-family comet. However, it is in a slightly elliptical orbit ($e \sim 0.38$), has a perihelion distance not far inside Saturn's orbit ($q \sim 8.5$ AU) and has a Tisserand parameter like Jupiter-family comets ($T \sim 3$). This is 95P/Chiron (1997 UB), which was assumed to be an asteroid when it was first discovered. Subsequently, it showed a weak coma. Chiron is discussed in more detail in section 8.2.2.

9.2 The origin of comets

In discussing the origin of comets, we assume that all comets have an essentially common physical origin. Until the 1990s, it was believed that the short-period comets are a natural consequence of repeated passages through the inner solar system by the long-period or nearly parabolic comets. As we described in section 9.1, evidence now points to the fact that separate processes are responsible for the origins of the short-period and long-period comets, and within the short-period comets separate processes are probably required for the origins of Jupiter-family and Halley-family comets.

9.2.1 The Oort-cloud concept

There is currently no serious question as to the existence of a peak in the frequency of $1/a$ values, falling between 10^{-4} and $10^{-3} \mathrm{AU}^{-1}$ (Fig. 9.3). This peak has been identified with the outer fringe of the Oort cloud of new comets. The nearly parabolic comets come from this cloud, with dimensions $\sim 10^4$ to 10^5 AU that are approaching the distances to the nearest stars. Because the orbital inclinations of long-period comets are distributed roughly at random, it is likely that at least the outer Oort cloud is spherical.

Oort postulated that stars passing, at random, through the outer regions of the cloud would perturb the comets located there. Some of these comets would enter the inner solar system. Once the comets enter the region of the planets, they are either ejected to interstellar space or are captured into more tightly bound orbits and become long-period comets. Since Oort's time, additional perturbers have been identified. Two perturbers, which have been shown to be reasonable, are giant molecular clouds and the tidal gravitational field of the galaxy. The combined perturbations on the outer Oort cloud would erase all evidence of the origin of the cloud.

Delsemme (1987) studied the original orbits of comets with $1/a \lesssim 0.002$, and found that their aphelia define an axisymmetric distribution in galactic coordinates, with a paucity of comets in the equatorial zone and polar caps (Fig. 9.4). He concluded that many of these comets have been perturbed into the inner solar system by tidal effects of the galactic disk.

Oort (1950) estimated the number of comets in the cloud, based on the known population of long-period comets, to be about 10^{11} comets. Other researchers have used alternate values for the population of long-period comets (Weissman 1991). Baily and Stagg (1988) have combined observations of long-period comets with the cratering record on earth to constrain models of the population in the Oort cloud. Since terrestrial craters arise from impacts of asteriods and short-period comets, as well as long-period comets, the authors carried out a separate analysis to evaluate

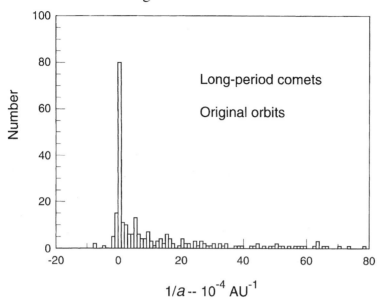

Fig. 9.3. Distribution of original $1/a$ values for the observed long-period comets. Dynamically new comets from the Oort cloud cause the spike near $1/a = 0$. (Courtesy of P. R. Weissman, NASA–Jet Propulsion Laboratory: From *Comets in the Post-Halley Era*, Vol. 1, 1991, pp. 463–86, ed. Newburn, Neugebauer, and Rahe, Dynamical history of the Oort cloud, by Weissman, Figure 1. Copyright © 1991, with kind permission of Kluwer Academic Publishers.)

the impact rates from the three sources. They find the total population in the inner and outer Oort cloud to be 9.5×10^{13} comets brighter than $H_{10} = 16$. Weissman (1991) summarizes the state of our understanding, as of about 1990, and states that the outer Oort cloud contains between 0.4×10^{12} and 1.3×10^{12} comets, and the inner Oort cloud contains between 2.0×10^{12} and 1.3×10^{13} comets. These numbers are consistent with the Levison *et al.* (2001) finding, based on the origin of Halley-family comets, that the inner Oort cloud is substantially denser that the outer Oort cloud. As the outer Oort cloud is depleted by comet ejections to the inner solar system and interstellar space, it is replenished by comets perturbed from the inner Oort cloud.

Weissman has derived a mass–brightness relationship for new comets,

$$\log M_c = 20.0 - 0.4\log H_{10}. \tag{9.2}$$

Everhart (1967) compiled the distribution of cometary magnitudes based on a group of 256 long-period comets. Weissman (1991) used Everhart's distribution and (9.2) to derive a mass distribution (Fig. 9.5). Weissman has summarized the state of our understanding of the mass of the Oort cloud by stating its mass is between 14 and

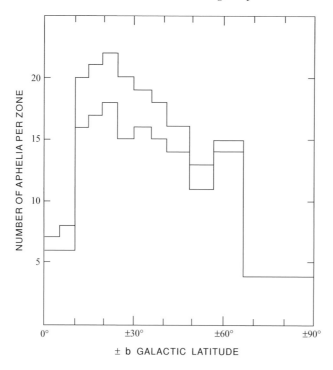

Fig. 9.4. Distribution of aphelia of long-period comets ($P > 10^4$ years) in 12 zones of equal areas, symmetric with respect to the plane of the galaxy. The upper curve has been corrected for the fact that the discoveries of long-period comets are affected by the geographical distribution of observers. The depletion of comets near the galactic pole and equator are caused by the influence of galactic tides on the Oort cloud. (Courtesy of *Astronomy and Astrophysics*: Delsemme, A. H., 1973. Galactic tides affect the Oort cloud: an observational confirmation. *Astron. Astrophys.* 187: 913–18.)

1000 M_e with a probable value around 46 M_e, of course with large uncertainty. Oort's original mass estimate was between 10^{-2} and 10^{-1} M_e.

Heisler (1990) has used a Monte Carlo simulation to model the evolution of the Oort cloud between 10 000 and 40 000 AU. The code was developed (Heisler *et al.* 1987) including only stellar perturbations, but was expanded for the 1990 paper to include the galactic tidal field. Rather than follow individual comets, Heisler and her co-workers followed groupings of a number of comets. The simulation followed more than 10^7 comets over a period of 270 million years. Figure 9.6 shows the model flux of new long-period comets (comets/yr) with $q < 2$ AU from the Oort cloud. The background flux predicted by the model is roughly 40 new comets per year. The figure also shows several comet showers – large numbers of simultaneous new comets – over the 270 million year time period covered by the simulation. It is interesting to note that the time interval between the two largest showers (10^8 years) is roughly the time interval between major extinction events on earth. Weissman (1990) briefly reviews the evidence for comet showers as the source of extinction level impacts, and concludes that there is currently no clustering of terrestrial crater ages that would indicate the effect of a major comet shower. Of course, erosion could easily have destroyed the evidence. Weissman suggests that a more detailed study of the ages of lunar craters – which are not subject to extreme weathering – might help resolve the question of the effects of comet showers.

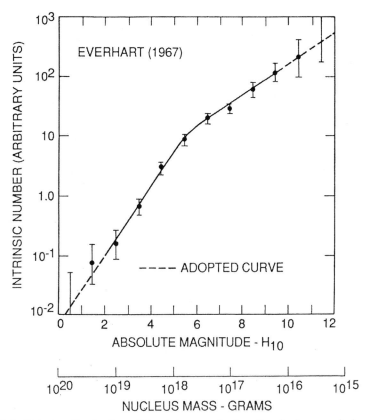

Fig. 9.5. Relative distribution of absolute magnitudes, H_{10}, for long-period comets (from Everhart 1967). The curve is corrected for observational selection effects. (Courtesy of P. R. Weissman, NASA-Jet Propulsion Laboratory: From *Comets in the Post-Halley Era*, Vol. 1, 1991, pp. 463–86, ed. Newburn, Neugebauer, and Rahe, Dynamical history of the Oort cloud, by Weissman, Figure 5. Copyright © 1991, with kind permission of Kluwer Academic Publishers.)

9.2.2 The Kuiper belt

We now know that the region of the solar system beyond Neptune's orbit is populated by small bodies. This region has come to be known as the Kuiper belt, and the bodies as Kuiper belt objects or KBOs. Jewitt (1999) has provided a detailed discussion of the history of the discovery of the Kuiper belt, and our current understanding of its nature. Between 1940 and 1990, there was considerable speculation about the existence of such a belt, strengthened by the fact that researchers demonstrated that the source of Jupiter-family comets could not be traced to the Oort cloud. In one of the earliest papers on the topic Edgeworth (1949) speculated that fledgling comets existed in the outer region of the solar system, beyond Pluto. Occasionally, clusters of them would be perturbed into the inner solar system. In honor of Edgeworth's contribution, the belt is sometimes called the Edgeworth–Kuiper belt.

10,000 < a < 40,000 AU, q < 2 AU

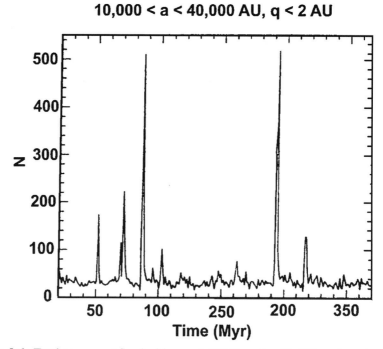

Fig. 9.6. Total new comet flux inside $q = 2$ AU. Comets with different semi-major axes were combined using the distribution predicted by Duncan *et al.* (1987). (Reprinted from *Icarus*, Vol. 88, by Heisler. Monte Carlo simulations of the Oort comet cloud, pp. 104–121. Copyright 1990, with permission from Elsevier.)

In the late 1980s, Luu and Jewitt began to search for KBOs, initially without success. However, in 1992 Jewitt and Luu (1993) discovered the object 1992 QB1, which was shown to lie in the Kuiper belt (Fig. 1.17). This discovery was quickly followed by the discovery of many more such objects.

As of mid 2001 there are approximately 370 known KBOs. Figure 9.7 is a plot of *e* vs *a* for known KBOs. These objects are divided into three distinct classes, based on their orbits: the classical KBOs, the Plutinos, and scattered objects. Classical KBOs are those objects which have semi-major axes $\sim$40 < $a \leq$ 47 AU, and which are not associated with resonances. The inclinations of the orbits of classical KBOs can be as large as $i = 30°$, while their eccentricities are small.

The Plutinos fall in the 3:2 mean motion resonance with Neptune (see section 2.6.1). The name Plutino was chosen for these bodies because Pluto, itself, also lies in the 3:2 resonance with Neptune. Jewitt (1999) points out that the large number of Plutinos relative to other KBOs is a selection effect, based on their relatively smaller heliocentric distances. Jewitt estimates that they may make up $\sim$30% of all KBOs. Figure 9.7 suggests that there is a small population of objects

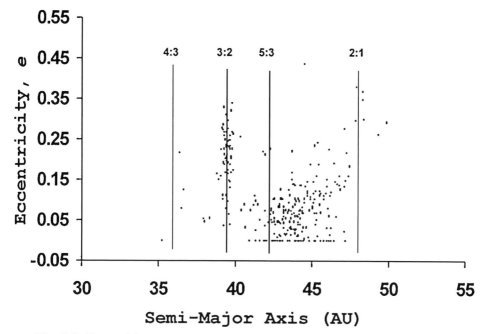

Fig. 9.7. Eccentricity versus semi-major axis plot for Kuiper belt objects. The resonances in the belt are indicated. (After Jewitt.)

in other resonances as well. The scattered KBOs are characterized by $q \sim 35$ AU, $a > 50$ AU and large eccentricities. As the name suggests, current thinking is that the scattered KBOs originate as a result of scatterings by Neptune. The first scattered KBO discovered was 1996 TL66.

How many KBOs are there? Jewitt *et al.* (1996, 1998) have made surveys specifically searching for KBOs, and estimate the number larger than 100 km with $30 < a < 50$ AU to be about 10^5. Extrapolating down to a size of 5 km, they estimate the number to be 8×10^8, with a large uncertainty.

The Kuiper belt is now believed by most comet researchers to be the source of short-period comets. This does not resolve the issue of the origin of short-period comets, however. Exactly how are the Kuiper belt objects scattered into the inner solar system? Duncan and Levison (1997) using their simulation models investigated the dynamical history of scattered KBOs. They found that the number of scattered KBOs decreased linearly on billion-year time scales, and after about 4 billon years 1% of the initial population remained. The orbital characteristics of those scattered KBOs that enter the inner solar system are similar to the characteristics of Jupiter-family objects. Trujillo *et al.* (2000) report on the discovery of three scattered KBOs, and suggest the number of large objects is sufficient for them to account for the short-period comets. Ip and Fernández (1997) investigated the dynamical behavior

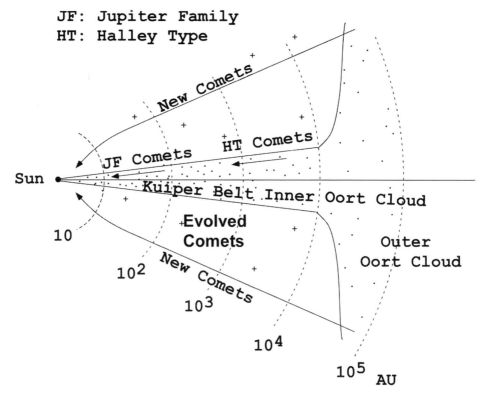

Fig. 9.8. Schematic of the Kuiper belt and inner and outer Oort cloud. (After Fernández.)

of KBOs stored in the 3 : 2 mean motion resonance with Neptune. They find that gravitational scattering can cause these bodies to move into chaotic orbits, from which they can then be injected into the inner solar system. At this point, we have quantitative evidence for the ability of two processes to feed comets into the Jupiter family. One or both may actually be at work. We must point out that all KBOs known so far are relatively large bodies, while short-period comets are generally small (median radius ~0.8 km). This is most certainly due to the fact that only the largest KBOs are observable at their distance; the KBO population is assumed to include comets of all sizes. Figure 9.8 is a summary illustration tying together our concept of the outer and inner Oort cloud and the Kuiper belt.

In mid 2002, Chad Trujillo and Mike Brown at CalTech discovered what is now the largest KBO, if we exclude Pluto from that group. The object, a classical KBO, has been named *Quaoar* after the creation force of the Native American Tongva tribe, which lived in the Los Angeles area. Its discovery designation is 2002 LM60. The KBO is roughly 1250 km in diameter, and moves in a nearly circular orbit 42 AU from the sun. The diameter is reasonably firm, since *HST* can resolve the

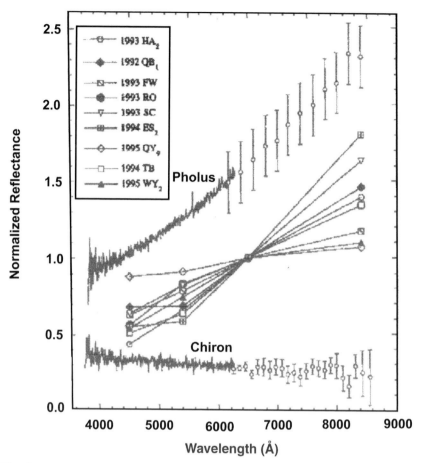

Fig. 9.9. Relative reflectivities of KBOs and Centaur objects, normalized to the R filter. The offset spectra of Chiron and Pholus are also shown. (Courtesy J. Luu & D. Jewitt: From Luu and Jewitt, 1996. Color diversity among the Centaurs and Kuiper belt objects. *Astron. J.* 112: 2310–18. Figure 3, Relative reflectivities of KBOs and Centaur objects. Reproduced by permission of the AAS.)

object's tiny disk. This discovery will add heat to the argument whether Pluto should be called a KBO instead of a planet.

KBOs accumulated in the outer region of the solar nebula at temperatures around 50 K. Consistent with this fact, they are thought to consist of molecular ices such as H_2O, CO_2, CO and N_2. Luu and Jewitt (1996) point out that long-term exposure to cosmic radiation leads to a carbon-rich solid mantle coating most KBOs. The authors point out that they have insufficient information on the time scale of the process to calculate instantaneous colors. To explore the surface of KBOs, the authors have carried out a program of broad-band optical photometry of Centaurs and KBOs. These objects show a wide spectral diversity as illustrated in Fig. 9.9.

The explanation of the diversity is not yet completely understood, but may be an indication of compositional diversity, arising from differences in the locations of formation of comets. For instance Mumma *et al.* (2001) point out that comets Halley, Hyakutake, Hale–Bopp, and Lee have measured compositions similar to those in the cores of dense interstellar clouds, consistent with formation from interstellar ices in the cold nebular region near Uranus and Neptune. On the other hand, the measured composition of comet LINEAR (C/1999 S4) is significantly different from that of those four comets, with the highly volatile species such as CO, CH_4, C_2H_6 and CH_3OH being significantly less abundant. Mumma *et al.* (2001) suggest that its nucleus condensed from material that had been processed in the warmer Jupiter-Saturn region. Kawakita *et al.* (2001) have obtained high resolution spectra of comet LINEAR (C/1999 S4) with the Subaru Telescope when the comet was at $r = 0.863$ AU. They derived an ortho-to-para ratio (OPR) of 3.33 ± 0.07 from NH_2 bands, which they assume reflects the OPR of ammonia (NH_3) in the nucleus, and indicates a spin temperature in the nucleus of 28 K. This suggests the nuclear ammonia originated in the Saturn–Uranus region of the solar nebula. These results are not fully consistent, and the differences will have to be resolved. We will discuss this topic further in section 9.3.1.

9.2.3 The fading problem

Comets arriving in the inner solar system from the Oort cloud are often referred to as nearly isotropic comets (NICs), because of the isotropic distribution of the directions of the major axes of their orbits. NICs include both comets on their first pass near the sun, and comets on later passes. Models of the orbital evolution of these comets predict far more returning comets than are actually observed. This effect has been called the "fading problem." Levison *et al.* (2002) point out that modeling shows this effect cannot be due to any dynamical effects not included in the model. The problem must be due to an intrinsic effect in the comet itself. One obvious suggestion is that the comets become dormant.

To test this hypothesis, Levison and his co-workers have found objects that appear to be asteroids, but with orbits that are consistent with them being dormant NICs. There are 11 such objects currently known. The team then posed the questions: "If these objects are dormant NICs, what do they tell us about the total numbers of such objects?" "How many dormant NICs would there be if our survey were to find them all?" The result of the numerical experiments is that there are too few dormant NICs (including too few active and dormant Halley-type comets) by two orders of magnitude.

If the objects becoming dormant is not the explanation, then what is? Levison *et al.* (2002) proposes that the NICs must disrupt into much smaller bodies (or even dust) after a few passes through the inner solar system. This suggests their nuclei must be very loosely bound together. Bailey (2002) points out the example of the over 400 known Kreutz group comets that must be the result of the disruption of a single body. The authors also point out that the numbers of Jupiter-family comets are more or less what is expected. Is it possible that the comets formed in the Kuiper belt, beyond Neptune, are more solidly held together than the NICs, which formed in the region of the giant planets and diffused out to the Oort cloud?

9.3 Formation of planetary systems

9.3.1 Theory of star and planetary system formation

The Milky Way galaxy, or Galaxy for short, is a type Sb spiral (Fig. 9.10). Binney and Merrifield (1998) is a good review of the current state of galactic and extragalactic astronomy. The spiral arms are density waves that move around the disk with periods on the order of 10^8 years. The wave compresses ambient interstellar gas clouds to the point where some become unstable to gravitational collapse, and stars are formed. The spiral arms of galaxies stand out from the remainder of the galaxian disks because of bright H II regions illuminated by ultraviolet radiation from the youngest, hottest of these stars, OB stars. Since the OB stars are observed in and near spiral arms, they must be born as a result of the density waves. This would seem to imply that, at any point in a galaxian disk, there is a burst of star formation every 10^8 years or so when a density wave passes, after which no new stars are formed until the next wave arrives. However, Cameron (1988), and others, suggest that there are other processes which can stimulate gravitational collapse.

Scattered throughout the Galactic disk are giant molecular clouds: high density, cold structures in the interstellar medium. These clouds are observed to contain a number of localized condensations, perhaps 0.1 pc in size, and a few times more massive than typical stars like the sun. These condensations have come to be known as cores. Plate 9.1 of the Eagle Nebula (M16), shows a region where there are numerous examples of stars in the early stages of formation. This *Hubble Space Telescope* picture shows embryonic stars at the tips of projecting columns of gas and dust.

Visible examples of such cores may be the so-called Bok globules. Shu *et al.* (1987) review the physical processes at work within the molecular clouds that lead to cores and their eventual collapse to stars. The ionization fraction in molecular clouds is very low, and a process called ambipolar diffusion (Mestel and Spitzer 1956) occurs, where the coupling between the plasma and the neutrals is small. The field and the attached plasma drift relative to the neutrals. The neutral component of the core will slowly contract across the field to form a core. The condensed core will collapse if it is given a "nudge." Cameron suggests that such a nudge could come

Fig. 9.10. The Sb spiral galaxy M81. (California Institute of Technology.)

from a nearby supernova explosion or a planetary nebula ejected by an asymptotic horizontal branch star. These two cases are supported by the existence of natural radioactivities such as ^{26}Al in the solar system. There are other possibilities as well. Once a protostar is formed, the core material with the highest angular momentum forms a disk around the star. There is a lot of interesting physics going on at this stage. The young stellar object (YSO) enters a bipolar outflow stage with inflow of core material onto the disk. YSOs have been observed in the Orion nebula, and the Herbig–Haro object HH30 (Plate 9.2).The outflow, for low mass objects, will reach the T Tauri stage. This brief discussion is summarized in Fig. 9.11. The Tarantula Nebula (Plate 9.3) in the Large Magellanic Cloud shows examples of stars at all stages of their life. The bright star cluster Hodge 301 in one corner of the image is an ancient cluster some of whose stars have already become supernovae. The supernova blasts have plowed into the main part of the nebula stimulating the birth of new stars.

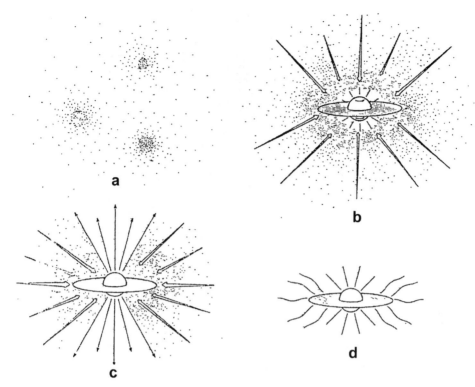

Fig. 9.11. Four stages of star formation. (a) Cores form in molecular clouds; (b) a protostar and surrounding nebular disk form in a core; (c) a stellar wind begins, creating a bipolar flow; (d) bipolar flow ceases, leaving a newly formed star with a circumstellar disk. (Courtesy F. Shu: With permission, from the *Annual Review of Astronomy and Astrophysics*, Volume 25 © 1987 by *Annual Reviews* www.annualreviews.org.)

Our brief description, so far, pictures a young stellar object surrounded by a flat nebular disk. Within that disk are small, fluffy clumps. Interstellar grains consist of a refractory core and a snowy mantle of atoms and molecules collected from the interstellar medium. As a molecular cloud core contracts, these grains will gently collide and build ever larger clumps as large as a centimeter. As the cloud contracts, its temperature will increase, and some of the icy mantles of interstellar grain clumps will evaporate, leaving fluffy grain clumps.

Safronov (1969) and Goldreich and Ward (1973) provided early discussions of the processes that lead to larger bodies. Their theory results in rocky planetesimals in the inner solar system and icy planetesimals in the outer solar system, consistent with the observed natures of the terrestrial versus the giant planets. The solid particles were assumed to drift toward the mid-plane of the solar nebula forming a thin sheet of solids, which would then fragment into objects with kilometer sizes.

Weidenschilling (1987) has found that the drift toward the mid-plane is inconsistent with the turbulent nature of the protoplanetary disk, especially in its inner regions. Instead, he pictures a process where bodies accumulate through collisions in the turbulence. Turbulence in the inner solar nebula is sufficient to keep the bodies small, while the turbulence in the outer solar nebula is weaker, and larger bodies will accumulate. As time goes on, gas and some small bodies in the inner solar nebula will spiral into the sun, and the flow in the disk will become laminar. Over a time period of roughly 10^4 years, kilometer-sized bodies will build up. Cameron (1988) discusses the physical processes in much more detail, and describes some additional hypotheses about planetary accumulation, which he now believes to be less plausible than the one we set forth here.

So far, we have discussed how kilometer-sized bodies may have accumulated in the solar nebula. The next step is the accumulation of planetary-sized bodies. Cameron (1988) discusses the issues associated with this step in the development of the solar system. From our point of view, the largest issue is the rapid accumulation of the giant planets. Earlier theories have the problem that the giant planets accumulate slowly. This is a problem, because we need the giant planets to provide the perturbations to send the excess kilometer-sized bodies in the outer solar system out to form the Oort cloud.

A widely accepted hypothesis is that the giant planets accumulated in a two-stage process. The first stage is the process described by Weidenschilling, where relatively large bodies accumulated in the low turbulence of the outer solar nebula. These bodies would have a relatively large mass of refractory material. Models of the giant planets show that they have rocky cores with masses between 10 and 20 earth masses. That is consistent with the object formed in stage one becoming the planetary core. In stage two, the solar nebula near the planetary cores first concentrates then collapses onto the cores, forming the outer portions of the giant planets today. As Cameron points out, if Jupiter had solar abundances, it would have only a one-earth-mass rocky core.

In section 9.2.2 we discuss the compositional differences between comets Halley, Hyakutake, Hale–Bopp and Lee on the one hand and comet LINEAR on the other. Mumma *et al.* (2001) suggests the differences are due to the location in the early solar system where the comets were formed. However, the recent paper by Kawakita *et al.* (2001) derives a nuclear temperature for NH_3 that is not significantly different from that of Halley. Yamamoto (1985) discusses the site of formation of comets based on their compositions and the condensation temperatures of CO and N_2. Mumma *et al.* (1993) discuss the formation of comets and state that the most likely site of formation was the Uranus–Neptune area, or perhaps just beyond. If S_2 is present at the time of condensation or is formed during condensation, temperatures less than or equal to 15 K are required. As we described in section 5.3.1, S_2 has

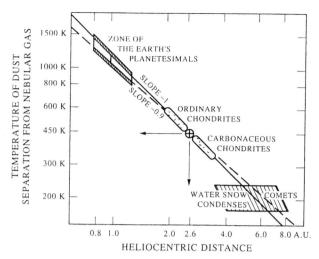

Fig. 9.12. Temperature distribution in the solar nebula. (Courtesy A. Delsemme. Reprinted from *Planetary and Space Science*, Vol. 46, by Delsemme, Recollections of a cometary scientist, pp. 111–24. Copyright 1998, with permission from Elsevier.)

been seen in several recent comets. Delsemme (1998) summarizes the accretion temperatures of comets formed in different regions of the solar nebula (Fig. 9.12). He states: "The comets from Jupiter's zone accreted at 225 K, those from Saturn's not far from 120 K, and those from Uranus and beyond not far from 60 K." Mekler and Podolak (1994) discuss the existence of amorphous water ice in the nuclei of comets. In order for the amorphous ice to be stable over the age of the solar system, it must remain at a temperature at or below 77 K. They examined models of the early solar system, and find that the comets must have formed at distances greater than 7 AU. Then, these comets would have been ejected into the Oort cloud by perturbations due to the proto giant planets. Fernández and Ip (1981) show that in such a case, most of the mass comes from ejections due to Uranus and Neptune. Ejections due to Jupiter can account for only about 3% of the mass in the Oort cloud. As you can see from this paragraph and section 9.2.2 the issue is not yet fully resolved, but cometary formation in the Uranus–Neptune region of the proto solar system is suggested.

Altweg *et al.* (1999a, b) present a series of review papers on the composition and origin of cometary materials, which should be of interest to the reader. The review by Fegley (1999) in that volume clearly shows that the temperatures at any point in the presolar nebula predicted by various theories agree only to about a factor of two.

A general understanding of the chemical composition of comets has been advanced by W. F. Huebner (Huebner and Benkhoff 1999a, b; Altweg *et al.* 1999a, 1999b). The starting point is the assumption of solar abundances of the elements, except for H and N. The abundance of H in comets is determined by its ability to bind chemically to available species. The abundance of N in comets is down by about a factor of 3 relative to the solar value; recall that N_2 is essentially inert. The solar abundances of C, O, Mg, Si, S, and Fe, along with available H and N, form the molecules that can condense at approximately 30 K. The Fe, Mg, Si, and O form the silicates Fe_2SiO_4 and Mg_2SiO_4. The remainder of O goes into H_2O and into HCO and CO-compounds. The remainder of the C, N, and S goes into HCNS-compounds.

When the bookkeeping is done, the ratio of total H to the H bound in H_2O is about 1.7. The ratio of total O to the O bound in H_2O is about 1.5. The relative abundance by mass of H_2O : silicates : carbonaceous molecules (CO, CO_2, and hydrocarbons) is approximately 1 : 1 : 1. And, the mass abundance of ices : dust (silicates and hydrocarbon polycondensates) is about 1 : 1.

We will leave the story of the formation of the planets at this point. The interested reader can consult the literature for additional details.

9.3.2 Interstellar comets

How many stars have comet clouds? Duncan *et al.* (1987) have simulated the origin and evolution of the Oort cloud. Their numerical model suggests that it may take more than 10^9 years for the icy bodies in the inner solar system to be perturbed out to the Oort cloud, and for the cloud to take on the proportions it has today. So, we would expect comet clouds around stars that have main sequence lifetimes longer than 10^9 years or so. The full picture of the origin of the solar system that we sketched above makes it seem unlikely that multiple stars would develop protoplanetary disks and, therefore, planetary systems. This means comet clouds should be expected around single stars with spectral types later than roughly A5. Stars of mid-M and later spectral types have very long pre-main-sequence lifetimes, and many such stars may not yet have reached the main sequence. Furthermore, only disk population stars will develop planetary systems. The older Population II stars are deficient in the heavy elements needed to form planets. So we suggest that single, disk population late A, F, G, K and early M stars are likely to have comet clouds. There are roughly 0.015 such stars pc^{-3} in the solar vicinity.

Calculations suggest that between 30 and 100 comets are lost to interstellar space for every comet that winds up in the Oort cloud. The estimate of the number of comets in the Oort cloud has grown in the last few years, but the number is from a few $\times$ 10^{12} to 10^{13} comets. This means the solar vicinity could contain 10^{13} comets pc^{-3}. McGlynn and Chapman (1989) have addressed the question of why we have not observed such comets. They suggest that at least six interstellar comets should have passed within Mars' distance of the sun in the last century and a half. Sekanina

(1976) has arrived at a somewhat smaller number that, given the uncertainties, is not significantly different from the McGlynn–Chapman number. We believe these comets should have clearly hyperbolic orbits, with eccentricities larger than any observed so far. The question of why they have not been observed remains open.

9.4 Summary

Our picture of the origin of comets has improved greatly in the last two decades. We now recognize three groups of comets: the Jupiter family with $P \leq 20$ years, the Halley family with $20 < P \leq 200$, and the long-period comets, with $P > 200$ years. We still refer to the Jupiter and Halley families as short-period comets. In a seminal paper, Duncan *et al.* (1988) have shown that the short-period comets cannot be explained by planetary scattering from a spherical population of comets (i.e., the Oort cloud). Instead, they hypothesized the existence of a comet belt in the outer solar system, now known as the Kuiper belt. In the late 1980s researchers searched for inhabitants of the Kuiper belt. In 1992 the object 1992 QB1 was discovered, and was shown to lie in the Kuiper belt. This discovery was quickly followed by the discovery of many more such objects. These bodies are now called Kuiper belt objects or KBOs.

In the 1950s, Oort postulated the existence of a cloud of comets surrounding the solar system. The nearly parabolic comets come from this cloud, with dimensions $\sim 10^4$ to 10^5 AU. Because the orbital inclinations of long-period comets are distributed roughly at random, it is likely that at least the outer part of the Oort cloud is spherical. Oort further postulated that stars passing, at random, through the outer regions of the cloud would perturb the comets located there. Some of these comets would enter the inner solar system. Once the comets enter the region of the planets, they are either ejected to interstellar space or are captured into more tightly bound orbits and become long-period comets. Since Oort's time, additional perturbers have been identified: giant molecular clouds and the tidal gravitational field of the Galaxy. More detailed studies of the Oort cloud suggest it is composed of two components: the spherical outer cloud and a more flattened inner cloud, with a boundary between the two at about 20 000 AU. The Halley-family comets may come from the inner cloud and Jupiter-family comets from the Kuiper belt. The total mass of the Oort cloud is around 45 M_e with a large uncertainty.

We briefly discuss the current theories of the formation of stars and planets. The process is a very complex one, which must explain the existence of both the terrestrial planets and the giant planets. Comets originated in the inner solar system early in the formation process, mainly between Uranus and Neptune. Then perturbations acting over the subsequent lifetime of the system sent some of the comets out into the Oort cloud. An equal number would have been sent into the inner solar system, where they would have decayed long ago.

We expect Oort clouds around most main sequence stars that have lifetimes greater than 10^9 years. The processes that send comets into the cloud and from the cloud back into the inner solar system are very inefficient, and many of the comets are sent into interstellar space. Have any of these comets found their way into the inner solar system? There is some disagreement about this point. However, none of the comets we have observed have eccentricities greater than a small fraction over 1.0. A truly interstellar comet would enter the solar system with a velocity of order of the sun's space velocity and would have an eccentricity significantly greater than 1.0. The fact that we have seen none remains a mystery.

10

Comets and the solar system

Comets have a variety of interactions in the solar system. The most dramatic are impacts – obvious ones such as the multiple impacts of the fragments of comet Shoemaker–Levy 9 on Jupiter and possible impacts such as ones that might produce damage on earth. The study of potential earth impactors, whether comets or asteroids, is now a major area of scientific activity.

Dust particles from comets fall into three general categories. First, the small particles experience strong radiation pressure and are blown out of the solar system. These are the particles that formed the dust tails as discussed in Chapter 4. Second, the intermediate-sized particles experience some radiation pressure and execute orbits around the sun under reduced gravity. These are the cometary contribution to the zodiacal dust cloud. Some of these particles have been collected by high flying aircraft in the earth's upper atmosphere (section 3.4.1). Third, the largest particles do not experience significant radiation pressure and continue in basically the same orbit as the comet to form meteor streams.

10.1 Meteoroid streams

The connection between comets and meteoroid streams was established long ago. In 1866, Schiaparelli noted that the orbit calculated for the Perseid meteor shower was similar to the orbit of comet P/Swift–Tuttle, which appeared in 1862.

The *meteor* is the light phenomenon in the earth's atmosphere. *Meteor showers* are a group of meteors that appear to come from a well-defined radiant point on the sky. The constellation in which the radiant point lies gives the shower its name, e.g., the radiant point for the Perseids is in the constellation Perseus. The particles producing the meteors are called *meteoroids* when they are still outside the earth's atmosphere and a group traveling in a similar orbit are called a *meteoroid stream*. The radiant point phenomenon implies that the particles producing the shower were traveling on essentially parallel paths in space. The direction of these paths allows

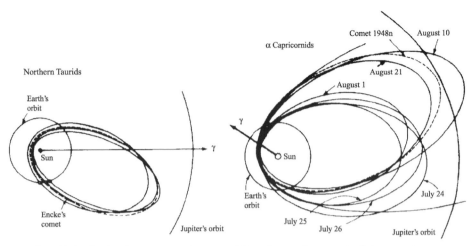

Fig. 10.1. Orbits for meteors (meteoroids) of two showers (solid curves) compared with the orbits of the associated comets (dashed curves). The agreement
is shown for the Northern Taurids and Encke's Comet. Contributions from other
comets are possible, and the Northern Taurids are not a major visual shower. The
possible agreement for the August stream of the α Capricornids with comet 1948n
is shown. The July stream of the α Capricornids probably has another origin. The
α Capricornids are also not a major visual shower. (After J. C. Brandt, University
of New Mexico, and P. W. Hodge, University of Washington.)

their orbit to be calculated. Some sample meteoroid stream orbits are shown in
Fig. 10.1. The basic idea for the comet/meteor shower connection was given by
Kirkwood in 1861 (Porter 1952: 67):

May not our periodic meteors be the debris of ancient but now disintegrated comets, whose
matter has become distributed round their orbits?

Table 10.1 gives details for several major meteor showers. For a detailed discussion
of the Leonid meteor shower, see Littmann (1999).

The meteor phenomenon – the bright streak in the sky – occurs when the particles
enter the earth's upper atmosphere at speeds of between 11 and 73 km s^{-1}. The
minimum speed of entry is the escape speed from earth or 11.2 km s^{-1}. This
is the case for prograde meteoroids just catching up to the earth; these produce
meteors seen on the evening side. Retrograde meteoroids encounter the earth head-
on at speeds up to 73 km s^{-1}; these produce meteors on the dawn side. With the
large difference in energy, meteor showers are generally expected to be fainter
on the evening side and brighter on the dawn side. Collisions with the earth's
atmospheric molecules heat the incoming meteoroid to incandescence and ablation
occurs. Meteors usually become visible at roughly 100 km altitude (Fig. 10.2)
and travel some tens of kilometers before burning out. Typical diameters for the
particles range from roughly 0.01 millimeter to about 10 meters. The lower end of

Table 10.1 *Principal nighttime meteor showers*

Shower	Period of activity	Comet
Quadrantids	January 1–5	96P/Macholz 1[a]
Lyrids	April 16–25	C/Thatcher (1861 G1)
η Aquarids	April 19 – May 28	1P/Halley
S. δ Aquarids	July 15 – August 28	96P/Macholz 1[a]
Perseids	July 17 – August 24	109P/Swift–Tuttle
Orionids	October 2 – November 7	1P/Halley
Draconids	October 6–10	21P/Giacobini–Zinner
Leonids	November 14–21	55P/Tempel–Tuttle
Geminids	December 7–17	(3200) Phaeton[b]
Ursids	December 17–26	8P/Tuttle

[a] An additional source may be present.
[b] An asteroid, possibly an extinct comet.

these sizes is consistent with our expectations for dust from comets. There can be little doubt of the cometary origin of this phenomenon with the direct observation by *IRAS* of material being laid down along the orbit for the case of comet Tempel 2 and others. The stream orbits can be altered by planetary perturbations, which also spread the particles along the comet's orbital path.

Particles smaller than the ones that produce meteors are slowed down in the atmosphere before they can vaporize. These are called *micrometeorites* and the particles eventually settle onto the earth's surface. The rate of accretion by earth of small meteoroids and dust is approximately 40×10^6 kg yr^{-1}.

The cometary debris that produces meteors is low-density and fragile. It should be distinguished from the higher density meteoroids that do not burn up completely and reach the ground to produce the iron, stony-iron, and stony *meteorites*; see Chapter 8. These can also produce the light phenomena, meteors, in the atmosphere. The brighter ones, approaching the brightness of Venus, are called *fireballs*, and the brightest ones, approaching the brightness of the full moon (particularly if they appear to explode), are called *bolides*. Meteorites are asteroidal in origin.

What do we learn from meteor showers? After all, the observations are recording pieces of comets burning up in the earth's atmosphere. The cometary meteoroids are fragile, perhaps porous, rocky bodies. Not even the brightest meteors associated with a meteor shower have produced a body recovered on the earth's surface. Meteor observations show a large range in density (and tensile strength) ranging from 0.01 to 1.06 g cm^{-3}. The lowest-density particles come from comet Giacobini–Zinner and the highest, not surprisingly, come from 3200 Phaeton. Most inferred densities, however, fall in the range 0.25 to 0.40 g cm^{-3}. These figures are certainly consistent with our knowledge of comets.

Fig. 10.2. Image of the Leonid meteor shower of November 17, 1966, showing many meteor trails. The star images are trailed by the time exposure. The two bright points near the center of the image are "head-on" meteors, close to the radiant of the shower. (Photograph by Dennis Milon and supplied courtesy of Scott and Betty Milon, Maynard, MA, USA.)

Some insight into the physics of the particles producing meteors is seen by considering some energies per gram (after Lewis 1997). Particles entering the earth's atmosphere have kinetic energies $\sim 10^{12}$–10^{13} ergs. To crush a gram of strong rock takes $\sim 10^9$ ergs. To vaporize a gram of rock takes $\sim 10^{11}$ ergs. Thus, meteor-producing particles should not survive their passage through the earth's atmosphere. For an extensive review of the meteor phenomena, see Ceplecha *et al.* (1998).

10.1.1 Comets and the zodiacal light

The zodiacal light is easily visible in a dark sky as a band of light approximately symmetrical around the plane of the ecliptic (Fig. 10.3). The brightness does not appear to change with the solar cycle.

The traditional explanation ascribes the source of dust to comets. Infrared images of comets (see Fig. 4.12 and Plate 4.2) clearly show that comets are a source, and comet Encke surely was a major contributor in the past. To sustain the zodiacal cloud, there must be a source because the zodiacal cloud loses dust particles from loss processes such as the Poynting–Robertson effect or collisions. In recent years, interplanetary dust has been directly measured by spacecraft (such as *Galileo* and *Ulysses*) and observed in the infrared (by *IRAS* and *COBE*).

The observations and measurements have revealed the complexity of the dust population in interplanetary space. Besides cometary and asteroidal sources, interstellar dust particles, dust streams from Jupiter, and a dust ring lying along the terrestrial orbit, shepherded by earth, have also been identified. For an overview of interplanetary dust, see Gustafson and Hanner (1996).

The particles in the zodiacal cloud are on bound Keplerian orbits. The masses are in the range 10^{-5} to 10^{-8} grams. For an assumed density of 2.5 g cm^{-3}, these particles are in the size range having radii from 10 μm to 100 μm. The total mass is estimated to be the equivalent of a comet or asteroid some tens of kilometers in diameter.

Losses in the zodiacal dust cloud could arise from the Poynting–Robertson effect and collisions. Collisions of the larger particles can produce smaller particles that experience strong radiation pressure and leave the solar system on hyperbolic orbits. If the particles are too small to experience strong radiation pressure, they are removed by electromagnetic forces because they should become charged and experience a $\mathbf{V} \times \mathbf{B}$ force with the solar-wind magnetic field.

The Poynting–Robertson effect arises when solar photons are scattered off the dust particles in the zodiacal cloud. Because of the orbital motion of the dust particle, the momentum imparted to the dust particle is not exactly in the antisolar direction. The angle with the antisolar direction is approximately $V_\perp/c$, where $V_\perp$ is the dust particle's speed perpendicular to the radial direction and c is the speed of light.

Fig. 10.3. Image of the inner zodiacal light taken from the *Clementine* spacecraft. (See Cooper *et al.* 1996.)

The motion of the dust particle is retarded. The dust orbits are circularized and the particles spiral in towards the sun. The time scale for removal of zodiacal dust is roughly 50 000 years for a 30 μm particle at 1 AU (Wyatt and Whipple 1950). This view could be altered if the dust particles are trapped in a planetary resonance.

The best estimates ascribe about 70% of the zodiacal cloud dust to comets and about 30% to asteroids. The cometary component probably accounts for the significant extent of the cloud in ecliptic latitude. The Themis and Koronis asteroid families (see Chapter 8) may account for 3% to 5% of the total.

Fig. 10.4. E. E. Barnard's drawing of Halley's Comet on May 18, 1910, showing the tail stretching across the morning sky. (Yerkes Observatory photograph.)

10.2 Impacts

Comets travel largely on highly elliptical orbits through the inner solar system. Collisions (and we include near misses) are rare, but they are also inevitable. Sections 10.3 and 10.4 explore specific impacts and the later sections discuss cumulative effects.

10.2.1 Comet Halley, May 1910

On May 18, 1910, the head of Halley's Comet passed directly between the earth and the sun at a distance from the earth of about 0.15 AU. For days before the close approach, the comet was a spectacular sight as it stretched across the sky (Figs. 10.4 and 10.5) and large angular lengths of about 100° were reported. This close approach of a famous comet was a major media event; see Chapter 12 for a discussion of the sociological aspects of this event.

The close approach distance of about 0.15 AU corresponds to 23 million kilometers. The dust and plasma tails of comets routinely exceed this length. The tails point away from the sun, as explained in Chapter 4, but not exactly in the antisolar direction. They lag behind in the sense of the comet's motion. Thus, depending on the exact amount of the lag and the movement of the comet out of the plane of the

Fig. 10.5. Image of comet Shoemaker–Levy 9 taken on March 30, 1993. The central "bar" consists of the multiple nuclei and dust trains extend from both ends of the "bar." Solar radiation pressure pushes the dust from the nuclei and the dust trains toward the upper right in the image. (Courtesy of James Scotti, Spacewatch Project, University of Arizona.)

ecliptic, the tails of comet Halley might have swept across earth on the night of May 18–19 or later.

Observations were hampered by bright moonlight and no significant phenomena could be attributed to passage through the tails. Specifically, no extra meteors were counted. The consensus conclusion was that the earth probably passed through the outer parts of the tails. Some dust and gases surely entered the earth's atmosphere. But, the amounts were small and no major effects were observed. As an impact, this event was just a near miss.

10.2.2 Comet Shoemaker–Levy 9

A dramatic example of cometary impacts came in 1994 when comet Shoemaker–Levy 9 (SL9) crashed into Jupiter. The story effectively begins on the afternoon of March 25, 1993 when Carolyn Shoemaker was scanning images taken with the 46-cm Schmidt telescope at Palomar Observatory. She noticed something near the center of the field about 4° away from Jupiter. She remarked:

I don't know what this is, but it looks like a squashed comet.

The appearance was highly unusual and cloudy weather was expected for the next few days. James Scotti at the Spacewatch Telescope at Kitt Peak in Arizona was contacted. He still had clear weather and agreed to take some CCD images that would confirm the unusual comet or show that the "image" was some kind of bizarre artifact. A few hours later Scotti had confirmed the discovery and is quoted as stating: "Have you got yourself a comet!" Figure 10.5 shows a Spacewatch telescope image of SL9 taken by Scotti on March 30, 1993. Images taken around this time showed that the central bar consisted of multiple nuclei each with a dust tail. By February 16, 1994, the comet was fainter and more spread out as shown in Fig. 10.6. On May 17, 1994, the Wide-Field Camera on the *Hubble Space Telescope* obtained an image showing the entire train of fragments. This image is shown in Plate 10.1 along with the nomenclature adopted for the fragments. For collections of papers on the comet SL9 phenomena, see Spencer and Mitton (1995) and Noll *et al.* (1996).

The discovery of SL9 near Jupiter as a comet that had broken into about 20 fragments immediately prompted calculations to determine the comet's past orbit. Observations were needed and taken; pre-discovery observations were also found. And, many earlier calculations were based on the idea of a "train center" and not on the orbits of the individual fragments. The calculations ultimately converged on a close approach to Jupiter at a distance of 90 000 km (0.0006 AU) on July 7, 1992. Several models were developed that involved the original nucleus splitting near perijove or somewhat after. Later splitting of some of the original fragments was also possible. The comet passed within the Roche limit of 0.001 AU for Jupiter. The understanding of SL9's orbit lead to the following conclusions: (1) the comet had passed quite close to Jupiter in July of 1992; (2) the comet was captured and orbiting Jupiter prior to 1960, and perhaps several decades earlier; and (3) the orbit would produce a comet crash on Jupiter with impacts from July 16 to 22, 1994.

The realization in May 1993 that SL9 would strike Jupiter in July 1994 galvanized the worldwide astronomical community into action. Clearly, this event, perceived as a once-in-a-millennium opportunity, demanded an extensive, coordinated observing campaign, involving both ground-based and space-based observations.

Fig. 10.6. Image of comet Shoemaker–Levy 9 taken on February 16, 1994. Dust tails are seen streaming away from each nucleus. The dust trains, visible in the March 30, 1993, image (Fig. 10.5), are no longer visible. (Courtesy of Robert Jedicke, Travis Metcalfe, and James Scotti, Spacewatch Project, University of Arizona.)

Electronic communications played a major role in the campaign. The players would include many observatories on the ground plus the *Galileo* spacecraft and the *Hubble Space Telescope*. Among the good news was the calculation by Yeomans and Chodas that the early impact points would fall only 5°–10° behind Jupiter's visible disk. This meant that they would be directly visible by *Galileo* and that they would rotate into view from earth in only 12 to 24 minutes. Thus, the early stages of the impact process would be visible. The later impacts would be closer to the limb.

The impacts would occur at 60 km s^{-1} and involve very large energies. If a fragment were 1.5 km or 1500 meters in radius and 0.5 g cm^{-3}, the kinetic energy released would be $\sim 10^7$ or 10^4 megatons, respectively. For comparison, the 1885 explosion of the Indonesian volcano Krakatoa was estimated at $\sim 10^2$ megatons. Thus, the SL9 impacts were major events.

The impacts took place essentially where and when predicted. Plate 10.2 shows the impact of the G fragment and subsequent evolution over a 5-day period. The

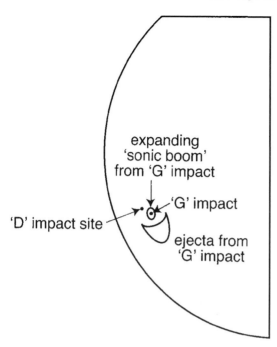

expanding
'sonic boom'
from 'G' impact

'G' impact

'D' impact site

ejecta from
'G' impact

Fig. 10.7. Schematic diagram for the impact features on Jupiter produced by the Shoemaker–Levy 9 impacts. (*Hubble Space Telescope* Comet Team; NASA.)

changes from the impacts on Jupiter's surface were quite spectacular. Figure 10.7 is a schematic showing some of the features seen on the surface. The impacts even stimulated aurora in the opposite hemisphere. This occurred some 45 minutes after the impact of the K fragment and well outside the normal regions for auroral activity on Jupiter. The events were also visible in x rays as recorded by the x-ray astronomy satellite *ROSAT*.

The material forming the dark scars on Jupiter would eventually settle out of the atmosphere and/or be dispersed by the winds of the Jovian stratosphere. The G site, for example, had been torn into tendrils and knots by late August 1994. But they were still easily visible. Faint remnant features would be visible for months, principally in the IR.

What did we learn from SL9 *about comets*? We found again that comet nuclei are exceptionally fragile. The tidal forces that split the nucleus near its closest approach produced stresses of less than 1% of the earth's atmospheric pressure. Recall the discussion of material strengths for cometary nuclei in Chapter 7. The mechanism responsible for the continued fragmentation is not clear although rotation is a possibility. The size of the nuclei has remained a mystery. Analysis of the *HST* images implies that the larger nuclei were about 3 to 4 km in diameter for an albedo of 0.04. But, significantly smaller sizes cannot be ruled out.

Are we sure that the nuclei in SL9 were comets? An asteroid that broke up would release dust and resemble a comet. The observation that the fragments had circular-appearing inner comas for most of the apparition supports continuous dust production rather than dust production associated with a break-up. Continuous dust production requires a sublimating ice and at 5 AU, this cannot be water ice. More volatile ices such as CO and CO_2 could have been exposed by the fragmentation, but they were not detected spectroscopically.

Spectroscopy of the plumes and impact sites did not appear to shed much light on the composition. The impacts are very energetic, the depths of penetration are variable, and material could be lifted from lower layers. Chemical reaction produced by the shock of the impact could remove evidence for trace materials. A convincing detailed model could change this conclusion in the future and indicate ices in the fragments. Lines from metals such as Li, Ca, K, Na, Mg and Fe were detected. These species may reasonably be presumed to come from the dust in the fragments because metallic compounds are not observed in the upper levels of the Jovian atmosphere. Recent models (e.g., Deming and Harrington 2001) point to major progress in modeling and rekindle hope that information about the comet can be extracted.

The orbital history of the object may help with its identification. Capture by Jupiter is easier if the body has a low initial velocity with respect to Jupiter. A plausible source is the Jupiter family of short-period comets. These comets are already captured into less eccentric orbits by Jupiter's gravity. Thus, the best conclusion is that SL9 was indeed a comet.

If split comets are an astronomically common event, others should have occurred in the Jovian system. The ones that impacted Jupiter would have produced surface features, perhaps similar to the SL9 impacts, but these have long since faded from view. However, if a split comet would have impacted one of Jupiter's satellites (instead of impacting Jupiter or passing out of the Jovian system), a chain of craters could be expected. Crater chains (Fig. 10.8) have been discovered on Callisto and Ganymede. The crater sizes imply impactors in the smaller part of the fragment size range suggested for SL9. They were typically a kilometer in diameter or smaller.

There is little doubt that impacts by comets are common and have occurred in the past. An earlier impact similar to the SL9 events may have been observed by J. Cassini on Jupiter in December of 1690. His drawings on December 5, 14, 15, 16, 19, and 23, 1690 are reminiscent of the dark marking produced by the SL9 fragments and the subsequent evolution.

10.3 Near-earth objects and the hazard to earth

Impact craters are ubiquitous in the solar system. The impacting bodies can be asteroids or comets. In some cases, the distinction between asteroids or comets can be made, but in many cases, the distinction cannot be made or is unimportant. In

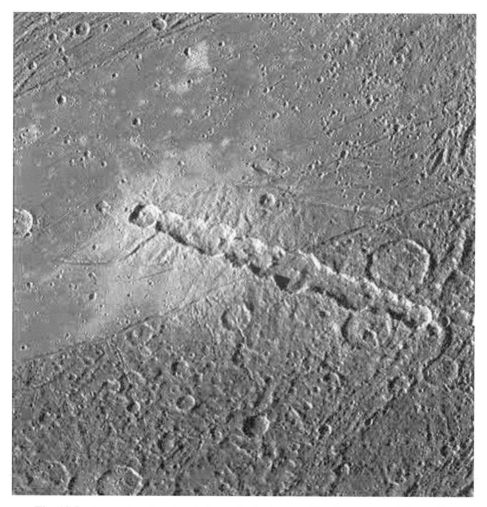

Fig. 10.8. A prominent crater chain on the Jovian satellite Ganymede. (NASA–Jet Propulsion Laboratory.)

either case, the impactor provides an extraordinary amount of energy and a crater is produced if the target area is solid.

The impacts on Jupiter from SL9 and the craters and crater chains on Callisto and Ganymede have been discussed. Some craters on the moon must have been produced by comet impacts. The impactor flux near earth has been estimated to be roughly 5–10% comets. Some unusual markings on the moon, mostly on the side facing away from the earth, might have been specifically produced by comets. These are bright swirled markings and a prominent one is near the crater Reiner Gamma. The infrared reflectivity is atypical. The cometary debris – gas and dust – could stir up the lunar soil. The lunar soil has been darkened by exposure to the solar wind for billions of years. The deeper soil has been protected from the solar wind and

Fig. 10.9. The Arizona Meteor Crater near Winslow, Arizona. (Yerkes Observatory photograph.)

should be lighter. If the comet impact occurred relatively recently, say 100 million years ago, the material exposed in the swirls could still be fairly light colored.

Mars has many obvious impact craters. Mercury is heavily cratered and radar observations strongly imply that Mercury has layers of ice at the north pole. Ices can be stable in Mercury's polar regions because there are areas that are permanently in shadow. Comet impacts over billions of years would bring more than enough water to Mercury. But would the water be retained by Mercury? Long-period comets would have high impact speeds and are not good candidates for the water source. But, short-period comets, particularly extinct comets with their ices insulated by a dust mantle, are plausible candidates.

With comets (and asteroids) impacting everything in the solar system (even the sun; see section 2.3.4.) and with the example of Halley's Comet coming close in 1910, has the earth been spared? The answer is a resounding no. See Gehrels (1994) for an extensive collection of papers on terrestrial impact phenomena. Also see Lewis (1996) and Verschuur (1996).

Figure 10.9 shows the Arizona Meteor Crater near Winslow, Arizona, USA. This impressive crater is approximately 1.2 km across and over 100 meters deep. It was produced by an iron meteoroid roughly 30 meters in diameter about 50 000 years ago. The impact speed was near 20 km s^{-1}.

For many years, a small comet was considered to be the cause of the Tunguska event in central Siberia. On June 30, 1908, a large explosion flattened an area of approximately 2000 km^2 of forest near the Tunguska River. Figure 10.10 shows an area of the downed forest. The measurements taken by expeditions to the area

Fig. 10.10. The Tunguska event. This photograph shows trees blown down approximately 8 km from ground zero. (Sovfoto.)

show that trees fell on average in a direction away from a single, central area. But, no single large crater was found at the location of "ground zero." The incoming object apparently did not survive to hit the ground, but exploded at an altitude of about 8 km. Microscopic nickel–iron fragments were found in the area. The contour of the devastated area resembles a "butterfly" and this contour is consistent with two shock waves producing the destruction: one from the projectile as it moved supersonically through the earth's atmosphere and one from the explosion. From barograms and seismograms the energy has been estimated at roughly 20 megatons.

The butterfly contour of the devastated area is consistent with a speed just before the explosion of about 8 km s^{-1}. Note that this value is below the escape speed of 11.2 km s^{-1} and this implies considerable deceleration in the earth's atmosphere. The object's mass was about 10^{12} grams and the diameter is often quoted as 60 meters.

Because of the lack of a major crater and other circumstantial evidence, the cometary explanation was accepted for years. But, Sekanina (1998) has reviewed the evidence and a cometary origin seems unlikely. The object has a long slant path through the atmosphere and was significantly decelerated. Recall (Chapter 7) that the nuclei of comets are extremely fragile. A comet-like body is likely to have disintegrated high in the atmosphere due to pressure forces. Also, the slow speed of about 8 km s^{-1} just before the explosion implies that the initial speed (outside the atmosphere) was close to the escape speed. This is not like most comet orbits, whereas some earth-crossing asteroids have such orbits.

In summary, the most likely explanation for the Tunguska event is the explosive disintegration of a small, stony-iron meteorite at an altitude of about 8 km. Thus, the ultimate origin is asteroidal. Different opinions are in the literature and new evidence is always a possibility.

In the 1980s, Luis Alvarez and his colleagues (Alvarez *et al.* 1980; Alvarez 1987) proposed that a fairly large asteroid or comet hit the earth about 65 million years ago. An important consequence of this impact was the extinction of many animal and plant species – some estimates go as high as 70% – including the dinosaurs. This time period marks the boundary between the Cretaceous and Tertiary geological periods (the K–T boundary). The extinction of the dinosaurs was particularly dramatic because they had flourished for the previous 140 million years.

A major piece of evidence for the impact theory was a thin layer of clay that separated the Cretaceous and Tertiary sediments; it is very rich in iridium (Ir). Iridium is quite rare in the earth's crust where its abundance is about 10^{-4} the meteoritic value. Iridium atoms have an affinity for iron. Early in the formation of the earth, the iridium was scrubbed out of the near-surface materials and carried down to the earth's core.

On the other hand, meteorites and possibly other solid bodies in the solar system have high abundances of iridium. The amount of iridium needed for the clay layer would be contained in a meteorite 10 kilometers in diameter. The impact speed is estimated to be approximately 20 km s^{-1} and the impact anywhere on earth would have catastrophic consequences. If this scenario is true, a large crater ~200 km in diameter should be somewhere on earth if the impact was on land.

Finding the crater took a while and the search was greatly hampered by the impact area finally implicated not showing the crater on the earth's surface; it was buried (see Verschuur 1996). The critical evidence came from petroleum geologists who were using gravimetric and magnetic field measurements to find areas likely to yield oil. The measurements indicated a huge buried ring some 200 km in diameter partially in the Gulf of Mexico and partially under the Yucatán Peninsula. The underground structure was named after the Chicxulub Pueblo on Yucatán's north coast. A major complication in the history of the discovery and understanding of

Fig. 10.11. Map showing the location of the 180 km Chicxulub Crater on the tip of the Yucatán Peninsula.

the Chicxulub Crater was that the surveys were funded by the Mexican national oil company, PEMEX, and were strictly proprietary. Plate 10.3 shows the gravity anomalies associated with the crater, and its location is shown in Fig. 10.11. Considerable evidence – including the dating of the sediments in the crater – supports the Chicxulub Crater as the site of the impact that produced the K–T event.

Major confirming evidence came from core samples also obtained by petroleum geologists. These cores contained some fascinating rocks, including breccias and shocked quartz. Breccias are created when rock fragments are cemented together in mineral matter. Temperature and pressure are needed. Certainly, the impact of a major asteroid or comet would be sufficient. Moreover, the shock waves from the impact would produce shocked quartz and this was found in the breccias. Plate 10.4 shows an example of shocked quartz from the Yucatán.

The general scenario is as follows: a ten-kilometer-diameter asteroid or comet plowed into the continental shelf off the north coast of the Yucatán Peninsula. It crashed through the shallow Gulf waters and penetrated some kilometers into the earth's crust. The devastation was catastrophic. Major tsunamis penetrated well into the interior of the North American continent. The conversion of the body's kinetic energy into an explosion threw a huge amount of material into the atmosphere and created a 180-km, red-hot crater. As the sea flowed back into the crater, more pyrotechnics occurred. The sky was undoubtedly black with soot.

Despite the convincing evidence that a major catastrophe occurred and that the high estimates for the percentages of extinction are real, the specific mechanism for the extinctions is debatable. The possibilities include the cessation of photosynthesis due to dust in the atmosphere, climatic cooling due to dust in the atmosphere, ozone depletion, burning, acid rain, the direct pressure of shock waves, drowning in the

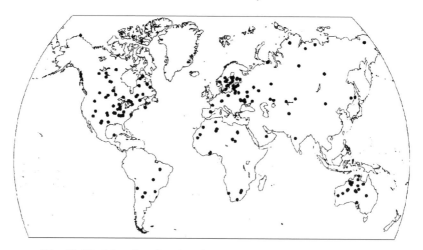

Fig. 10.12. Map showing the global distribution of impact craters.

tsunamis, and poisoning due to heavy metals and other chemicals thrown into the atmosphere. Toon *et al.* (1994) have noted:

During the past decade there has been a creative burst in finding ways that impacts and explosions might damage the global environment.

Toon *et al.* conclude that the specifics depend on the energy of the event. The K–T event is estimated to have had an energy of roughly 10^8 megatons. The principal extinction mechanisms would be produced by the lack of sunlight caused by dust in the atmosphere and by global fires.

The examples presented show the scales of destruction from local to regional to global. These are not isolated events. The geological record shows at least 140 significant terrestrial impact craters; see Fig. 10.12 for a map showing the global distribution of known impact craters. But, the number of known craters is almost surely a lower limit because discoveries continue, because many have been eroded away, and because many, like Chicxulub, are not exposed at the surface. In short, the record, although extensive, is incomplete. Nevertheless, Hughes (2000) has used the cratering record on the earth's surface to estimate the rate at which small craters are produced. The rates derived from this approach are much smaller than the rates described below which are based primarily on lunar crater data.

Estimates on the frequency of significant impacts are summarized in Fig. 10.13. Reasonable assumptions have been used to convert the megaton yield to an asteroid diameter. An event similar to the Arizona Meteor Crater on average occurs every 100 years. An event similar to Tunguska occurs on average every few hundred years. A globally catastrophic event such as the Chicxulub/K–T impact occurs on average

Log[Impact Energy, MT]

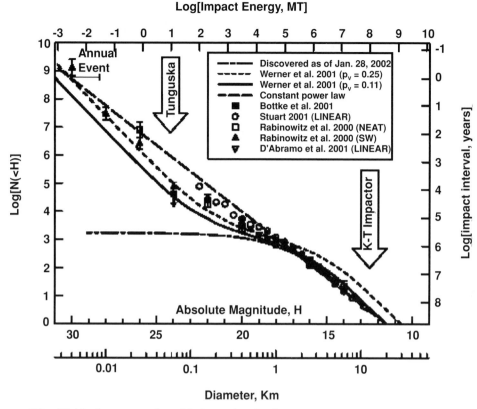

Fig. 10.13. Summary plot of information for impacts on earth. The plot shows the number of impactors greater than a given absolute magnitude or diameter, the energy of the impact, and the time interval between impacts for a given size. (Courtesy of D. Morrison, NASA–Ames Research Center, and C. Chapman, Southwest Research Institute; see Morrison *et al.* 2003. Dealing with the impact hazard. In *Asteroids III*, ed. W. Bottke, A. Cellino, P. Paolicchi, and R.P. Binzel.)

every 100 million years. These figures illustrate the great problem in explaining the hazard to the public and to political leaders. The catastrophic events simply do not occur very often and no one alive has experienced one.

Nevertheless, efforts are underway to quantitatively evaluate the threat. The study of NEOs, near-earth objects, is now a major scientific effort. The NEOs are principally asteroids and short-period comets with orbits that bring them close to earth. Specifically, an NEO is any asteroid or comet that has a perihelion distance $q \leq 1.3$ AU. A more stringent designation is a PHA or Potentially Hazardous Asteroid, defined as an asteroid with an absolute magnitude of 22 or less and whose orbit brings it 0.05 AU or less from the earth's orbit. The current emphasis is on search and discovery. While both asteroids and comets are being discovered, the emphasis is on asteroids because of their regular orbits in the inner solar system.

Comets are estimated to supply approximately 5–10% of the objects that may pose a hazard. A blur in this distinction is the growing evidence that extinct comets could comprise a significant fraction of near-earth asteroids (NEAs).

Astronomical realization of the NEA hazard goes back to the discovery of 433 Eros in 1898 and the determination that its orbit came close to earth. The first asteroid determined to be earth-crossing was 1862 Apollo in 1932. The first dedicated searches were carried out at Palomar Observatory. There were two teams. One was led by Eleanor Helin of the Jet Propulsion Laboratory and the other by Eugene and Carolyn Shoemaker of the U.S. Geological Survey. These were photographic surveys. Around 1980, Tom Gehrels and Robert McMillan at the University of Arizona began the Spacewatch program using a 0.9-meter telescope on Kitt Peak. This program used CCD detectors.

The NASA Spaceguard Survey Report (Morrison 1992) stimulated considerable public and legislative interest in the USA. The following quote is from the U.S. House of Representatives Committee for Science and Technology (1994):

To the extent practicable, the National Aeronautics and Space Administration, in coordination with the Department of Defense and the space agencies of all other countries, shall identify and catalog within 10 years the orbital characteristics of all comets and asteroids that are greater than 1 km in diameter and are in an orbit around the Sun that crosses the orbit of the Earth.

In Congressional Hearings in 1998, NASA formally began the search by adopting the objective of finding 90% of all NEOs 1 km in diameter or greater.

The process is complicated by the fact that diameters are not usually measured directly. Rather, a brightness is measured, converted to an absolute brightness, and an assumed albedo (taken to be 0.10) gives the projected surface area. If the object, usually an asteroid, is roughly spherical, the diameter is obtained.

In 2001, there are several major search efforts. Spacewatch continues its efforts (R. McMillan and T. Gehrels). In Arizona, USA, we also have the Catalina Sky Survey (S. Larson) and the Lowell Observatory Near-Earth Object Survey (LONEOS) (T. Bowell). The Near-Earth Asteroid Tracking (NEAT) effort (E. Helin and S. Pravdo) uses telescopes on Palomar Mountain and in Maui, Hawaii. Currently, 70% of near-earth-object discoveries are made by the Lincoln Near-Earth Asteroid Research (LINEAR) program (G. Stokes) near Socorro, New Mexico. The LINEAR program is a major discoverer of comets as we have seen in Chapter 7. We note that a Japanese Spaceguard effort (S. Isobe) is scheduled to begin full-time operations.

Estimates of the size of the population of NEOs/NEAs larger than 1 km in diameter are in the range 700 to 1300. Figure 10.14 shows the progress toward meeting the goal. Progress toward the goal depends somewhat on the final estimate

Known Kilometer–Size Near–Earth Asteroids

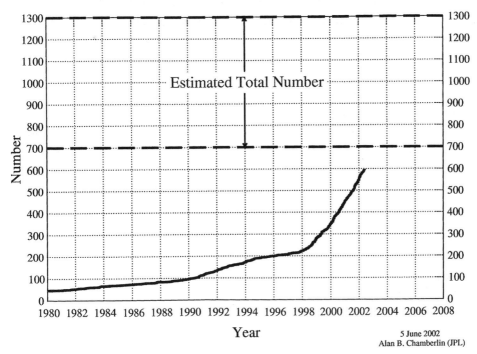

Fig. 10.14. Cumulative number of known 1-kilometer-diameter near-earth asteroids as of June 5, 2002. The dashed lines at 700 and 1300 illustrate the uncertainty in the estimated total number. (Courtesy of Alan B. Chamberlin, NASA–Jet
Propulsion Laboratory.)

of the population to be surveyed. This number should become clearer as the survey
continues. In sum, reasonable progress is being made, but there is concern that the
goal may not be met.

While the main threat may come from asteroids, comets can pose a major
threat too. Verschuur (1996) has noted the description of comet Swift–Tuttle
(109P) as "…the single most dangerous object known to humanity." The comet
has a period of about 135 years and last passed perihelion in 1992. In August
2126 A.D., the calculated orbit has the comet missing the earth by about 23 million
kilometers. However, nongravitational forces add considerable uncertainly to the
prediction.

If comet Swift–Tuttle scored a direct hit on earth, the impact energy could be as
high as roughly 50 times the impact energy of the comet (or asteroid) responsible
for the K–T event. We do not want to think about the consequences of such an
impact. We do want to monitor the comet carefully and be prepared for mitigation,
if necessary.

The astronomical side of the NEO problem is just the first step. If a "potentially hazardous asteroid" is identified, then the trajectory and probability of impact would be carefully evaluated by independent groups. If the impact probability is high and the object is large enough, mitigation would be attempted by destruction or by deflection. Note that a relatively small deflection impulse applied far enough in advance could be sufficient in many cases.

The issue of impacts and their possible mitigation contains problems far transcending straightforward scientific and technical areas of endeavor. As experiences with the communication of potential disasters – floods, hurricanes, etc. – show, the enterprise is fraught with pitfalls. To announce possible catastrophic collisions of asteroids or comets with earth, as noted by R. Binzel, presents:

> ...a topic so provocative and so prone to sensationalism that great care must be taken to assess and publicly communicate the realistic hazard (or non hazard) posed by such events. At the heart of this risk communication challenge resides the fact that low probability/high consequence events are by their very nature not within the realm of common human experience.

To address this problem, a scale, now called the Torino Scale, was developed at a workshop held in Torino, Italy on June 1–4, 1999, entitled, "International Monitoring Programs for Asteroid and Comet Threat (IMPACT)." The Torino Scale (Binzel 2000) is presented schematically in Plate 10.5 and summarized in Plate 10.6. The scale ranges from 0 for events having no likely consequences to 8 to 10 for certain collisions with consequences ranging from local to global.

The Torino Scale is analogous to the diagnosis of a patient's illness by a physician. It is only the beginning, and, as emphasized by the proposers, has limitations. Should a dangerous event be likely anytime in the near future, scenes of major panic as often depicted in science-fiction movies could become reality.

10.4 Comets, terrestrial water, and the origin of life

The origin of terrestrial water is a major area of research and controversy in solar system science. Planetesimals forming in the region of earth's orbit are believed to have a very low water content. In addition, a significant fraction of water in these planetesimals was probably lost in the accretion process. Therefore, there has been considerable interest in a cosmic source for terrestrial water.

One view of the cosmic source of terrestrial water is called the "cometary late veneer scenario." Basically, the earth had formed and the earth's water, including the oceans, was added by a bombardment of bodies rich in water ice, i.e., comets. This view was bolstered by the initial value for the ratio of deuterium to hydrogen, D : H, for Halley's Comet determined by the mass spectrometer on *Giotto*. Within the substantial error bars, the value for Halley's Comet was the same as the oceans (SMOW or Standard Mean Ocean Water).

However, detailed analysis of the *Giotto* measurements produced a value for D : H with a much smaller error bar. The value of D : H was also determined spectroscopically for comets Hyakutake and Hale–Bopp. The cometary values are in substantial agreement with a D : H of 3.1×10^{-4} and an error bar of less than 10%. This value is approximately twice the value for ocean water. Thus, if the value of D : H for these comets is typical, comets cannot be the sole source of terrestrial water.

The solution to this problem requires a dynamical capability to deliver material to earth and a source with essentially the correct D : H value. Morbidelli *et al.* (2000) have investigated the idea that the water source is massive planetary embryos in the asteroid belt that delivered the water as part of the process of forming the earth. The earth should have received about 10^{-2} earth masses of water on this approach. Other possible sources such as primitive asteroids and Jupiter–Saturn region comets do not supply enough water. The asteroid embryos are expected to have a D : H ratio similar to the water in carbonaceous chrondites; this D : H value is nearly identical to the D : H value for ocean water. And, the delivery from asteroid embryos occurs late in the earth's formation and, hence, could be retained. This scenario is currently the accepted one.

Another possible cometary source is from trans-Uranian-region comets. Even under very optimistic assumptions, these comets could deliver at most 10% of the water presently on earth. This value or a smaller value would not greatly alter the D : H value from the value for carbonaceous chrondites.

However, a spirited defense of the cometary late veneer scenario has been presented by Delsemme (2000, 2001). There are many details in the scientific arguments, but basically Delsemme suggests that the comets that contributed the earth's seawater are different than comets Halley, Hyakutake, and Hale–Bopp. Specifically, comets formed in the Jupiter–Saturn region would be formed at higher temperatures and acquire more water and less deuterium than comets formed in the Uranus–Neptune region. Hence, the comets supplying the oceans would have a lower D : H value.

Thus, comets may have provided only a minor component of the earth's water. They are a rich source of complex hydrocarbons (see Chapter 3) and may still be important in terms of the source of seed chemicals for pre-biotic evolution. See Thomas *et al.* (1996) for an overview of these possibilities. As with any developing and fascinating topic, important developments are expected in the near future.

One of these developments involves the study of amorphous ice as a carrier for organic molecules (Ehrenfreund and Charnley 2000; Blake and Jenniskens 2001). The amorphous ice has some properties of water. Even when the ice warms and crystallizes, some remains in the amorphous form and can preserve the organic materials. These new results on phase transitions in ice are important for the discussion of ice in comets (Chapter 7). Comets are thought to be the most likely objects to deliver the organics-laced water to earth.

Hoyle and Wickramasinghe (1985) proposed that comets could be composed largely of organic material, including bacteria, and that comets are not composed largely of water ice. The evidence presented in this monograph is overwhelmingly in favor of water ice as the major constituent of comet nuclei. But, comets can certainly be a source of organic material for earth. Hoyle and Wickramasinghe suggest that some bacteria could take about two years to descend through the earth's atmosphere and that they could be responsible for the periodic outbreak of some diseases such as whooping cough. This view was bolstered by the report by Wickramasinghe and Allen (1986) of organic grains in comet Halley based on a broad absorption band at 3.4 μm. While the fit to this band from biological sources is good, the fit is widely regarded as not unique. The contentious nature of the debate is captured by the discussion following the paper by Meadows (1987). The approach by Hoyle and Wickramasinghe is not widely accepted.

A highly controversial source for terrestrial water has been proposed by L. Frank. The suggestion was originally proposed on the basis of ultraviolet observations obtained by the *Dynamics Explorer* satellite that appeared to show dark spots in the airglow (Frank *et al.* 1986a). The proposed explanation was house-sized chunks of water ice that were below the satellite and were suppressing the airglow. The objects have a mass of some 30 tons and on average 20 of them hit the earth per minute (Frank *et al.* 1986b). This source could supply the earth's oceans if it operated over the age of the earth.

However, these objects have extraordinary and some say unbelievable properties. They are nearly pure ice or snow and easily pull apart in the earth's atmosphere. But, they do not sublimate because they are cloaked with a thin layer of dark material that prevents sublimation and the dark surface makes them hard to detect. The objects must be nearly pure ice or snow to prevent, in essence, the sky from being bright in the meteors they would otherwise produce. And, they seem to strike only the earth and not, for example the Moon where they would be detected by seismographs. Finally, where and how are these objects formed?

The debate was rekindled by data obtained from the *Polar Orbiter* satellite. Basically, the new results by Frank and Sigwarth (1997) supported the view described above. Most scientists believe that the explanation for the dark spots cannot be the one proposed. But what about the dark spots themselves? Are they real and do they still require an explanation? Independent analysis of part of the raw data finds that the dark spots are consistent with instrumental noise. As with many subjects steeped in controversy, future developments are likely.

The authors of this comet text have a minor, side issue with the use of the words "small comet" or "mini-comet" to describe the proposed objects. These objects are not comets, which are a mixture of ices and dust, and which sublimate at heliocentric distances ~1 AU to form comas or tails. This misleading use of a well-known term can only further muddy a contentious debate.

10.5 Summary

We often see the fiery death of cometary dust particles in meteor showers. Of course, these particles come from comets, and comets are a major contributor to the population of zodiacal light particles as well.

Comets impact everything in the solar system. The spectacular impacts of comet Shoemaker–Levy 9 on Jupiter during July 1994 reinforced this point. On earth, we have had some near misses and several well-known impacts by comets or asteroids. The impact hazard has generated major efforts to discover objects (asteroids or comets) that might pose a hazard to earth. The Torino Scale was developed to help categorize the extent of the danger.

Comet are known to carry volatiles such as water and complex organic compounds. In some scenarios, they are an important source of terrestrial water and may have supplied seed chemicals for the origin of life. Currently, there is no consensus on the details and further study is needed.

11

The future of cometary research

The beginning of the third millennium is a time of great promise for cometary research. After a drought of over 15 years since the Halley Armada in 1986, several space missions to comets should take place in the near future. The bright comets Hyakutake in 1996 and Hale–Bopp in 1997 have also stimulated major scientific interest.

In addition to these developments, we also perceive changes in attitudes and thinking that auger well for the future. In this chapter, we attempt the risky enterprise of describing cometary research for the near future.

11.1 Space missions

The space missions to comets described below offer the prospect of a steady stream of new results being announced until 2015 (Schilling 2002). Of course, these prospects should be tempered by the reality that comet missions can encounter technical difficulties or can be delayed or cancelled for political reasons.

Scientists interested in missions to comets have experienced an emotional "roller coaster" ride through the years. In the 1980s, NASA had great difficulty mounting a mission to Halley's Comet and finally cancelled the *International Comet Mission* (*ICM*). The next hope was the *Comet Rendezvous/Asteroid Flyby* (*CRAF*) mission that was cancelled in 1992. Then, we come to NASA's New Millennium Program. While the prime goal of the program is to develop and test new technologies for exploration of the solar system, two missions, *Deep Space 1* and *Deep Space 4, Champollion*, offered the possibility of significant comet science return. *Champollion* was to have been a comet rendezvous and lander mission to comet 9P/Tempel 1. Citing budget contraints, NASA cancelled this mission in 1999. *Deep Space 1* was launched on October 25, 1998 and it flew by the asteroid Braille (1992 KD) on July 28, 1999. The primary mission ended in September 1999, but an extended mission was planned which could encounter comet 19P/Borrelly in September 2001.

355

Technical problems had developed with the spacecraft, but work-around solutions were found. The situation in early September 2001 was that *Deep Space 1* – an ailing spacecraft with no dust protection and nearing the end of its fuel supply – was poised to encounter comet Borrelly on September 22, 2001. Some of the expectations were dire, but the encounter was a great success! Imaging results are discussed in section 7.2 and plasma results in section 6.3.2.3. But, this pleasant situation was followed by the loss of *CONTOUR* in 2002.

11.1.1 Stardust

The *Stardust* spacecraft was successfully launched on February 7, 1999 (see section 3.6.2). The trajectory is complex, including earth swingbys. The spacecraft passes through the inner coma of comet 81P/Wild 2 in January 2004. The goals of the *Stardust* mission are ambitious. In addition to relatively straightforward science involving imaging of the coma and nucleus, and the compositional analysis of dust particles and volatiles, the *Stardust* mission proposes to bring samples of cometary dust back to earth. The Principal Investigator for *Stardust* is D. Brownlee, University of Washington.

The mission trajectory involves three loops around the sun. This trajectory was chosen to minimize the fuel requirement, particularly after launch, to encounter the comet at the low relative speed of 6.1 km s^{-1}, and to bring the spacecraft back to earth. The first loop with the earth gravity assist is shown in Fig. 11.1. Figure 11.2 shows the second loop and the encounter with the asteroid Annefrank. The plan for the subsequent encounter with comet Wild 2 is to pass through the coma at a distance of 100 to 150 km on the sunward side as shown schematically in Fig. 11.3. At the time of closest approach, comet Wild 2 is 97.5 days past perihelion and at a heliocentric distance of 1.86 AU. The encounter period spans about 150 days. The approach of the spacecraft (with the aerogel collector deployed; see below) to comet Wild 2 is shown in an artist's conception in Fig. 11.4.

Figure 11.5 shows the encounter period and the return trajectory to earth. As the spacecraft approaches earth, the Sample Return Capsule (SRC) is released and it enters the atmosphere initially at 12.8 km s^{-1}. Aerobraking slows the SRC and the heat shield protects it. The entry is shown schematically in Fig. 11.6. The final stage of the SRC with parachute is shown in Fig. 11.7. The SRC is recovered in the desert south-west of Salt Lake City, Utah, USA and transported to the planetary curatorial facility at NASA's Johnson Space Center, Houston, Texas, USA. Laboratory analyses are then performed on the material (dust and volatiles) trapped in the aerogel (described below).

The objective for cometary particles is to recover more than 1000 particles larger than 15 μm as well as volatile molecules. The objective for fresh interstellar particles

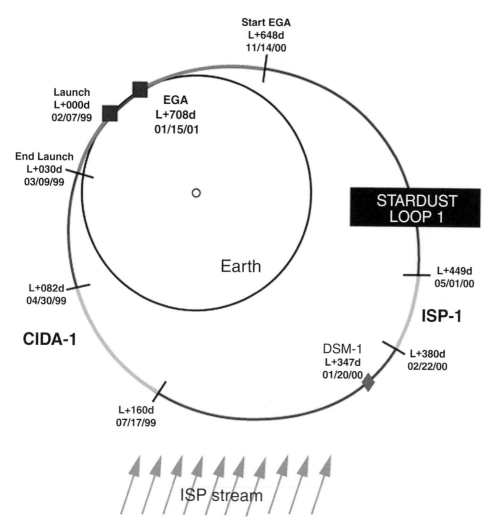

Fig. 11.1. *Stardust* Loop 1, showing the launch and the return to earth for the Earth Gravity Assist (EGA). The arrows labeled "ISP stream" indicate the direction of the stream of interstellar dust particles. The periods of ISP collection and CIDA operation (see text) are indicated. (Courtesy of NASA/JPL/Caltech.)

is to collect a significant sample (over 100 particles) in the size range of 0.1 to 1.0 μm. The goal is to gather all particles in a manner designed to preserve, at least, the elemental and isotopic composition for major elements in individual particles. This involves gently slowing down the particles for capture using a substance called *aerogel*.

In terms of everyday experience, aerogel is an extraordinary substance. It consists of colloidal silica units that form bubbles or foam. Some 99.8% of the volume

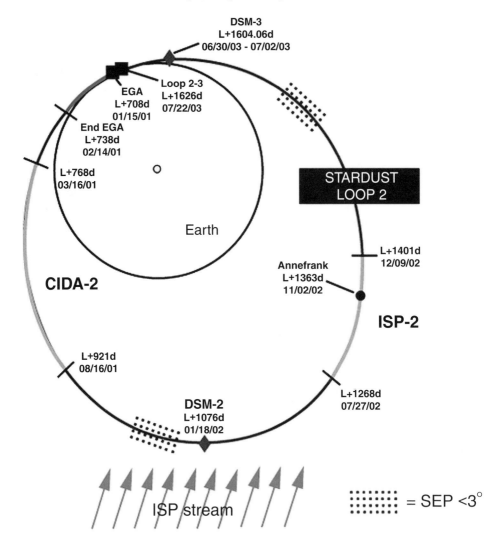

DSM-3
L+1604.06d
06/30/03 - 07/02/03

Loop 2-3
L+1626d
07/22/03

EGA
L+708d
01/15/01

End EGA
L+738d
02/14/01

L+768d
03/16/01

Earth

CIDA-2

STARDUST
LOOP 2

L+1401d
12/09/02

Annefrank
L+1363d
11/02/02

ISP-2

L+921d
08/16/01

DSM-2
L+1076d
01/18/02

L+1268d
07/27/02

ISP stream

⋮⋮⋮ = SEP <3°

Fig. 11.2. *Stardust* Loop 2, showing the flyby of the asteroid Annefrank and the second return to earth. The arrows labeled "ISP stream" indicate the direction of the stream of interstellar dust particles. The periods of ISP collection and CIDA operation (see text) are indicated. The stippled area (coded with SEP < 3°; SEP = sun–earth–probe angle) indicates a time when the spacecraft is effectively behind the sun and mission operations are not performed. (Courtesy of NASA/JPL/Caltech.)

is empty space. While aerogel has a distinctive smoky-blue cast, objects can be seen through it, as shown in Fig. 11.8. Particles striking the aerogel bury themselves in the foam and form a carrot-shaped track up to 200 times their size, as shown in Fig. 11.9. Thus, the particles are captured in the aerogel and their impact direction determined. This fact allows single slabs of aerogel to collect interstellar

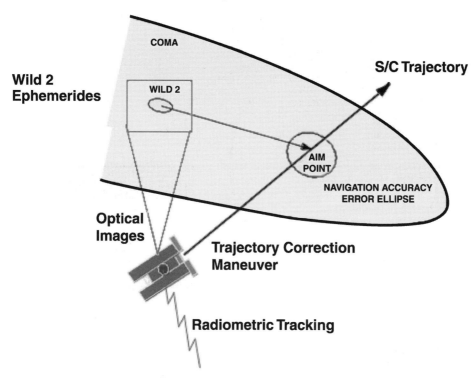

Fig. 11.3. Schematic illustration of the targeting strategy and encounter circumstances for the flyby of comet Wild 2. (Courtesy of NASA/JPL/Caltech.)

dust particles (ISPs) on one side during cruise phase and cometary dust particles during the encounter. Periods of ISP collection are marked on Figs. 11.1 and 11.2. The slabs are mounted into a collector shaped like a tennis racket. The spacecraft configuration at comet encounter is shown in Fig. 11.10. After the encounter with Comet Wild 2, the aerogel collector is retracted into the Sample Return Capsule (SRC).

During the encounter itself, the rate of impacts is measured by the Dust Flux Monitor. The composition of the dust particles and volatiles is determined by the Comet and Interstellar Dust Analyzer (CIDA), a mass spectrometer derived from instruments flown on the *Giotto* and *VEGA* Halley missions. Periods of CIDA operation for ISPs are indicated on Figs. 11.1 and 11.2.

The imaging system serves two purposes. First, wide-band filter images are used for navigation. Second, images through blue, yellow, red and infrared filters are used to study the coma and nucleus. Around the closest approach, the nucleus subtends 60×60 pixels. This corresponds to determining the topology of the nucleus at 10 times better resolution than the *Giotto* images of comet Halley.

Fig. 11.4. Artist's depiction of the *Stardust* spacecraft approaching comet Wild 2. Note that the aerogel collector is deployed. (Courtesy of NASA/JPL/Caltech.)

Comet Wild 2 is the choice for the *Stardust* mission because it has not been exposed to the sun during many passes through the inner solar system. Thus, its composition is probably little altered from its original condition. The comet's orbit was changed in 1974 by a close approach to Jupiter. When *Stardust* encounters it in 2004, Comet Wild 2 will be on its fifth pass through the inner solar system.

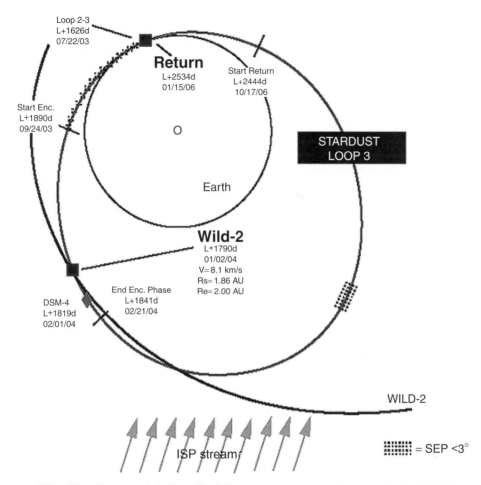

Fig. 11.5. *Stardust* Loop 3, showing the encounter with comet Wild 2 and the spacecraft's final return to earth. The arrows labeled "ISP stream" indicate the direction of the stream of interstellar dust particles. The stippled area (coded with SEP < 3°; SEP = sun–earth–probe angle) indicates a time when the spacecraft is effectively behind the sun and mission operations are not performed. (Courtesy of NASA/JPL/Caltech.)

11.1.2 CONTOUR

The *CONTOUR* mission has been lost. The spacecraft was in earth orbit and an on-board engine firing on August 15, 2002 should have put it on a trajectory to intercept comets Encke and Schwassmann–Wachmann 3. The firing took place when the spacecraft was not in contact with the ground. No signal was received at the scheduled time for re-contact. Subsequent optical imaging showed several fragments at approximately the location expected for the spacecraft. The spacecraft is presumed to have exploded during the rocket firing.

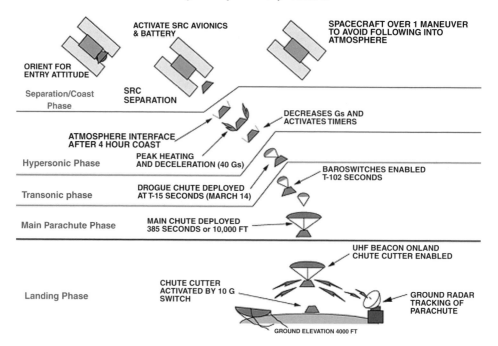

Fig. 11.6. Schematic showing the spacecraft entering the earth's atmosphere, separation of the sample return capsule (SRC), and the SRC's return to the earth's surface. (Courtesy of NASA/JPL/Caltech.)

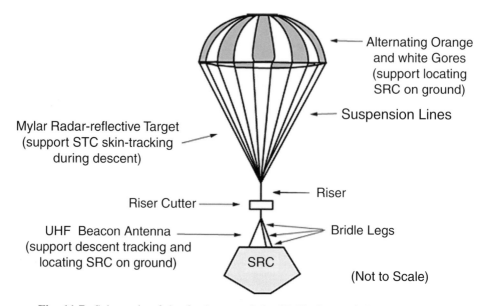

Fig. 11.7. Schematic of the final stage of the SRC's descent with the parachute deployed. (Courtesy of NASA/JPL/Caltech.)

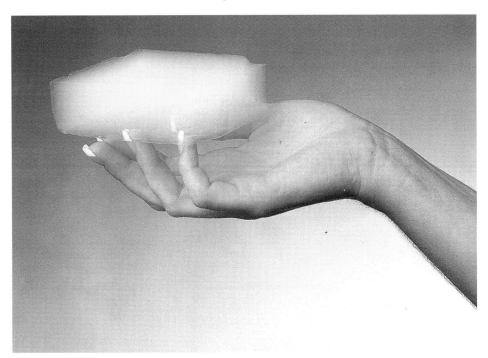

Fig. 11.8. A slab of aerogel being held in a hand. The tips of several fingers are visible through the aerogel. (Courtesy of NASA/JPL/Caltech.)

Fig. 11.9. Particle tracks in aerogel. Many of the dust particles are visible at the tip of the cone-shaped tracks. (Courtesy of NASA/JPL/Caltech.)

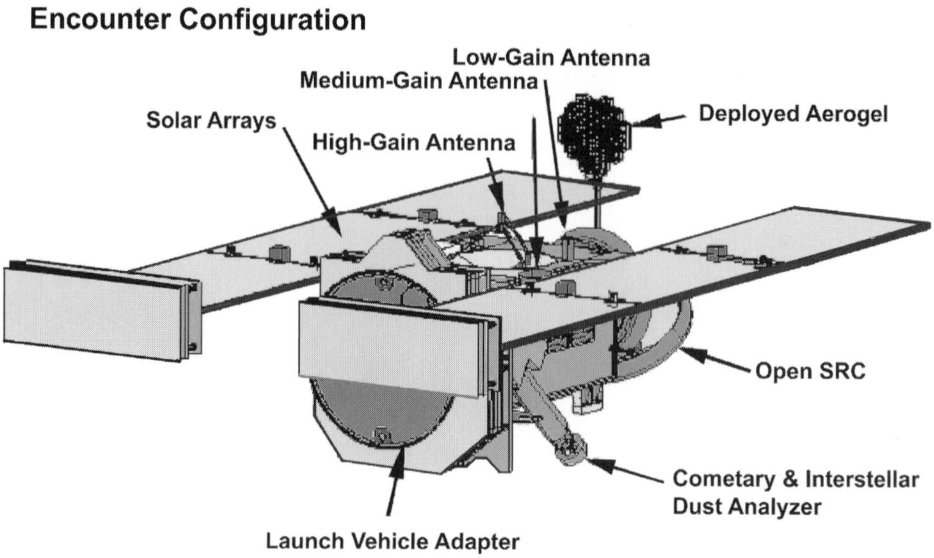

Encounter Configuration

Fig. 11.10. The Encounter Configuration of the *Stardust* spacecraft with the aerogel collector deployed. (Courtesy of NASA/JPL/Caltech.)

We present the description of the *CONTOUR* mission essentially as written before the spacecraft was lost for two reasons. First, the mission was well-designed and nicely illustrates the cometary science accessible with current technology. Second, there is a reasonably high likelihood that a similar mission, say *CONTOUR 2*, will take place.

The Comet Nucleus Tour or *CONTOUR* addresses the diversity of comets and attempts to improve our knowledge of comet nuclei by studying at least two comets and possibly a third. The original targets include: a highly evolved comet, 2P/Encke, which has made many orbits around the sun (period = 3.3 years); a less evolved (in the sense of time spent near the sun) comet, 73P/Schwassmann–Wachmann 3, (but, this comet was observed to split in late 1995 and in late 2000 (see Fig. 7.21) and relatively fresh material could be observed by *CONTOUR*); and a comet with apparently stable rotation and vent structure, 6P/d'Arrest. Note that comet d'Arrest has been officially dropped from the mission, but there is hope that it will be restored later. The Principal Investigator of *CONTOUR* is J. Veverka, Cornell University.

The comets' orbits are shown in Fig. 11.11. The spacecraft trajectory uses multiple earth swingbys and this allows the spacecraft to be retargeted if an interesting comet, like Hale–Bopp, should appear during the mission. Launch was scheduled for July 2002. Each encounter – November 2003 for comet Encke, June 2006 for comet Schwassmann–Wachmann 3 and August 2008 for comet d'Arrest – takes place at a distance of about 100 km and near the comet's peak activity. The comet

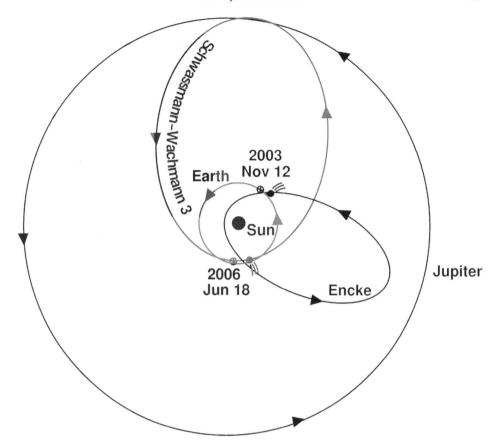

Fig. 11.11. Orbits of comets Encke and Schwassmann–Wachmann 3, the nominal targets for the *CONTOUR* mission. (Courtesy of the *CONTOUR* Project/NASA.)

distances from earth are roughly one-third AU and the geometry for earth-based observations is good.

CONTOUR carries two camera systems. The *CONTOUR* Remote Imager Spectrograph (CRISP) with a field of view of $1.2° \times 1.2°$ has the capability of imaging the nuclei with a resolution of 4 meters, an improvement of 25 times over the images of the comet Halley nucleus obtained by *Giotto*. A capability for spectral mapping of the nuclei in the wavelength range 8000 Å to 25 500 Å is also provided. CRISP is behind the dust shield and uses a scanning mirror to track the comet. Because of the geometry, the scanning mirror could be hit by a dust particle when the spacecraft is far from the comet. The scanning mirror is two-sided and the pristine side is used for observations near the comet when the mirror is protected by the dust shield.

The *CONTOUR* Forward Imager (CFI) has a field of view of $2.5° \times 2.5°$ and looks through a small opening in the dust shield. To avoid damage from the stream of cometary dust particles that are encountered near the nuclei, CFI does not look directly at the nucleus. The images are obtained by looking at one of four mirrors mounted on a rotatable cube. Only one mirror is used per comet and a fresh mirror is rotated into place for the next comet. It provides a wide-field capability and redundancy for imaging. The images are taken when the spacecraft is more than 2000 km from the nucleus. A mass spectrometer provides compositional data for the neutral gases and ions. And, a dust analyzer provides compositional data for the dust. The intensive data gathering time is a few days near closest approach. When not involved in an encounter, the spacecraft is placed in a spin-stabilized hibernation mode much of the time.

11.1.3 Deep impact

Knowledge of cometary nuclei is restricted to the surface layers. Short-period comets have been exposed repeatedly to the thermal consequences of intense solar radiation. These effects are estimated to extend more than 1 meter beneath the surface of the nucleus. All comets are exposed to the background flux of cosmic rays that could modify the composition to depths of several tens of meters.

Deep Impact proposes to sample a comet's interior by delivering an impactor that punches through the crust to expose the relatively fresh material below (Belton and A'Hearn 1999). An instrumented spacecraft records the crater-forming process and the expected enhanced activity of the newly exposed volatile layer. The Principal Investigator for *Deep Impact* is M. A'Hearn, University of Maryland.

Launch of *Deep Impact* is scheduled to occur on January 2, 2004, and the trajectory is shown in Fig. 11.12. The spacecraft swings by the earth on December 31, 2004, and the impact occurs on Comet 9P/Tempel 1 on July 4, 2005. These are nominal dates and small changes are possible.

The spacecraft at launch consists of the combined impactor and flyby spacecraft. Approximately 24 hours before impact, the flyby spacecraft releases the impactor on a course to hit the comet's sun-facing side, as shown in Fig. 11.13. The impactor carries out its own navigation using on-board cameras. These cameras also provide images of the nucleus and the impact area up to a few seconds before impact.

The impactor has a mass of 350 kg and an approach speed of 10.2 km s^{-1}. The impactor is mostly copper and is not expected to react with the cometary material. Calculations indicate that impact should create a crater some 100 meters in diameter and 25 meters deep. The impact should produce a splash of solid and vaporized ejecta. Then, the crater, containing freshly exposed material, should be visible. The investigators note that predictions for the size of the crater are highly uncertain and depend on assumptions about the density and strength of the nucleus.

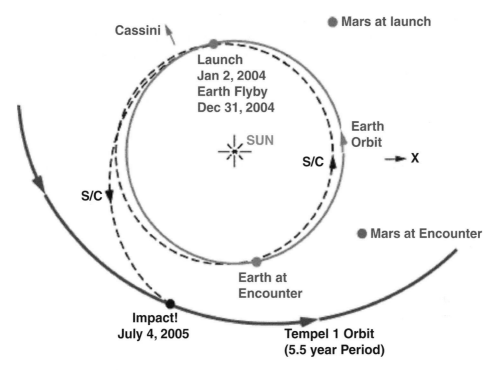

Fig. 11.12. The trajectory of the *Deep Impact* spacecraft to comet 9/P Tempel 1, including the earth flyby in January 2005. (Courtesy of the *Deep Impact* Project/University of Maryland/NASA.)

These events are documented by instruments on the flyby spacecraft. Specifically, these are a high-resolution imager and an infrared spectrometer. Data are collected for a few days after impact.

The science goals of the mission are to observe the formation of the crater, measure the crater's dimensions, determine the composition of the ejecta and the crater interior, and characterize the changes in sublimation/outgassing produced by the impact. The goals are ambitious. As mentioned in Chapter 7, our knowledge of the interiors of comet nuclei is meager and some surprises might be expected. In any event, July and August of 2005 promise to be an exciting time.

Comet 9P/Tempel 1 was discovered in 1867. The comet's orbital history is complex. Its perihelion distance has moved inward and outward through time because of close approaches to Jupiter. Currently, the comet's orbital period is 5.5 years. This is roughly one-half of Jupiter's orbital period. Hence, the comet's orbit is in resonance with Jupiter. But, there are no close approaches to Jupiter between 1967 and 2022 and this resonance is currently somewhat stable. Estimates of the diameter of the nucleus are around 5.2 km. The impulse delivered by the impactor to

Fig. 11.13. Artist's rendition of the *Deep Impact* impactor after release from the flyby spacecraft on its way to 9P/Tempel 1. (Courtesy of the *Deep Impact* Project/University of Maryland/NASA.)

comet Tempel 1 is quite small. The change in the orbit is negligible, particularly when compared with the large orbital changes produced by close encounters with Jupiter.

11.1.4 Rosetta

Missions to comets received a major setback when the European Space Agency (ESA) cancelled the January 2003 launch of *Rosetta* and placed the spacecraft in storage. The reason was the failure of a launch vehicle (similar to the one scheduled to be used) in December 2002. ESA has announced that *Rosetta* will rendezvous with comet 67P/Churyumov-Gerasimenko and the launch is scheduled for 26 February 2004. The details are still being worked out. Because *Rosetta* superbly illustrates the future of mission to comets, we present the description of the *Rosetta* mission essentially as written before the postponement.

The *International Rosetta Mission* was scheduled for launch on January 20, 2003. This highly ambitious mission is the European Space Agency's premier

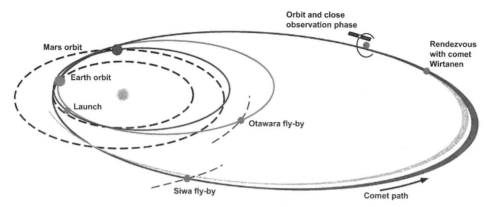

Fig. 11.14. The trajectory of the *Rosetta* spacecraft to comet 46P/Wirtanen. The flybys of the asteroids Otawara and Siwa are also shown. (Copyright © 2001 European Space Agency.)

comet mission; its scope is extensive and the spacecraft carries an extraordinary suite of instruments (Schwehm and Schulz 1999). The Principal Investigator for *Rosetta* is G. Schwehm, European Space Agency. Just getting to the target, comet 46P/Wirtanen, requires an 8-year journey that includes one Mars and two earth gravity assists. The orbit is shown in Fig. 11.14. Two asteroid flybys are planned; these are of the asteroids 4979 Otawara and 140 Siwa. A mass determination should be possible for Siwa, but not for Otawara.

The mission is named after the Rosetta Stone, which was the key to modern knowledge of ancient Egyptian hieroglyphics, as deduced by J. F. Champollion. The expectation is that the results from the *Rosetta* mission will provide keys to the modern understanding of comets. The measurements planned to achieve this goal include a complete characterization of the nucleus of a comet; a thorough determination of the chemical, isotopic, and mineralogical composition; and studying the activity (and other surface layer processes) produced by the cometary volatiles.

The spacecraft will spend about 2 years near comet Wirtanen. Its orbital period is 5.5 years and, if the albedo is 0.04, the nuclear radius is about 600 meters (section 7.3.1). See Schulz and Schwehm (1999) for a review of the properties of comet Wirtanen. The comet was chosen for its intermediate range of activity – active enough to show typical phenomena, but not so active as to damage the spacecraft.

The mission design includes extended hibernation periods for the spacecraft to help deal with the long length of the mission. The timeline is shown in Fig. 11.15. After rendezvous with the comet, the detailed study of the nucleus begins at a heliocentric distance of about 3.25 AU. The measurements and observations begin

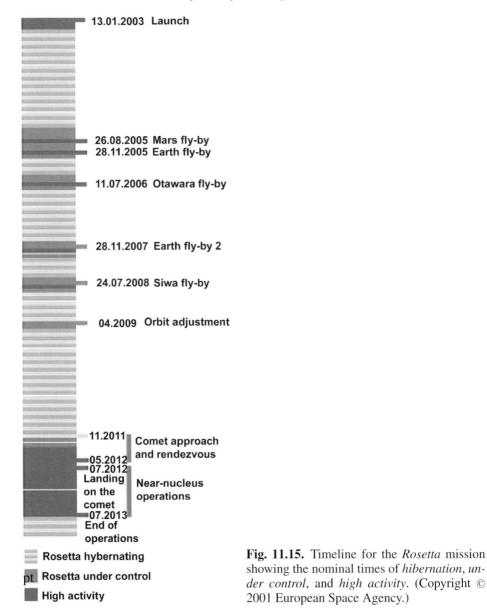

13.01.2003 Launch

26.08.2005 Mars fly-by
28.11.2005 Earth fly-by

11.07.2006 Otawara fly-by

28.11.2007 Earth fly-by 2

24.07.2008 Siwa fly-by

04.2009 Orbit adjustment

11.2011 Comet approach
and rendezvous
05.2012
07.2012
Landing   Near-nucleus
on the   operations
comet
07.2013
End of
operations

Rosetta hybernating

pt. Rosetta under control

High activity

Fig. 11.15. Timeline for the *Rosetta* mission showing the nominal times of *hibernation, under control*, and *high activity*. (Copyright © 2001 European Space Agency.)

at substantial distances from the comet and range down to close observations at about 1 kilometer. The spacecraft will be put into orbit around the comet, thus allowing the mass of the nucleus to be determined. The surface of the nucleus will be thoroughly mapped and the site for the *Rosetta* Lander determined. After the Lander has been deployed, the main spacecraft remains near the comet for at least 200 days to monitor activity on the surface and to provide a communication relay service for the Lander. The nominal end of mission date is the time of perihelion passage for comet Wirtanen in July 2013. The spacecraft orbiter is shown schematically in Fig. 11.16. Figure 11.17 shows the orbiter spacecraft escorting the comet. Figure 11.18 is an artist's rendition of the Lander anchored to the comet's surface.

Rosetta carries an extensive suite of instruments optimized for remote sensing and *in situ* coma measurements. These include: imaging; spectroscopy from the ultraviolet through the microwave region; mass spectroscopy of dust, neutral gas, and ions; plasma measurements for particles and magnetic fields; and a radio sounder for tomography of the nucleus. The latter experiment is carried on both the Main Spacecraft and the Lander.

The Lander carries an extensive suite of instruments optimized for lander science and a sample acquisition system. The instruments include: imaging; material and gas analyzers; x-ray spectroscopy; and plasma instruments for particles and magnetic fields.

The *Rosetta* mission is exceptionally ambitious and comprehensive. It is also challenging, difficult, and demanding. The community of scientists interested in comets can hope for a successful mission and the fascinating results from *Rosetta* starting circa 2012.

11.1.5 Pluto/Kuiper belt mission

The solar system beyond Neptune is a largely unexplored region. The planet Pluto is a tempting target and has attracted the attention of scientists for years. Pluto is both an icy planet and the brightest and largest-known Kuiper belt object (see Chapter 9). Besides the intrinsic interest in Pluto, including its satellite Charon, Pluto and the Kuiper belt objects are distant remnants of planet-building processes that took place during the formation of the solar system. See Stern (2002) for a summary of the possible scientific return and approach for a mission to Pluto and the Kuiper belt region.

NASA concluded a competitive selection in November 2001 for a team to carry out the Pluto–Kuiper belt mission. They selected a team from Southwest Research Institute (SWRI) and several other institutions. The Principal Investigator was S. A. Stern of SWRI. If the spacecraft were launched in 2006, it would fly by the Pluto–Charon system in 2015 and go on to study several KBOs.

Unfortunately, funds to carry out this mission were removed from the US Federal Budget for fiscal year 2003. If this decision stands, the mission, if done at a later time, would suffer from missing the correct alignment of Jupiter for the slingshot maneuver to send the spacecraft to Pluto and run the risk of Pluto's atmosphere freezing out if the delay is too long.

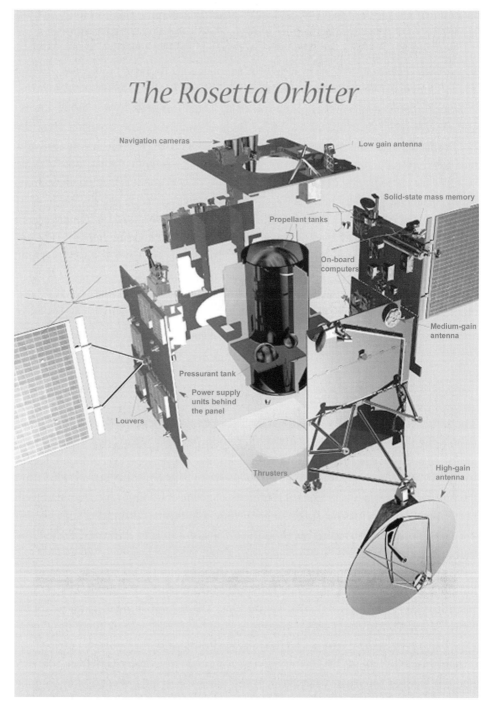

Fig. 11.16. Schematic of the *Rosetta* orbiter. (Copyright © 2001 European Space Agency.)

Fig. 11.17. Artist's rendition of the *Rosetta* orbiter spacecraft escorting the comet, July 2012 to July 2013. (Copyright © 2001 European Space Agency.)

Fig. 11.18. Artist's rendition of the *Rosetta* lander spacecraft anchored to the surface of comet Wirtanen. (Copyright © 2001 European Space Agency.)

The situation has not been settled and there is hope that the mission may yet be launched on schedule. The editors of *Scientific American* (2002) have noted that "...it can take longer for a space mission to escape from Washington, DC, than to cross the solar system." They also pointed out a straightforward solution to the fiscal and managerial problems that involves a reasonably simple reshuffling of some funding. As in any political/managerial situation, we can only hope that sensible things will be done.

11.2 Trends in observational capability and theoretical work

New observational techniques are coming online, using both observatories in space and telescopes on the ground. A major area of expected advances is in the infrared. The *Stratospheric Observatory for Infrared Astronomy* (*SOFIA*), a facility which is an integral part of an airplane, and the *Space Infrared Telescope Facility* (*SIRTF*), the last of NASA's "Great Observatories" should be in operation soon. In the future, the successor to the *Hubble Space Telescope* called the *James Webb Space Telescope* (*JWST*) should provide a major infrared capability.

New ground-based telescopes are in various stages of development. The twin 10-meter segmented-mirror Keck Telescopes on Mauna Kea in Hawaii are now operational. These telescopes have achieved exceptional resolution using adaptive optics. The 8-meter Gemini Telescopes on Mauna Kea, Hawaii, and Cerro Pachon, Chile, have both started operating, and became fully available for science in late 2001. These two telescopes are outfitted with high-resolution infrared instrumentation. Finally, we have the four 8.2-meter telescopes of the Very Large Telescope (VLT) at the ESO Paranel Observatory, on Cerro Paranel, Chile. All these modern telescopes should provide excellent facilities for cometary research if sufficient time is available. Another promising possibility is the Large Binocular Telescope (LBT) that offers excellent sensitivity and spatial resolution (see Campins *et al.* 2001). New radio telescope arrays such as the Atacama Large Millimeter Array (ALMA) project located at a high (16 400 feet), dry site in Chile should provide remarkable resolution and sensitivity in the millimeter and submillimeter wavelength regions. While new instruments are becoming available, some scientists feel that progress in comet research as well as other solar system small bodies will ultimately require dedicated telescopes.

The thrust of progress due to recent bright, well-studied comets, new observatories or observing techniques, and space missions is expected and somewhat inevitable. Currently, we find a renaissance in attitudes toward and thinking about comet research.

Anticipating the next theoretical breakthrough or unifying concept is very difficult, but a look back at the last decade gives an idea of the *kind* of theoretical advance that might be expected. New computing techniques and the development of symplectic integrators (Saha and Tremaine 1992) have enabled the calculation of long-term situations in solar system dynamics (Chapter 9). Computing techniques utilizing variable grid sizes allow treatment of cometary plasma physics problems involving a very large range of scales (Chapter 6). And, we have been made aware of the importance of chemistry in comets. Observers cannot always use simple models (Chapter 3) to represent their data. Understanding the species observed from earth requires an understanding of the chemistry. Finally, we strongly suspect that the species seen in the coma do not have abundances representative of the interior (Chapter 7).

An unexpected development has been the upsurge in the discovery rate for comets largely due to the search and discovery effort for NEOs. This high rate, particularly from the LINEAR project (section 10.4), provides the benefits of the richness effect, i.e., the more comets that are studied, the greater the opportunity for unusual phenomena that provide insight into the overall population. Complete discovery statistics have been published for 1998 and 1999, and the first 5 months of 2000. In that time frame, LINEAR discovered 43 comets, 399 NEOs and over 59 000 new asteroids.

All of these new investigations can be expected to test current theories, require revisions, and provide major progress into the understanding of comets.

11.3 Summary

The prognosis is excellent for comet research in the years up to 2015. If a major fraction of the projects described above are carried through to completion, comet scientists and historians will remember these as halcyon years.

As the history of cometary science clearly shows, Halley's Comet has been a stimulus to research over many returns. The return of 1759 demonstrated that Halley's Comet, and by extension all comets, are members of the solar system, orbiting the sun. The return of 1835 marked the beginning of research on the physical nature of comets. At the time of the return of 1910, cometary research was progressing steadily. When the head of the comet transited the sun, the lack of a silhouette put a firm upper limit on the size of the nucleus. Progress between the previous two returns was remarkable. Theories published in the 1950s – Whipple's dirty snowball model of the nucleus and Oort's theory of the origin of comets – began to revolutionize the field. Scientific studies at the comet Halley return of 1986 took a spectacular turn when multiple spacecraft flew through the comet and collected a wealth of *in situ* data, including images of the nucleus, furthering the revolution. For the first time, we had a firm direct measurement of the size of a nucleus, and strong evidence that at least one nucleus is monolithic. But, Halley is only one comet. In 2001, the *Deep Space 1* spacecraft flew by comet Borrelly and collected a second image of a nucleus. Since the last return of comet Halley, we have added much to our information base on the composition of comet nuclei. However, this information is all indirect. We have observed material after it has sublimated from the nucleus and been exposed to solar ultraviolet radiation, and changed by chemical reactions.

So here is where we stand today. We know a lot about comets. As this volume clearly shows, we can predict with increasing precision the physical phenomena observed in the gas, plasma and dust of comets and their interactions with the solar-wind plasma and fields. But, our knowledge of nuclei is still inadequate. We cannot safely generalize the nature of all nuclei based on direct images of only two nuclear surfaces, and on the basis of studies of processed nuclear material in the inner comas of comets. We know what we have to do: collect substantially more image data; collect and return samples of pristine nuclear material; and study sub-surface nuclear material. There are missions on the way to comets, today, that will do most of these things for a few comets. The missing mission is one to provide a sample return of pristine cometary material. Such a mission, admittedly technically demanding and expensive, should provide a phenomenal increase in our understanding of comets. The process must continue beyond those missions. This is not the collection of data for the data's sake. Sound scientific theories are based firmly on statistically sound collection of observational data. For another view of future research, see Ball *et al.* (1999).

In our 1981 technical treatise we summed up the discussion of planned missions at that time by saying, "It is tremendously exciting to contemplate the fascinating discoveries awaiting completion of cometary missions of almost any description." We have described the discoveries made by missions between then and now. We think you will agree that they are tremendously exciting. We have no doubt that the next two decades will push even further the frontiers of comet science. Missions alone are not enough! Scientists must carry on with remote studies of comets to continue progress on our understanding of other cometary phenomena. Young scientists interested in comets should not be concerned about their futures, the best is yet to come.

12

Comet lore

A very bright comet is a beautiful sight to behold. Because a comet is brightest when it is simultaneously near perihelion and near the earth, naked-eye comets are most often seen in the western sky just after sunset or in the eastern sky just before sunrise. Then, a comet can be the most obvious object in the sky. Naturally enough, an impressive comet will excite considerable public interest. Both of the authors have spent considerable time speaking to general audiences and answering questions from them, as well as from the press. People are genuinely interested in the nature of comets – probably because those bright enough to catch the public interest are infrequent.

Recent comets have brought home an unfortunate truth: Most people will not be able to see even bright comets from their home sites in the future. Comet Kohoutek (C/1973 E1) was discovered many months before it passed perihelion. Preliminary calculations indicated that it would be very bright during the Christmas season of 1973–4, and considerable excitement was built up beforehand. After the fact, the press dubbed comet Kohoutek a dud; most people never saw it. Yet to astronomers on their mountaintop observatories, it was a conspicuous naked-eye object for almost a month. Part of the difference was due to bad weather conditions in the populous northeastern United States during the time Kohoutek was visible. But a big problem was (and is) pollution. Filthy air and bright city lights conspire to overwhelm the subtle light of a comet. Let us hope that this situation will change in the future.

Although much interest in comets comes from a public seeking knowledge about a fascinating phenomenon, not all interest is so nobly motivated. Superstition, fear, and charlatanism rear their ugly heads, too. Unhappily, the idea that comets are omens is widespread. As we saw in Chapter 10, a real danger from comets does exist, though it is very small. We have already discussed some of the early conceptions of comets in Chapter 1. Here we will look more closely at some of the superstitions.

12.1 Comets and superstition

Will Durant (1954:80) points out that "astrology antedated – and perhaps will survive – astronomy; simple souls are more interested in telling futures than in telling time." The belief that comets are bad omens is an integral part of Aristotle's philosophy. However, it is not difficult to believe that the concept is far more ancient. Memorable events in human existence occur constantly. A bright comet might have appeared to Stone Age people at the time of a great drought when food was scarce, or at the time of a particularly disastrous hunt when tribesmen were injured or killed. The next bright comet would strike fear in their hearts. And the fear would be borne out by another setback in the people's hard existence. By extension, it is not difficult to imagine why similar superstitions continue in regions where civilization has barely intruded.

That superstitions regarding comets are still touted in the modern civilized world is more difficult to understand. It is clear that some writers are exploiting the belief for their own purposes. It is difficult to imagine a physical mechanism whereby a passing comet could cause specific events such as the death of a king, and scientists do not take the idea seriously.

The Roman Catholic Church was one of the bastions of cometary superstition during the Middle Ages, and the idea persisted in teachings through the Reformation period. A scriptural basis for the belief was the book of Joel, Chapter 2, verses 30 and 31: "And I will shew wonders in the heavens and in the earth, blood, and fire, and pillars of smoke. The sun shall be turned into darkness, and the moon into blood, before the great and terrible day of the Lord come." Bright comets were warnings to keep us on our toes. St. Thomas Aquinas' great *Summa Theologica*, written in the thirteenth century, is full of references to the influences of heavenly bodies on human events. With scripture and the church's greatest thinkers behind the concept, it is no surprise that the idea has persisted.

Modern superstition is not a high point of humanity's intellectual development, and it should not be dignified by extensive discussion. The spirit of the historical attitude is captured in Figs. 12.1, 12.2, and 12.3, and by the following quotation from Sir Stanislaus Lubeinietski in his *Theatrum Cometrium* published in 1668: "Known to God and man ... never had there been a disaster without a comet or a comet without a disaster ... had there not been a comet overhead which had been responsible for the epidemic of sneezing sickness among the cats of the Rhenish areas of Westphalia?" (Haney 1965: 177). If human misfortune truly were connected with the appearance of comets, the sky should be bright with them!

It is relatively easy to forgive the colorful folklore of previous centuries, and we can be amused with Mark Twain's statement that he came (born 1835) and would go (died 1910) with Halley's Comet. However, it is difficult to consider

AN
ALLARM
TO
EUROPE:

By a Late Prodigious
COMET
feen November and December, 1680.

With a Predictive Difcourfe. Together with fome preceding and fome fucceeding Caufes of its fad Effects to the *Eaft* and *Noth Eaftern* parts of the World.

Namely, *ENGLAND*, *SCOTLAND*, *IRELAND*, *FRANCE*, *SPAIN*, *HOLLAND*, *GERMANY*, *ITALY*, and many other places.

By *John Hill* Phyfitian and Aftrologer.

The Form of the *COMET* with its Blaze or Stream as it was feen *December* the 24*th*. Anno 1680. In the Evening.

London Printed by *H. Brugis* for *William Thackry* at the Angel in Duck-Lane.
8 *Jan.* 168⁵⁄₁

Fig. 12.1. Title page of John Hill's *An Alarm to Europe*. (By permission of the Houghton Library, Harvard University.)

Fig. 12.2. French cartoon reflecting fears that a comet would collide with earth in 1857.

seriously the following statement published in this decade: "Every comet in history has been associated with war, catastrophe, and extraordinary events…Like it or not, Kohoutek will be, too." (Goodavage 1973:1). To be fair, the author of the same book warns the reader not to "be fooled by weighty scientific pronouncements."

In 1811, a series of powerful earthquakes shook the eastern half of North America. The quakes were probably the largest quakes in North America in recorded history. The epicenter of the earthquakes was near a small Mississippi river town – New Madrid, Missouri. Shocks were felt as far away as New York and Florida. It was reported that church bells sounded in Richmond, Virginia. The Mississippi River was observed to flow backward at one point. The great naturalist John James Audubon was in Kentucky when the first tremor struck. Initially, he thought the roar was a tornado and headed for shelter. Audubon and others reported strange darkenings and brightenings in the sky.

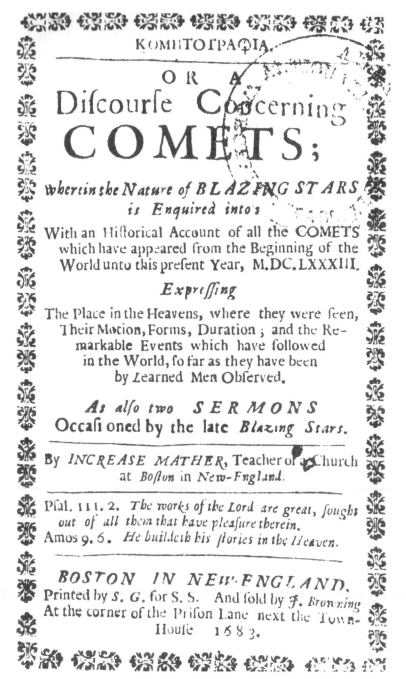

KOMHTOГPAФIA.

OR A

Difcourfe Concerning

COMETS;

wherein the Nature of BLAZING STARS
is Enquired into:

With an Hiftorical Account of all the COMETS
which have appeared from the Beginning of the
World unto this prefent Year, M.DC.LXXXIII.

Expreffing

The Place in the Heavens, where they were feen,
Their Motion, Forms, Duration ; and the Remarkable Events which have followed
in the World, fo far as they have been
by Learned Men Obferved.

As alfo two SERMONS
Occafioned by the late *Blazing Stars.*

By *INCREASE MATHER*, Teacher of a Church
at *Bofton* in *New-England.*

Pfal. 111. 2. *The works of the Lord are great, fought
out of all them that have pleafure therein.*
Amos 9. 6. *He buildeth his ftories in the Heaven.*

BOSTON IN NEW-ENGLAND.
Printed by *S. G.* for *S. S.* And fold by *J. Browning*
At the corner of the Prifon Lane next the Town-
Houfe 1683.

Fig. 12.3. Title page of Increase Mather's *Discourse Concerning Comets.*

The great New Madrid Earthquakes occurred before the age of seismographs. Charles Richter, the inventor of the Richter scale, examined the great mass of anecdotal evidence from the event, and estimated that at least three of the most severe shocks exceeded magnitude eight on his scale. The New Madrid Earthquakes thus stand as one of the most severe, if not the most severe, series of quakes recorded in US history. The Good Friday Earthquake of 1964 in Alaska is certainly the largest quake for which we have quantitative scientific observations. (There is much information on the New Madrid Earthquakes on the web; a good place to start is www.ceri.memphis.edu.)

The New Madrid Earthquake was just one of the events of late 1811 and 1812 that occurred after the Great Comet of 1811. In December, 1811, a fire broke out in a packed Richmond, Virginia theater. The governor of the state, George Smith, and scores of others died in the flames. Elsewhere, events were leading to the War of 1812. In contrast, 1811 was a particularly good year for wine. In honor of the Great Comet, the wine was dubbed *vin de la comète*. Thus, the Great Comet of 1811 seems to have been the cause of both unfortunate and good events. In fact, on average, there is a naked-eye comet almost every year, so comets and events on earth come and go independently of one another.

12.2 Comets and literature

Most references to comets in literature are based on the folkloric view of them as omens or producers of disasters. Examples are William Shakespeare in *Caesar:*

> When beggars die, there are no comets seen,
> The Heavens themselves blaze for the death of Princes.

John Milton in *Paradise Lost:*

> Satan stood
> Unterrified, and like a Comet burn'd
> That fires the length of Ophiuchus huge
> In th' Artick sky, and from its horried hair
> Shakes pestilence and war.

Shakespeare in *Henry IV* wrote these words, which sound much better to our ears:

> By being, seldom seen, I could not stir,
> Buts like a comet, I was wondered at.

Some departures from the folklore mold were made in the prose of two famous science fiction writers. *Off on a Comet* was published by Jules Verne in the 1860s. It is a good adventure story (given a few liberties), and it has an interesting scientific curiosity. The comet of the story has a satellite, and this fact allows the astronomers to calculate the mass of the comet. Recall that no real comet has had its mass directly determined as of this writing.

H. G. Wells has an interesting twist on the folkloric approach. In *The Days of the Comet* he describes a comet with a sunward tail and an unprecedented band in the green. The fictional comet approaches the earth, but the influences on earth and its inhabitants are all favorable.

A fairly recent entry in the literature of comets is the book *Lucifer's Hammer* by Larry Niven and Jerry Pournelle (1977). It is a fictional account of a collision between the earth and a very large comet. It is particularly noteworthy here because the authors have taken care to be as scientifically accurate as they can, consistent with their tale.

If you will forgive us for mentioning popular motion pictures in a section on literature, we will do so. The movie *Deep Impact* was released by Paramount Pictures in 1998. It is about fictional comet Wolf–Biederman that was discovered on a collision path to the earth over a year before the impact. The long lead time allowed astronauts to be sent on a mission to destroy the comet. Herein is the main message we should take away from the movie; don't blow up comets on a collision course to the earth. After the first attempt to destroy the comet, there were two smaller bodies, each on collision course with our planet. In the end, the larger of the two bodies was shattered by a bomb, but all its pieces still hit the atmosphere. This would be more or less all right if the pieces were so small that they burned up in the atmosphere. Even so, they could add some dust to the atmosphere, and decrease the amount of sunlight reaching the ground for some period of time. So, what is the best solution? We think it is hinted at by the non-gravitational forces. These forces are quite small, but they act for the entire time a comet is near perihelion. The impulse – i.e., force times time acting – can be quite significant. For instance, these forces can change the time comet Halley reaches perihelion by as much as 4 days, without modifying its orbit. Four days is quite enough for a comfortable miss. So, faced with a collision, we should send a robotic mission to the comet or asteroid, as far in advance as possible, with an appropriate engine to supply an impulse to change the orbital speed of the body for a long enough time to ensure that it will miss the earth. Remember the asteroid 1997 XF11? It missed the earth by several hundred thousand kilometers. But, it actually crossed the earth's orbit when the earth wasn't there. We had no clue it was coming until it had passed us by. This is a good argument for monitoring the sky for near-earth objects! Back to the movie: the special effects of impact of the smaller body were not bad, in our opinion. If you are interested in a more detailed discussion of the good and bad science in the movie, you might want to consult Phil Plait's Bad-Astronomy web page (www.badastronomy.com). That website also discusses the movie *Armageddon* (Touchstone Pictures, 1998), with a plot based on an asteroid heading for the earth.

Comets have often been a favorite topic for humorous cartoons. Two of our favorites appeared in *Punch*. One was associated with the 1910 apparition of Halley's Comet (see Fig. 12.4a). A masterpiece of the cartoonist's art is shown in Fig. 12.4b.

Burglar (with sudden enthusiasm for astronomy). "'SCUSE ME, GUV'NER, CAN YOU TELL ME WHERE I CAN GET A VIEW OF THIS 'ERE COMET?"

(a)

Fig. 12.4. Comet related cartoons from issues of *Punch*; (a) June 1, 1910; (b) December 5, 1906.

OUR UNTRUSTWORTHY ARTIST IN LONDON.
DISCOVERY OF A COMET AT GREENWICH OBSERVATORY.

(b)

Fig. 12.4. (*Cont.*)

The Greenwich Observatory buildings are faithfully represented, but almost all else is delightful fantasy.

12.3 Comets in art

Comet Halley has been observed at almost every return starting in 240 B.C. The only return with no reported observations is the one in 164 B.C. Then, starting in 88 B.C., it has been seen at every return. This phenomenal string of returns covers a time period of over 2000 years. It is easy to imagine that the comet was visible during some very significant events over that time period. For instance, comet Halley was visible throughout the spring of 1066. In January of that year, Harold was crowned King of England, after the death of Edward. However, Edward had promised the throne to William in return for William's support. This broken promise, and a number of other intrigues led to the famous battle of Hastings in October, 1066; Harold was killed by the Norman invaders, and William became the King of England.

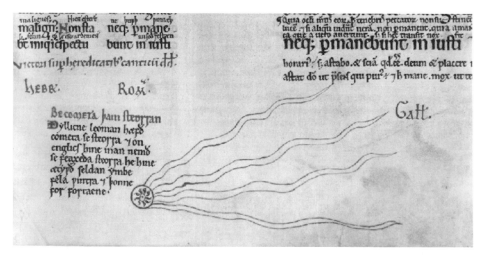

Fig. 12.5. Halley's Comet in its 1145 apparition was probably the model for the comet shown on the *Eadwine Psalter*. (Courtesy of the Master and Fellows of Trinity College, Cambridge University.)

The Bayeux Tapestry, completed in the eleventh century, was reportedly embroidered by William's queen, Matilda, and her ladies in waiting. It consists of sixty panels showing events beginning with the battle preparations and ending with the death of Harold. A comet is shown in the sky in one panel (Plate 1.1). Given the fact that comet Halley had been visible early in the year, the depicted comet is most certainly comet Halley. Depending on your point of view, you can regard the apparition as a bad omen for Harold – as the tapestry does – or as a good omen for William the Conqueror.

Halley's Comet reached perihelion in October 1301. A fresco painted by Giotto di Bondone (1267–1337) around 1303 shows a typical nativity scene, but the star of Bethlehem looks remarkably comet-like. Was Giotto so impressed with Halley's Comet that he used it to represent the Christmas star? Some art historians think so. Certainly the rendition of the comet agrees well with several written descriptions. It does look surprisingly realistic.

Art historian Roberta J. M. Olson (Olson 1979; Olson and Pasachoff 1987) has studied this and other portraits of Halley's Comet produced over the years. Her work led to an exhibit of comet art at the Smithsonian's National Air and Space Museum in 1985–86. The catalog for that exhibit, *Fire and Ice* (Olson 1985), contains 118 illustrations of comets, including many of comet Halley. The earliest that she has found represents the A.D. 684 apparition, and it appears in the *Nuremberg Chronicles* published in 1493 (Fig. 1.1). The charming woodcut is supported by a text that recounts all the disasters brought on by the comet. Comet Halley, at its 1145 return, was drawn by the monk Eadwine (Fig. 12.5), and appeared in the *Eadwine Psalter*. Olson and Pasachoff (1987) point out that the drawing is similar

Fig. 12.6. Albrecht Dürer's *Melencolia I* of 1514. (Courtesy of the Metropolitan Museum of Art, Harris Brisbane Dick Fund, 1941 [43.106.1].)

to the appearance of the comet in April 1986. It is interesting to note the appearances of comets in many of the old drawings and woodcuts. In several cases, the comets are represented as if they were swords.

Albrecht Dürer, in a beautiful engraving entitled *Melencolia I*, dated 1514, presented (Fig. 12.6) a "luminous and dynamic comet" that was "much less naturalistic than Giotto's." Dürer's picture seems to express the unhappy side of cometary lore. The return of comet Halley in 1759, as we have discussed in Chapter 1, was the first predicted return of any comet, and established these objects as members of the solar system, orbiting the sun. Plate 12.1 is a painting by Samuel Scott, showing comet Halley as seen over the Thames. Westminster Cathedral can be seen in the background, and the royal barge in the foreground.

Many representations of comets dwell on the disasters they are thought to foretell. We should remember Giotto's fresco. There, Halley's Comet is associated with an event that many people see as one of great joy and promise.

12.4 Recent comets

12.4.1 Comet Halley

Long before comet Halley was recovered, interest began to grow in the minds of both the public and the scientific community. The public was prepared for the worse-case scenario by Don Yeomans, who pointed out that the comet would be near superior conjunction with the sun when it was at perihelion; it would be nearly impossible to observe when it was near its absolute brightest. When it was visible from the earth, it was not spectacularly bright. In fact, it was difficult to see from our largest cities, because of the light pollution. Our personal experiences are typical. RDC was temporarily living in Houston, Texas, at the time. He had to drive out of the city to see the comet, but it was visible to the unaided eye. JCB did not get a good view from the continental USA. But, on a trip to Puerto Rico the comet was easily seen although the experience was not spectacular.

The scientific community was intensely interested in studying comet Halley. They began to formulate plans for a mission to the comet as early as the 1970s, because of the long lead time needed for such cutting-edge technologies as Solar Electric Propulsion (SEP). SEP is a low-thrust engine that is capable of operating for long periods of time. A one-pound thrust engine operating for one year provides 3×10^7 pound-seconds of impulse, and a corresponding change in momentum, not greatly different from the first-stage performance of a Delta rocket lifting a payload toward low earth orbit. The SEP could place a spacecraft in a parallel orbit with comet Halley, and cause it to fly in formation with the comet, for extended *in situ* observations. NASA built a prototype SEP engine, and successfully tested it in the late 1970s. Unfortunately, in 1979, the U.S. Office of Management and Budget (OMB) removed the appropriation for the construction of the SEP from the NASA budget request. This was a time of economic woes in the USA, with rampant inflation, and NASA had more than it could afford on its plate. National advisory committees gave the Gamma-Ray Observatory higher priority than SEP.

Meet Halley's Comet

GEORGE F. WILL

In 1910, the last time Halley's comet came by, an Oklahoma sheriff had to stop some citizens from sacrificing a virgin to the comet. The comet is coming again in 1986, so Oklahomans should lock up their daughters. And David Stockman should stop sacrificing science on the altar of parsimony.

I shall use my dying breath to whisper praise of Stockman, but he should not have killed NASA's, plan to send a satellite to intercept the comet. It would have cost $300 million over five years, 25 cents per person a year, and it should have been an occasion for the Administration to leaven its frugality with a farsighted exception.

Comets, and especially Halley's, have excited superstition far from Oklahoma. The historian Josephus, said a comet resembling a sword (Halley's. in A.D. 66) foretold the destruction of Jerusalem (A.D. 70). The visit of Halley's comet in 1066 was thought to have been a portent of the unpleasantness that befell King Harold at Hastings. Shakespeare said: "When beggars die, there are no comets seen; the heavens themselves blaze forth the death of princes." The day Edward VII died (May 6,1910) Halley's comet was especially vivid (more vivid than that particular prince merited). Mark Twain, born during the comet's 1935 visit, said he would be disappointed if he didn't depart when it came again. He died April 21, 1910, just before the comet's "tail" brushed earth and as (Twain would have loved this) people were selling anti-comet pills to a public panicky about gases in the tail.

In "The Comet is Coming! The Feverish Legacy of Mr. Halley," Nigel Calder says Halley's comet is, as most comets probably are, "sky pollution." a "dirty snowball that comet stumbling out of the freezer of twilight space. (There are an estimated 100 billion comets in our itsy-bitsy solar system.)

Flu Machines: These cosmic jaywalkers rarely bump into anything because space is even more vacant than Wyoming. (If there were just three bees in America, the air would be more congested with bees than space is with stars.) But there is a constant rain onto earth of meteoritic debris, and an occasional "thump." Calder writes: "Early in the morning of 30, January 1908 the driver of the trans-Siberian express board loud bangs and imagined that his train had exploded... his wide-eyed passengers said they had seen a bright

blue ball of fire ..." A small comet had leveled a 70 -mile-long strip of forest.

But some collisions may have been constructive. One theory is that a comet brought to earth the first bacteria or whatever it was that started the ol' ball of life rolling 4 billion years ago (fortunately, before governments demanded environmental impact statements). Another theory is that comets are "flu machines," bringing viruses to earth. Ask now what caused the fall or Rome and the rise of Christianity. Calder, says some theorists argue: "During the period from A.D. 400 to 1400 the earthlings bad a particularly nasty time with the clouds of diseases spun off from comets. A

> *Conservatives cannot turn space exploration over to their two loves, federalism or capitalism.*

'disease-filled' millennium...forced people to live farther apart and thus to 'uncivilise' themselves; it also...moved the Europeans to adopt the 'sombre' religion of Christianity."

But if that is true, Calder asks, why not now? "If a millionth part of the meteoric debris falling to earth from comets consists of viruses, a small garden could collect millions of viruses every day, ready to assail plants, pets and humans."

Calder finds a bit more plausible the theory that a comet killed the dinosaurs; they did die out suddenly, and folks used to think they were just too big for Noah's ark. Today some scientists think a big comet, perhaps 6 miles in diameter, struck earth, throwing up a hundred times its weight it dust--much more dust than was sent up by the eruption in 1883 of the Krakatau volcano, which produced "glorious sunsets" around the world for two years. The theory is that the cloud produced by the comet collision blocked out sunlight, and in the four-year "night" much vegetation and most dinosaurs died.

Why, then, is there no crater? Well, there is a suspicious ring-shaped something the seabed north of Australia. (Oceans and continents have

him meandering around during the last 65 million years.) And in geological formations around the world are thin layers of clay with a chemical composition that suggests that long ago the earth was suddenly swamped with a particular element (iridium) in an amount that seems unlikely to have come from a source on earth. If this theory is true, then if, 65 million years ago, the comet bad come by an how earlier or later, it would have missed and dinosaurs might still rule the earth. So a comet may have been a benefactor.

Mysteries: Anyway, comets are owed the respect due the elderly. Most comets in out solar system spend most of their time loitering (relatively speaking) beyond the outer planets, So they are among the "oldest," meaning least changed, objects: they experience less of the erosion and evolution that erases the imprint of the birth of the solar system. A rendezvous with one might reveal evidence about the origins of the universe, the human race and Oklahoma.

If our curiosity about such things atrophies, so will our humanity. That is why the Halley's comet intercept program, which can still be saved, should be used by the Administration as an opportunity to practice "creative exceptionalism." The country wants conservatism, but needs conservatism subtle enough to make exceptions to the principle of parsimony. Conservatives cannot turn space exploration over to their two loves, federalism or capitalism--to the states or the private sector. Neither Utah nor Exxon, can do it. Only Big Government--only our government--can do it

Conservatives are supposed to take the long view, and to take intangibles seriously. They should went to look back toward the creation of the universe that has produced as its crowning glory, the Reagan Administration. And they should look far into the future and imagine a future in which mankind is not curious about the wondrous mysteries of its situation.

We know next to nothing about virtually everything. It is not necessary to know the origins of the universe; it is necessary to want to know. Civilization depends not on any particular knowledge, but on the disposition to crave knowledge.

Fig. 12.7. George Will editorial from the August 3, 1981 issue of *Newsweek*. (Courtesy of George Will. Reprinted by permission.)

A rendezvous mission with comet Halley was effectively scrapped. The scientific community began to plan for a flyby mission. But, once again, the national boards gave another mission higher priority than a Halley mission. Venus Orbiting Imaging Radar – which became known as Magellan – won the priority battle. The significance of the loss of a major U.S. role in space studies of comet Halley was not lost on non-scientists. Figure 12.7 is a supportive editorial by George F. Will that appeared in *Newsweek* in 1981.

The European Space Agency (ESA), the Japanese Institute of Space and Astronomical Sciences (ISAS), and the former Soviet Union all had approved flyby missions. The USA was left "standing at the altar" until Robert Farquhar came up with a brilliant idea. The *Third International Sun–Earth Explorer* (*ISEE 3*) was sitting at the Lagrangian point between the earth and the sun, monitoring the solar wind and other inputs to the earth's environment. *ISEE 3* had a powerful engine that had placed it at the Lagrangian point and plenty of fuel, if needed, to maintain its position. Farquhar showed that he could fire the engine, and direct the spacecraft on a complex series of maneuvers involving five looping passes around the earth and moon. The final pass, only 120 km above the lunar surface, would propel the spacecraft to comet Giacobini–Zinner. It would get there before the Halley flybys, and could do some excellent observations and provide a test of some of the observing techniques of the Halley Armada. NASA approved Farquhar's idea, and *ISEE 3* became the *International Comet Explorer* (*ICE*). Incidentally, Farquhar was the individual who oversaw the flight of the *NEAR* spacecraft to orbit the asteroid Eros.

Comet Halley was recovered on October 16, 1982, by two scientists using a cutting-edge detector system and the 200-inch Hale telescope at Palomar Observatory. This was not a chance discovery; David C. Jewitt and G. Edward Danielson were looking exactly where the orbit calculations told them it should be (Fig. 12.8). It is interesting to compare the four historical recoveries of the comet. In 1759 it was recovered on Christmas night, less than three months before perihelion passage; in 1835 it was recovered about four months before perihelion; in 1910 it was discovered nine months before perihelion; and, in 1982 it was discovered a full three years before perihelion (see Fig. 1.17). This sequence of recoveries shows how technology had progressed in nearly two and a quarter centuries.

In early 1986, NASA decided to test a spectrograph designed to observe comet Halley while it was near the sun. It was to be taken into orbit on the Space Shuttle *Challenger* where the crew would assist with the observations. As we know, *Challenger* blew up in a disastrous explosion. This not only destroyed the spectrograph, but grounded the Shuttle fleet. The planned *Astro I* mission could not fly to study the comet, and the launch of the *Hubble Space Telescope* was delayed until long after the comet passed perihelion.

The 1986 perihelion passage of comet Halley occurred at a low point for the U.S. space program. Nonetheless, the scientific return of comet Halley was enormous. The U.S. scientific community fought long and hard for a mission to comet Halley, but they were thwarted by bad economic times, committees that gave higher priority to other missions, and a really disastrous event in the space program. In the end, the *ICE* mission was extended, and the spacecraft did fly through the tail of comet Giacobini–Zinner. Hall (1990) gives an extensive history of this time period for anyone interested in more than our thumbnail sketch. The history of the GZ encounter is detailed in Yeomans (1991).

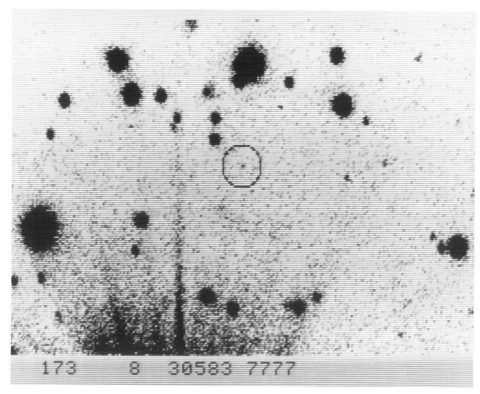

173 8 30583 7777

Fig. 12.8. The recovery image of Halley's Comet obtained by G. E. Danielson and D. C. Jewitt on October 16, 1982 with an advanced electronic camera and the 200-inch Hale telescope at Palomar Observatory. (California Institute of Technology.)

Comet Halley caused considerable public interest as it passed through the inner solar system. It garnered more interest than the other two comets we will discuss, even though it was never as bright at the last return as they were. A number of books – both technical and non-technical – were inspired by the comet. One science fiction novel that attracted interest was *The Further Adventures of Halley's Comet*, a first novel by John Calvin Batchelor. The story concerns a family who comes up with the idea of launching a private probe to Halley's Comet in order to lay claim to it, and by extension to all solar system objects. This is not too far-fetched an idea. After the U.S. Halley missions were cancelled, a group calling themselves The Halley Fund attempted to raise funds for a private mission to the comet. Figure 12.9 is a reproduction of an advertisement for The Halley Fund.

In the 1980s, Halley the man or Halley the comet, or both, were popular topics for postage stamps and first day covers. Plate 12.2 shows stamps issued by Britain in honor of the comet and in commemoration of Halley's work on the Atlantic Ocean island of St. Helena. As late as 2001, a one man show entitled "Halley's Comet" was touring the USA. The author and performer, John Amos, presents a humorous version

Fig. 12.9. Public support for a mission to comet Halley was supported by an advertisement by The Halley Fund.

of the story of an 87-year-old man who saw the comet when he was an 11-year-old boy. The performance describes the effect the incident had on his later life.

12.4.2 Comet Hyakutake

Comet Hyakutake was discovered the morning of January 30, 1996, by the Japanese amateur comet hunter Yuji Hyakutake. He was using a pair of giant Fujinon

25 × 150 binoculars, which is his comet-seeking tool of choice. Hyakutake wrote that he left home for the site where he usually searched for comets about 3:30 A.M., about the time the moon was due to set. He had discovered comet Hyakutake (C/1995 Y1) about a month earlier. At this time, he found it was still visible and about 9th magnitude. Then he began scanning the pre-dawn sky. Soon, he found an object that looked like a comet, not far from the place where he discovered the earlier comet. This object was compact and roughly 11th magnitude. He sketched the position, then had to quit observing, because of morning twilight. Unfortunately, he was not able to confirm the object's motion. He sent a report of his observation to the New Astronomical Findings Information Department at the National Observatory.

The next night was not auspicious. The sky was overcast, and it was raining lightly. Just as he decided that he would have to give up the search, a fax arrived confirming his discovery. He had discovered his second comet Hyakutake (C/1996 B2) in as many months. The preliminary orbit from the Central Bureau for Astronomical Telegrams showed that the comet had a 17 000-year period. So, it had been in the inner solar system earlier.

From mid-March to mid-April comet Hyakutake zipped past the earth with a minimum distance of only 0.1 AU. We both remember being impressed by how quickly it moved across the sky. *Sky & Telescope* magazine called it the finest comet in 20 years and featured spectacular wide-field images on its cover for July 1996.

On March 27, 1996, the comet was first magnitude, and it moved the full length of Ursa Minor in a 24-hour period. Visual observers described interesting telescopic features, including a brilliant tail spike.

12.4.3 Comet Hale–Bopp

As comet Hyakutake faded from view, comet Hale–Bopp (C/1995 O1) began to brighten. The comet was discovered by Alan Hale and Thomas Bopp in July 1995. Alan Hale is a comet scientist who searches for comets as a hobby; Thomas Bopp is an amateur comet hunter.

While comet Hyakutake became spectacular a matter of three months after discovery, comet Hale–Bopp did not reach maximum brightness until the summer and fall of 1996, more than a year after discovery. It actually reached naked-eye brightness twice, once before conjunction with the sun, and once after conjunction, in the late winter of 1996–7.

Comet Hale–Bopp was associated with one of the more bizarre incidents in history. A New-Age California-based group that was known as Heaven's Gate shared beliefs based on a mixture of their interpretation of Christianity and space-age technology. They decided that a UFO was trailing behind Comet Hale–Bopp, and if they could shed their "containers" they would be transported to the UFO and be carried to a higher level of existence. (See, for instance, the *Washington Post*, March 28, 1997.) So the group committed mass suicide by "ingesting a homemade recipe of drugs, applesauce and vodka." In all, 21 women and 18 men died. Their bodies were found by deputies from the San Diego County Sheriff's department.

12.5 Summary

Comets are fascinating objects. We have presented a survey of the history of cometary science in this volume. It should be pretty clear that we have only scratched the surface of the cometary literature. There is much more known now, and much more to learn. In this chapter, we have tried to show how comets affect people's views of the world and themselves. Comets have encouraged the creation of great art and great literature, as well as unfortunate acts of self destruction.

The study of comets has inspired us, and we are sure it can inspire you, too. We fervently hope that this volume will encourage some of its readers to invest their life's work in cometary research.

13

Summary

The science of comets is quite diverse. It draws results from many other fields to achieve physical understanding of the phenomena we observe in comets. We have discussed a complex series of topics in the preceding 12 chapters. The topics have included a long list of observations, a host of different objects in the solar system, and a variety of theoretical subjects. In writing this final chapter, we have attempted to tie all these topics together with a short overview based on a few summary illustrations.

The first illustration (Fig. 13.1) shows a summary plot of effective diameters versus visual geometric albedos as presented by Fernández for cometary nuclei, Centaurs, and some near-earth asteroids. The cometary nuclei are generally found in the low albedo part of the plot and cover a large range in sizes, particularly if the Centaurs are considered as comets. Note that some near-earth asteroids fall into the "cometary" part of the plot.

The dark icy-dust cometary nuclei, with a wide range of sizes, produce the phenomena associated with comets as they pass through the inner solar system. Figure 13.2 shows the appearance of comet Halley for approximately $8^1/_2$ months during 1985–1986 as documented in *The International Halley Watch Atlas of Large-Scale Phenomena*. The comet developed starting in late 1985, was spectacular in March and April 1986, and waned in the time period of the later images. Recall that the comet passed perihelion on February 9, 1986, and was near superior conjunction with the sun. The images show the dust tail, synchronic bands, the plasma tail, disconnection events, and the neck-line structure. Figure 13.3 is a summary diagram of comet features.

The phenomena are produced by many processes that begin with the sublimation of ices by solar visible and infrared radiation as the nucleus approaches the sun. Figure 13.4 is an attempt to summarize these processes in one schematic diagram.

395

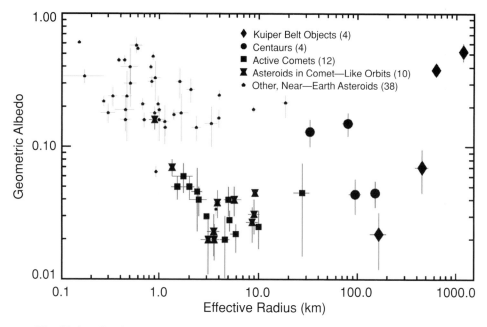

Fig. 13.1. Effective diameter versus visual geometric albedo for comets, Centaurs, and some NEAs. (Courtesy Y. Fernández, University of Hawaii.)

Where did these icy nuclei form and what has been their orbital evolution? Figure 13.5 is a schematic summary of these processes. Note the complexity of this diagram and that Figure 13.4 only applies to the area on Fig. 13.5 marked "Many orbits with sublimation."

We hope that these summary diagrams help unify our discussion of comets. However, the results from cometary studies over the next few years, including the missions, could dictate major changes. And, that would be great.

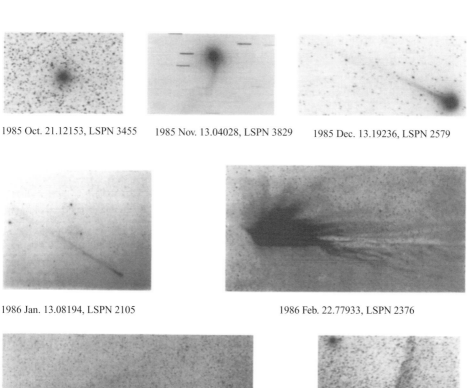

1985 Oct. 21.12153, LSPN 3455   1985 Nov. 13.04028, LSPN 3829   1985 Dec. 13.19236, LSPN 2579

1986 Jan. 13.08194, LSPN 2105   1986 Feb. 22.77933, LSPN 2376

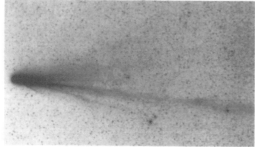

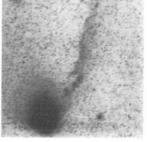

1986 Mar. 16.48611, LSPN 895   1986 Apr. 12.10417, LSPN 1194

1986 May 2.00660, LSPN 2923   1986 June 1.06250, LSPN 759   1986 July 6.39137, LSPN 2453

Fig. 13.2. Monthly sequence of comet Halley in 1985–1986. Top row, left to right: October 21, 1985. Coma only. November 13, 1985. Plasma activity begins. December 13, 1985. Plasma tail develops. Second row, left to right: January 13, 1986. Prominent plasma tail. February 22, 1986. Shows an antitail (toward left), synchronic bands in the dust tail (above), and a disconnection event (lower right). Third row, left to right: March 16, 1986. Shows a dust tail (above) and a plasma tail. April 12, 1986. Shows a spectacular disconnection event. Bottom row, left to right: May 2, 1986. Last vestige of the plasma tail (above) and the neck-line structure (below). June 1, 1986. The neck-line structure. July 6, 1986. The neck-line structure. (After Figure 858 of *The International Halley Watch Atlas of Large-Scale Phenomena* 1992.)

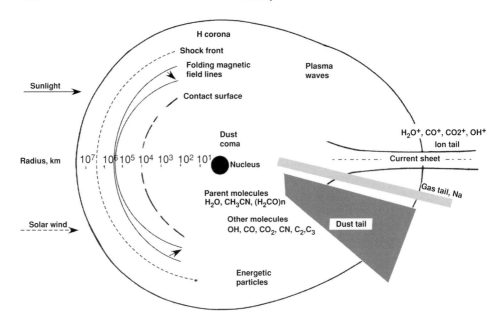

Fig. 13.3. Summary of comet features and phenomena.

Grand Summary of Physical Processes in Comets

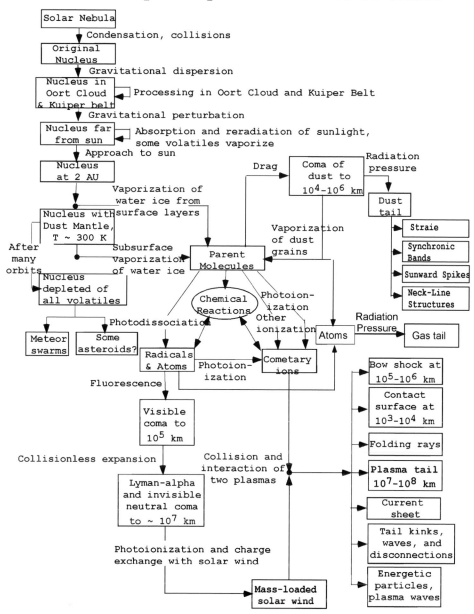

Fig. 13.4. Grand summary of physical processes in comets.

Grand Summary of Orbital Evolution and Terminal States

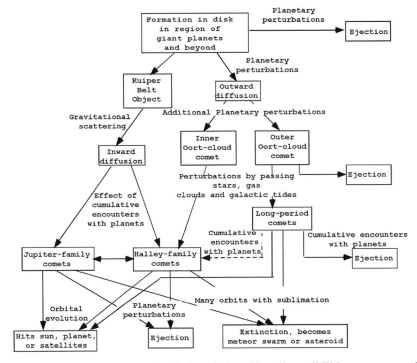

Fig. 13.5. Grand summary of orbital evolution. Not all possibilities are captured by this diagram. Note that ejection from the solar system is a likely fate for comets.

References

Chapter 1

Alfvén, H. 1957. On the theory of comet tails. *Tellus 9*: 96–6.

Aristotle. 1952. Meteorology. *Great Books of the Western World 8*: 445. Chicago: Encyclopaedia Britannica.

Armitage, A. 1966. *Edmond Halley*. London: Thomas Nelson.

Biermann, L. 1951. Kometenschweife und Korpuskularstrahlung. *Z. Astrophys. 29*: 274–86.

    1968. *On the Emission of Atomic Hydrogen in Comets*. J1LA Report No. 93. Boulder, CO: Joint Institute for Laboratory Astrophysics.

Brandt, J. C., Niedner, M. B., Jr., and Rahe, J. 1992. *The International Halley Watch Atlas of Large-Scale Phenomena*. Boulder, CO: LASP-University of Colorado.

    1997. *The International Halley Watch Atlas of Large-Scale Phenomena:Supplement 1997*. Boulder, CO: LASP-University of Colorado.

Delsemme, A. H., and Miller, D. C. 1970. Physico-chemical phenomena in comets – II. Gas adsorption in the snows of the nucleus. *Planet. Space Sci. 18*: 717–30.

Delsemme, A. H., and Miller, D. C. 1971a. Physico-chemical phenomena in comets – III. The continuum of comet Burnham (1960 II). *Planet. Space Sci. 19*: 1229–57.

Delsemme, A. H., and Miller, D. C. 1971b. Physico-chemical phenomena in comets – IV. The C_2 Emission of comet Burnham (1960 II). *Planet. Space Sci. 19*: 1259–74.

Delsemme, A. H., and Swings, P. 1952. Hydrates de gaz dans les noyaux comètaires et les grains interstellaires. *Ann. Astrophys. 15*: 1–6.

Delsemme, A. H., and Wenger, A. 1970. Physico-chemical phenomena in comets – I. Experimental studies of snows in a cometary environment. *Planet. Space Sci. 18*: 709–15.

Donn, B., and Urey, H. C. 1956. On the mechanism of comet outbursts and the chemical composition of comets. *Astrophys. J. 123*: 339.

Drake, S., and O'Malley, C. D. (transl.). 1960. *The Controversy on the Comet of 1618*, Philadelphia: University of Pennsylvania Press. (Writings by Galileo, Kepler, Griassi, and Guiducci.)

Dreyer, J. L. E. 1953, *A History of Astronomy from Thales to Kepler*. New York: Dover.

Everhart, E. 1972. Origin of short-period comets. *Astrophys. Lett. 10*: 131–5.

Fernandez, J. A. 1980. On the existence of a comet belt beyond Neptune. *Mon. Not. R. Astron. Soc. 192*: 481–91.

FitzGerald, G. F. 1892. *The Electrician 30*: 481.

Galileo, 1623. The assayer. In *Discoveries and Opinions of Galileo*, transl. S. Drake, pp. 217–80. Garden City, NY: Doubleday, 1957.

Harwit, M., and Hoyle, F. 1962. Plasma dynamics in comets. II. Influence of magnetic fields. *Astrophys J. 135*: 875–82.

Hellman, C. D. 1944. *The Comet of 1577: Its Place in the History of Astronomy*. New York: AMS Press. (A very good reference for the history of cometary thought through the sixteenth century.)

Herschel, J. F. W. 1871. *Outlines of Astronomy*. London: Longmans, Green.

Hertzberg, G., and Lew, H. 1974. Tentative identification of the H_2O^+ ion in comet Kohoutek. *Astron. Astrophys. 31*: 1232–4.

Hoffmeister, C. 1943. Physikalische Untersuchungen an Kometen. I. Die Beziehungen des primdren Schweifstrahls zum Radiusvector. *Z. Astrophys. 22*: 265–85.

Hoffmeister, C. 1944. Physikalische Untersuchungen an Kometen. II. Die Bewegung der Schweifmaterie und die Repulsivkraft der Sonne beirn Kometen 1942g. *Z. Astrophys. 23*: 118.

Hoyle, F., and Harwit, M. 1962. Plasma dynamics in comets. I. Plasma instability. *Astrophys. J. 135*: 867–74.

Kouchi, A., Yamamoto, T., Kuroda, T., and Greenberg, J. M. 1994. Conditions for condensation and preservation of amorphous ice and crystalline astrophysical ices. *Astron. Astrophys. 290*: 1009–18.

Lewis, John S. 1997. *Physics and Chemistry of the Solar System*. San Diego: Academic Press.

Lyttleton, R. A. 1953. *The Comets and their Origin*. Cambridge: Cambridge University Press.

Mekler, Y., and Podolak, M. 1994. Formation of amorphous ice in the protoplanetary nebula. *Planet. Space Sci. 42*: 865–70.

Newton, I. 1686. *Principia*. Transl. by A. Mott. Berkeley: University of California Press, 1962.

Olivier, C. P. 1930. *Comets*. Baltimore: Williams & Wilkins.

Oort, J. H. 1950. The structure of the cloud of comets surrounding the solar system and a hypothesis concerning its origin. *Bull. Astron. Inst. Neth. 11*: 91–110.

Oort, J. H., and Schmidt, M. 1951. Differences between new and old comets. *Bull. Astron. Inst. Neth.* 11: 259–70.

Pannekoek, A. 1961. *A History of Astronomy*. New York: Interscience.

Proctor, M., and Crommelin, A. C. D. 1937. *Comets: Their Nature, Origin and Place in the Science of Astronomy*. London: Technical Press.

Roemer, E. 1963. Comets: discovery, orbits, astrometric observations. In *The Moon, Meteorites, and Comets*, ed. B. M. Middlehurst and G. P. Kuiper. Chicago: University of Chicago Press.

Russell, H. N. Dugan, R. S., and Stuart, J. Q. 1945. *Astronomy. I. The Solar System*. Bosten: Ginn.

Schwarzschild, K., and Kron, E. 1911. On the distribution of brightness in the tail of Halley's comet. *Astrophys. J. 34*: 342–52.

van Woerkom, A. J. J. 1948. On the origin of comets. *Bull. Astron. Inst. Neth. 10*: 445–72.

Vsekhsvyatskij, S. K. 1977. Comets and the cosmogony of the solar system. In *Comets, Asteroids, Meteorites: Interrelations, Evolution and Origins*, ed. A. H. Delsemme, pp. 469–74. Toledo: University of Toledo Press.

Whipple, F. L. 1950a. A comet model. I. The acceleration of comet Encke. *Astrophys. J. 111*: 375–94.

1950b. A comet model. II. Physical relations for comets and meteors. *Astrophys. J. 113*: 464–74.

1955. A comet model. III. The zodiacal light. *Astrophys. J. 121*: 750–70.

Whipple, F. L. 1985. *The Mystery of Comets.* Washington, DC: Smithsonian Institution Press.

Zanstra, H. 1929. The excitation of line- and band-spectra in comets by sunlight. *Mon. Not. R. Astron. Soc. 89*: 178–97.

Additional reading

Brandt, J. C. (ed.) 1981. *Comets: Readings from Scientific American.* San Francisco: W. H. Freeman and Co.

Brown, P. L. 1974. *Comets, Meteorites and Man.* New York: Taplinger.

Chambers, G. F. 1909. *The Story of Comets.* Oxford: Clarendon Press.

Delsemme, A. H. 1998. Recollections of a cometary scientist. *Planet. Space. Sci. 46*: 111–24.

Lodge, O. 1900. *The Electrician 46*: 249.

Opik, E. 1932. Note on stellar perturbations of nearly parabolic orbits . *Proc. Am. Acad. Arts and Sciences 67*: 169–82.

Chapter 2

Boulet, D. L. 1991. *Methods of Orbit Determination with the Micro Computer.* Richmond: Willmann-Bell, Inc.

Brouwer, D., and Clemence, G. M. 1961. *Methods of Celestial Mechanics.* New York: Academic Press.

Brouwer, D., and van Woerkom, A. J. J. 1950. The secular variations in the orbital elements of the principal planets. *Astron. Papers U.S. Naval Obs. 13*: 85–105.

Danby, J. M. A. 1988. *Fundamentals of Celestial Mechanics.* Richmond: Willmann-Bell, Inc.

Dubyago, A. D. 1961. *The Determination of Orbits.* New York: Macmillan.

Escobal, P. R. 1965. *Methods of Orbit Determination.* New York: John Wiley.

Froeschlé, C., and Greenberg, R. 1989. Mean motion resonances. In *Asteroids II*, ed. Binzel, R. P., Gehrels, T., and Matthews, M. S., pp. 827–44. Tucson: University of Arizona Press.

Froeschlé, C., and Scholl, H. 1986. The secular resonance ν_6 in the asteroidal belt. *Astron. Astrophys. 166*: 326–332.

Goldreich, P., and Tremaine, S. 1982. The dynamics of planetary rings. *Ann. Rev. Astron. Astrophys. 20*: 249–83.

Goldreich, P., Rappaport, N., and Sicardy, B. 1995. Single sided shepherding. *Icarus 118*: 414–7.

Greenberg, R., and Scholl, H. 1979. Resonances in the asteroid belt. In *Asteroids*, ed. T. Gehrels, pp. 310–33. Tucson: University of Arizona Press.

Herget, P. 1948. *The Computation of Orbits.* Cincinnati: University of Cincinnati Press.

Jackson, A. A., and Zook, H. A. 1989 Solar system dust ring with earth as its shepherd. *Nature 337*: 629–31.

Kresák, L. 1992. Are there any comets coming from interstellar space? *Astron. Astrophys. 259*: 682–91.

Marsden, B. G. 1968. Comets and Nongravitational Forces. I. *Astron. J. 73*: 367–79.

1969. Comets and Nongravitational Forces. II. *Astron. J. 74*: 720–34.

1973. Comets in 1972. *Quart. J. R. Astron. Soc. 14*: 389–406.

1985. Initial orbit determination: The pragmatist's point of view. *Astron. J. 90*: 1541–7.

1995. Cfa-www.harvard.edu/iau/lists/CometResolution.html

1999. *Catalogue of Cometary Orbits*, 13th Edn. Cambridge, MA: IAU. Central Bureau for Astronomical Telegrams.

2001. *Catalogue of Cometary Orbits*, 14th Edn. Cambridge, MA: IAU. Central Bureau for Astronomical Telegrams.

Marsden, B. G., and Sekanina, Z. 1971. Comets and nongravitational forces. IV. *Astron. J. 76*: 1135–51.

Marsden, B. G., Sekanina, Z., and Yeomans, D. K. 1973. Comets and nongravitational forces. V. *Astron. J. 78*: 211–25.

Moulton, F. R. 1914. *An Introduction to Celestial Mechanics*. New York: The Macmillan Company.

Peale, S. J. 1976. Orbital resonances in the solar system. *Ann. Rev. Astron. Astrophys. 14*: 215–46.

Rickman, H. 1986. *Uppsala reprints in Astronomy*, No. 8.

Scholl, H., Froeschlé, C. Kinoshita, H., and Yoshikawa, M. 1989. Secular resonances. In *Asteroids II*, ed. Binzel, R. P., Gehrels, T., and Matthews, M. S., pp. 845–61. Tucson: University of Arizona Press.

Sekanina, Z. 1988. Outgassing asymmetry of periodic Comet Encke II. Apparitions 1868–1918 and a study of the nucleus evolution. *Astrophys. J. 96*: 1455–75.

Sekanina, Z. 2002. Statistical investigation and modeling of sungrazing comets discovered with the *Solar and Heliospheric Observatory*. *Astrophys. J. 566*: 577–98.

Sitarski, G. 1994. On perihelion asymmetry for investigations of nongravitational motion of comets. *Acta Astron. 44*: 91–8.

Smart, W. M., and Green, R. M. 1977. *Textbook on Spherical Astronomy*, 6th Edn. Cambridge: Cambridge University Press.

Szebehely, V. G., and Mark, H. 1998. *Adventures in Celestial Mechanics*. New York: John Wiley and Sons, Inc.

Whipple, F. L. 1950. A comet model. I. The acceleration of Comet Encke. *Astrophys. J. 111*: 375–94.

Williams, J. G. 1969. Secular perturbations in the solar system. Ph.D. Thesis, University of California at Los Angeles.

Wisdom, J. 1982. The origin of the Kirkwood Gaps: A mapping for asteroidal motion near the 3/1 commensurability. *Astron. J. 87*: 577–93.

1983. Chaotic behavior and the origin of the 3/1 Kirkwood gap. *Icarus 56*: 51–74.

1985. A perturbative treatment of motion near the 3/1 commensurability. *Icarus 63*: 272–89.

Yabushita, S. 1996. On the effect of non-gravitational processes on the dynamics of nearly parabolic comets. *Mon. Not. R. Astron. Soc. 283*: 347–52.

Yao, K., Yeomans, D., and Weissman, P. 1994. The past and future motion of Comet P/Swift–Tuttle. *Mon. Not. R. Astron. Soc. 266*: 305–16.

Yeomans, D. K. 1971. Nongravitational forces affecting the motions of periodic comets Giacobini–Zinner and Borrelly. *Astron. J. 76*: 83–6.

1984. The orbits of comets Halley and Giacobini–Zinner. In *Cometary Astrometry*, pp. 167–85. Pasadena: JPL Report.

Yeomans, D. K., and Chodas, P. W. 1989. An asymmetric outgassing model for cometary nongravitational accelerations. *Astron. J. 98*: 1083–93.

Yu, Qingjuan and Tremaine, S. 1999. The dynamics of plutinos. *Astron. J. 118*: 1873–81.

Additional reading

Chebotarev, G. A., Kazimirchaka-Polonskaya, E. L., and Marsden, B. G. (eds.). 1972. *The Motion, Evolution of Orbits and Origins of Comets.* IAU Symposium No. 45. Dordrecht: D. Reidel.

Marsden, B. G. 1970. Comets and Nongravitational Forces. III. *Astron. J. 75*: 75–84.

  1972. Comets in 1971. *Quart. J. R. Astron. Soc. 13*: 415–34.

  1974a. Comets. *Ann. Rev. Astron. Astrophys. 12*: 1–21. (Contains studies of orbital statistics.)

  1974b. Comets in 1973. *Quart. J. R. Astron. Soc. 15*: 433–60.

Moons, M., Morbidelli, A., and Migliorini, F. 1998. Dynamical structure of the 2/1 commensurability with Jupiter and the origin of the resonant asteroids. *Icarus 135*: 458–68.

Morais, M. H. M. 1999. A secular theory of Trojan-type motion. *Astron. Astrophys. 350*: 318–26.

  2001. Hamiltonian formulation of the secular theory for Trojan-type asteroids. *Astron. Astrophys. 369*: 677–89.

Yeomans, D. K. 1974. The nongravitational motion of comet Kopff. *Pub. Astron. Soc. Pacific 86*: 125–7.

Chapter 3

A'Hearn, M. F., Schleicher, D. G., Feldman, P. D., Millis, R. L., and Thompson, D. T. 1984. Comet Bowell 1980b. *Astron. J. 89*: 579–91.

Altenhoff, W. J., Huchtmeier, W. K., Schmidt, J., Schraml, J. B., Stumpff, P., and Thum, C. 1986. Radio continuum observations of comet Halley. *Astron. Astrophys. 164*: 227–30.

Altenhoff, W. J., Bieging, J. H., Butler, B., Butner, H. M., Chini, R., Haslam, C. G. T., Kreysa, E., Martin, R. N., Mauersberger, R., McMullin, J. Muders, D. Peters, W. L., Schmidt, J., Schraml, J. B., Sievers, A., Stumpff, P., Thum, C., von Kap-Herr, A., Wiesmeyer, H., Wink, J. E., and Zylka, R. 1999. Coordinated radio continuum observations of comets Hyakutake and Hale-Bopp from 22 to 660 GHz. *Astron. Astrophys. 348*: 1020–34.

Balsiger, H., Altwegg, K., Bühler, F., Fischer, J., Geiss, J., Meier, A., Rettenmund, U., Rosenbauer, H., Schwenn, R., Benson, J., Hemmerich, P., Säger, K., Kulzer, G., Neugebauer, M., Goldstein, B. E., Goldstein, R., Shelley, E. G., Sanders, T., Simpson, D., Lazarus, A. J., and Young, D. T. 1986. The Giotto Ion Mass Spectrometer. In *The Giotto Mission – Its Scientific Investigations*, ed. R. Reinhard, and B. Battrick, pp. 129–48. ESA SP-1077.

Biver, N., Bockelée-Morvan, D., Crovisier, J., Henry, F., Davies, J. K., Matthews, H. F., Colom, P., Gérard, E., Lis, D. C., Phillips, T. G., Rantakyrö, F., Haikala, L., and Weaver, H. A. 2000. Spectroscopic observations of comet C/1999 H1 (Lee) with the SEST, JCMT, CSO, IRAM and Nançay radio telescopes. *Astrophys. J. 120*: 1554–70.

Bowyer, S., and Malina, R. F. 1991. The extreme ultraviolet explorer mission. In *Extreme Ultraviolet Astronomy*, ed. R. F. Malina and S. Bowyer. Elmsford, NY: Pergamon.

Brandt, J. C., Niedner, M. B., Jr., and Rahe, J. 1992. *The International Halley Watch Atlas of Large-Scale Phenomena.* Boulder: University of Colorado.

Brandt, J. C., Niedner, M. B., Jr., and Rahe, J. 1997. *The International Halley Watch Atlas of Large-Scale Phenomena – Supplement 1997*. Boulder, CO: LASP-University of Colorado.

Brownlee, D. E., Rajan, R. S., and Tomandl, D. A. 1977. A chemical and textural comparison between carbonaceous chondrites and interplanetary dust. In *Comets, Asteroids, Meteorites*, ed. A. H. Delsemme, pp. 137–42. Toledo: University of Toledo Press.

Combs, M., *et al.* 1988. The 2.5–12 μm spectrum of comet Halley from the IKS-Vega experiment. *Icarus 76*: 404–36.

Crovisier, J., and Schloerb, F. P. 1991. The study of comets at radio wavelengths. In *Comets in the Post-Halley Era*. ed. R. L. Newburn, Jr., M. Neugebauer, and J. Rahe. Dordrecht: Kluwer Academic Publishers.

Delsemme, A. H., and Miller, D. C. 1971. Physico-chemical phenomena in comets – III. The continuum of comet Burnham (1960 II). *Planet. Space Sci. 19*: 1229–57.

Edberg, S. J. and Levy, D. H. 1994. *Observing Comets, Asteroids, Meteors and the Zodiacal Light*. Cambridge: Cambridge University Press.

Edenhofer, P., Bird, M. K., Buschert, H., Esposito, P. B., Porsche, H., and Volland, H. 1986. The Giotto radio-science experiment. In *The Giotto Mission – Its Scientific Investigations*, ed. R. Reinhard, and B. Battrick, p. 173. ESA SP-1077.

Hirao, K. and Itoh, T. 1987. The Sakigake/Suisei encounter with comet P/Halley. *Astron. Astrophys. 187*: 39–46.

Hobbs, R. W., Maran, S. P., Brandt, J. C., Webster, W. J., and Krishna-Swamy, K. S. 1975. Microwave continuum radiation from Comet Kohoutek 1973f. Emission from the icy grain halo. *Astrophys. J. 201*: 749–55.

Jewitt, D. 1991. Cometary photometry. In *Comets in the Post-Halley Era*, ed. R. L. Newburn, Jr., M. Neugebauer, and J. Rahe, pp. 19–66. Dordrecht: Kluwer Academic Publishers.

Johnstone, A. D., Bowles, J. A., Coates, A. J., Coker, A. J., Kellock, S. J., Raymont, J., Wilken, B., Studemann, W., Weiss, W., Cerulli Irelli, R., Formisano, V., de Giorgi, E., Perani, P., de Bernardi, M., Borg, H., Olsen, S., Winningham, J. D., and Bryant, D. A. 1986. The Giotto three-dimensional positive ion analyser. In *The Giotto Mission – Its Scientific Investigations*, ed. R. Reinhard, and B. Battrick, pp. 15–32. ESA SP-1077.

Kissel, J., Sagdeev, R. Z., Bertaux, J. L., Angarov, V. N., Audouze, J., Blamont, J. E., Buchler, K., Evlanov, E. N., Fechtig, H., Fomenkova, M. N., von Hoerner, H., Inogamov, N. A., Khromov, V. N., Knabe, W., Krueger, F. R., Langevin, Y., Leonasv, B., Levasseur-Regourd, A. C., Managadze, G. G., Podkolzin, S. N., Shapiro, V. D., Tabaldyev, S. R., and Zubkov, B. V. 1986. Composition of comet Halley dust particles from VEGA observations. *Nature 321*: 280–2.

Korth, A., Marconi, M. L., Mendis, D. A., Krueger, F. R., Richter, A. K., Lin, R. P., Mitchell, D. L., Anderson, K. A., Carlson, C. W., Rème, H., Sauvaud, J. A., and d'Uston, C. 1989. Probable detection of organic-dust-borne aromatic $C_3H_3^+$ ions in the coma of comet Halley. *Nature 337*: 53–5.

Krankowsky, D., Lämmerzahl, P., Dörflinger, D., Herrwerth, I., Stubbemann, U., Woweries, J., Eberhardt, P., Dolder, U., Fischer, J., Herrmann, U., Hofstetter, H., Jungck, M., Meier, F. O., Schulte, W., Berthelier, J. J., Illiano, J. M., Godefroy, M., Gogly, G., Thévenet, P., Hoffman, J. H., Hodges, R. R., and Wright, W. W. 1986. The Giotto neutral mass spectrometer. In *The Giotto Mission – Its Scientific Investigations*, ed. R. Reinhard, and B. Battrick, pp. 109–280. ESA SP-1077.

Lamy, F. L. and Toth, I. 1995. Direct detection of a cometary nucleus with the Hubble Space Telescope. *Astron. Astrophys. 293*: L43–5.

Lamy, P. L., Toth, I., and Weaver, H. A. 1998. Hubble Space Telescope observations of the nucleus and inner coma of comet 19P/1904 Y2 (Borrelly). *Astron. Astrophys. 337*: 945–54.

Levasseur-Regourd, A. C., Bertaux, J. L., Dumont, R., Festou, M., Giese, R. H., Giovane, F., Lamy, P., Llebaria, A., and Weinberg, J. L. 1986. The Giotto optical probe experiment. In *The Giotto Mission – Its Scientific Investigations*, ed. R. Reinhard, and B., Battrick, pp. 187–206. ESA SP-1077.

Lisse, C. M., and 11 colleagues, 1996. Discovery of x-ray and extreme ultraviolet emission from comet Hyakutake C/1996 B2. *Science 274*: 205–9.

Maas, R. W., Ney, E. P., and Woolf, N. J. 1970. The 10-micron Emission Peak of Comet Bennett 1969i. *Astrophys. J. 160*: 101–4.

Mason, J. W. (ed.) 1990. *Comet Halley. Vol. 1, Investigations, Results and Interpretations.* Chichester: Ellis Horwood Limited.

McDonnell, J. A. M., Alexander, W. M., Burton, W. M., Bussoletti, E., Clark, D, H., Evans, G. C., Evans, S. T., Firth, J. G., Grard, R. J. L., Grün, E., Hanner, M. S., Hughes, D. W., Igenbergs, E., Kuczera, H., Lindblad, B. A., Mandeville, J.-C., Minafra, A., Reading, D., Ridgeley, A., Schwehm, G. H., Stevenson, T. J., Sekanina, Z., Turner, R. F., Wallis, M. K., and Zarnecki, J. C. 1986. The Giotto dust impact detection system. In *The Giotto Mission – Its Scientific Investigations*, ed. R. Reinhard, and B. Battrick, pp. 85–108. ESA SP-1077.

McKenna-Lawlor, S., Thompson, A., O'Sullivan, D., Kirsh, E., Melrose, D., and Wenzel, K.-P. 1986. The Giotto Energetic Particle Experiment, In *The Giotto Mission – Its Scientific Investigations*. ed. R. Reinhard, and B. Battrick, pp. 53–66. ESA SP-1077.

Meisel, D. D., and Morris, C. S. 1976. Comet brightness parameters: Definition, determination and correlations. In *The Study of Comets*, ed. B. Donn, M. Mumma, W. Jackson, M. A'Hearn, and R. Harrington, pp. 413–32. NASA SP-393. Washington, DC: GPO.

  1982. Comet head photometry: past, present and future. In *Comets*, ed. L. L. Wilkening, Tucson: The University of Arizona Press.

Mendis, D. A. 1988. Postencounter view of comets. *Ann. Rev. Astron. Astrophys. 26*: 11–49.

Moroz, V. I. *et al.* 1987. Detection of parent molecules in comet P/Halley from the IKS-VEGA experiment. *Astron. Astrophys. 187*: 513–18.

Mumma, M. J., Krasnopolsky, V. A., and Abbott, M. J. 1997. Soft x-rays from four comets observed by EUVE. *Astrophys. J. 491*: L125–8.

Neubauer, F. M., Acuna, M. H., Burlaga, L. F., Franke, B., Gramkow, B., Mariani, F., Musmann, G., Ness, N. F., Schmudt, H. U., Terenzi, T., Ungstrup, E., and Wallis, M. 1986. The Giotto Magnetic-Field Investigation. In *The Giotto Mission – Its Scientific Investigations*, ed. R. Reinhard, and B. Battrick, pp. 1–14. ESA SP-1077.

Ney, E. P. 1974. Multiband photometry of comets Kohoutek, Bennett, Bradfield and Encke. *Icarus 23*: 551–60.

Ó Ceallaigh, D. P., Fitzsimmons, A., and Williams, I. P. 1995. CCD photometry of comet 109P/Swift–Tuttle. *Astron. Astrophys. 297*: L17–20.

Opal, C. B., and Carruthers, G. R. 1978. Lyman-α observations of comet West. *Icarus 31*: 503–9.

de Pater, I., Palmer, P., and Snyder, L. E. 1991. A review of radio interferometric imaging of comets. In *Comets in the Post-Halley Era*, ed. R. L. Newburn, Jr., M. Neugebauer, and J. Rahe, pp. 175–208. Dordrecht: Kluwer Academic Publishers.

Reinhard, R. 1990. Space missions to Halley's comet and international cooperation. In *Comet Halley. Investigations, Results and Interpretations. Vol. 1, Organization,*

Plasma, Gas, ed. J. W. Mason, pp. 1–32. Chichester: Ellis Horwood Limited.

Rème, H., Cotin, F., Cros, A., Médale, J. L., Sauvaud, J. A., d'Uston, C., Korth, A., Richter, A. K., Loidl, A., Anderson, K. A., Carlson, C. W., Curtis, D. W., Lin, R. P., and Mendis, D. A. 1986. The Giotto RPA-Copernic Plasma Experiment. In *The Giotto Mission – Its Scientific Investigations*, ed. R. Reinhard, and B. Battrick, pp. 33–52. ESA SP-1077.

Richter, K., Combi, M. R., Keller, H. U., and Meier, R. R. 2000. Multiple scattering of hydrogen Lyα: Radiation in the coma of comet Hyakutake (C/1996 B2). *Astrophys. J., 531*: 599–611.

Russell, H. N. 1916. On the albedo of the planets and their satellites. *Astrophys. J. 43*: 173–96.

Schmidt, W. K. H., Keller, H. U., Wilhelm, K., Arpigny, C., Barbieri, C., Biermann, L., Bonnet, R. M., Cazes, S., Cosmovici, C. B., Delamere, W. A., Heubner, W. F., Hughes, D. W., Jamar, C., Malaise, D., Reitsema, H., Seige, P., and Whipple, F. L. 1986. The Giotto Multicolor Camera. In *The Giotto Mission – Its Scientific Investigations*, ed. R. Reinhard, and B. Battrick, ESA SP-1077.

Usher, P. D. 1990. Photometric theory for wide-angle phenomena. *Icarus 86*: 93–9.

Weaver, H. A., Mumma, M. J., Larson, H. P., and Davis, D. S. 1986a. Airborne infrared investigations of water in the coma of Halley's Comet. *Proceedings of the 20th ESLAB Symposium on the exploration of Halley's Comet. Vol. 1 Plasma and Gas*, pp. 329–34.

1986b. Post perihelion observations of water in comet Halley. *Nature 324*: 441–4.

Weaver, H. A., Feldman, P. D., A'Hearn, M. F., Arpigny, C., Brandt, J. C., and Stern, S. A. 1999. Post-perihelion HST observations of comet Hale–Bopp (C/1995 O1). *Icarus, 141*: 1–12.

Additional reading

Arpigny, C. 1965. Spectra of comets and their interpretation. *Ann. Rev. Astron. Astrophys. 3*: 351–76.

1977. On the nature of comets. *Proceedings of the Robert A. Welch Foundation Conferences on Chemical Research, XXI. Cosmochemistry.* Houston.

Brockelée-Morvan, D., Lis, D. C., Wink, J. E., Despois, D., Crovieier, J., Bachiller, R., Benford, D. J., Biver, N., Colom, P., Davies, J. K., Gérard, E., Germain, B., Houde, M., Mehringer, D., Moreno, R., Paubert, G., Phillips, T. G., and Rauer, H. 2000. New molecules found in comet C/1995 O1 (Hale–Bopp). Investigating the link between cometary and interstellar material. *Astron. Astrophys. 353*: 1101–14.

Bobrovnikoff, N. T. 1941. Investigations of the brightness of comets. *Popular Astronomy 49*: 467–79.

Brownlee, D. E., Tomandl, D. A., and Hodge, P. W. 1976. Extraterrestrial particles in the stratosphere. In *Interplanetary Dust and Zodiacal Light*, ed. H. Elsasser and H. Fechtig, pp. 279–83. New York: Springer-Verlag.

Bruston, P., Coron, N., Dambier, G., Laurent, C., Leblanc, J., Lena, P., Rather, J. D. G., and Vidal-Madjar, A. 1974. Observations of comet Kohoutek at 1.4 mm. *Nature 252*: 665–6.

Dennerl, K., Englhauser, J., and Trümper, J. 1997. X-ray emission from comets detected in the Röntgen X-ray satellite all-sky survey. *Science 277*: 1625–9.

Greenstein, J. L. 1958. High-resolution spectra of comet Mrkos (1957d). *Astrophys. J. 128*: 106–13. (This paper originates the Greenstein effect.)

Hanner, M. S., and Tokunaga, A. T. 1991. Infrared techniques for comet observations. In *Comets in the Post-Halley Era*, ed. R. L. Newburn, Jr., M. Neugebauer, and J. Rahe, pp. 67–92. Dordrecht: Kluwer Academic Publishers.

Krasnopolsky, V. A., Mumma, M. J., and Abbott, M. J. 2000. EUVE Search for X-rays from comets Encke, Mueller (C/1993 A1), Borrelly, and postperihelion Hale–Bopp. *Icarus 146*: 152–60.

Lamy, P. L., Toth, I., A'Hearn, M. F., and Weaver, H. A. 1999. Hubble Space Telescope observations of the nucleus of comet 45P/Honda–Mrkos–Pajdusakova and its inner coma. *Icarus 140*: 424–38.

Lisse, C. M., Christian, D., Dennerl, K., Englhauser, J. Trümper, J., Desch, M., Maeshall, F. E., Petre, R., and Snowder, S. 1999. X-Ray and Extreme Ultraviolet Emission from Comet P/Encke 1997. *Icarus 141*: 316–30.

Marsden, B. G. 1974. Comets. *Ann. Rev. Astron. Astrophys. 12*: 1–21.

Mason, J. W. (ed.) 1990. *Comet Halley. Vol. 2, Dust, Nucleus, Evolution*. Chichester: Ellis Horwood Limited.

Rahe, J. 1980. Ultraviolet spectroscopy of comets. In *Second European IUE Conference*, ed. B. Battrick and J. Mort, p. 12. ESA SP-157.

Rayman, M. D., Varghese, P., Lehman, D. H., and Livesay, L. L. 2000. Results from the Deep Space 1 technology validation mission. *Acta Astronautica 47*: 475.

Weaver, H. 1998. Comets. In *The Scientific Impact of the Goddard High Resolution Spectrograph*, ed. J. C. Brandt, T. B. Ake, III, and C. C. Petersen, p. 213. Astronomical Society of the Pacific Conference Series Volume 143.

Weaver, H. A., Mumma, M. J., and Larson, H. P. 1991. Infrared spectroscopy of cometary parent molecules. In *Comets in the Post-Halley Era*, ed. R. L. Newburn, Jr., M. Neugebauer, and J. Rahe, pp. 93–106. Dordrecht: Kluwer Academic Publishers.

Report of the Comet Science Working Group, Executive Summary. August 1979. NASA TM 80542.

Websites

Websites with information on CCDs:
http://zebu.uoregon.edu/ccd.html
http://www.not.iac.es/CCD-world/
http://www.ing.iac.es/~smt/CCD_Primer/CCD_Primer.htm

Other web sites of interest:
http://www.jpl.nasa.gov/
http://www.jpl.nasa.gov/galileo

Chapter 4

Alfvén, H. 1957. On the theory of comet tails, *Tellus 9*: 92–6.

Barnard. E. E. 1920. On Comet 1919b and on the rejection of a comet's tail. *Astrophys. J. 51*: 102–6.

Belton, M. J. S. 1965. Some characteristics of Type II comet tails and the problem of distant comets. *Astron. J. 70*: 451–65.

Biermann, L. 1951. Kometenschweife und solare Korpuskularstrahlung. *Zeit. für Astrophysik 29*: 274–86.

Brandt, J. C., Hawley, J. D., and Niedner, Jr., M. B. 1980. A very rapid turning of the plasma-tail axis of comet Bradfield 1979l on 1980 February 6. *Astrophys. J. Lett. 241*: L151–4.

Cremonese, G. 1999. Hale–Bopp and its sodium tails. *Space Sci. Rev. 90*: 83–9.

Cremonese, G. and Fulle, M. 1989. Photometric analysis of the neck-line structure of comet Halley. *Icarus 80*: 267–79.

Cremonese, G., Boehnhardt, H., Crovisier, J., Fitzsimmons, A., Fulle, M., Licandro, J., Pollacco, D., Rauer, H., Tozzi, G. P., and West, R. M. 1997. Neutral sodium from comet Hale–Bopp: a third type of tail. *Astrophys. J. Lett. 490*: L199–L202.

Finson. M. L., and Probstein, R. F. 1968a. A theory of dust comets. I. Model and equations. *Astrophys. J. 154*: 327–52.

1968b. A theory of dust comets. II. Results for comet Arend–Roland. *Astrophys. J. 154*: 353–80.

Fulle, M. 1987. A possible neck-line structure in the dust tail of comet Halley. *Astron. Astrophys. 181*: L13–14.

1989. Evaluation of cometary dust parameters from numerical simulations: comparison with an analytical approach and the role of anisotropic emissions. *Astron. Astrophys. 217*: 283–97.

Gloeckler, G., Geiss, J., Schwandron, N. A., Fisk, L. A., Zurbuchen, T. H., Ipavich, F. M., von Steiger, R., Balsiger, H., and Wilken, B. 2000. Interception of comet Hyakutake's ion tail at a distance of 500 million kilometers. *Nature 404*: 576–8.

Hoffmeister, C. 1943. Physicalische Untersuchungen an Kometen. I. Bie Beziehungen des primären Schweifstrahls zum Radiusvector. *Z. Astrophys. 22*: 265–85.

Jambor, B. J. 1973. The split tail of comet Seki-Lines. *Astrophys. J. 185*: 727–34.

Jones, G. H., Balogh, A., and Horbury, T. S. 2000. Identification of comet Hyakutake's extremely long ion tail from magnetic signatures. *Nature 404*: 574–6.

Koutchmy, S., Coupiac, P., Elmore, D., Lamy, P., and Sèvre, F. 1979. Comet West 1975n. I. Observations near and after perihelion passage. *Astron. Astrophys. 72*: 45–9.

Lamy, P. L., and Koutchmy, S. 1979. Comet West 1975a. Part II: Study of the striated tail. *Astron. Astrophys. 72*: 50–4.

Larsson-Leander, G. 1957. The anomalous tail of comet Arend–Roland. *Observatory 77*: 132–5.

McDonnell, J. A. M., Alexander, W. M., Burton, W. M., Bussoletti, E., Evans, G. C., Evans, S. T., Firth, J. G., Grard, R. J. L., Green, S. F., Grün, E., Hanner, M. S., Hughes, D. W., Igenbergs, E., Kissel, J., Kuczera, H., Lindblad, B. A., Langevin, Y., Mandeville, J.-C., Nappo, S., Pankiewicz, G. S. A., Perry, C. H., Schwehm, G. H., Sekanina, Z., Stevenson, T. J., Turner, R. F., Weishaupt, U., Wallis, M. K., and Zarnecki, J. C. 1987. The dust distribution within the inner coma of Comet P/Halley 1982i: Encounter by Giotto's impact detectors. *Astron. Astrophys. 187*: 719–41.

Mendillo, M., Wilson, J. K., Baumgardner, J., Cremonese, G., and Barbieri, C. 1998. Imaging studies of sodium tails in comets. *Bull. Am. Astron. Soc. 30*: 1062.

Moore, E. P. 1991. Cometary ray closing rates: comet Kobayashi–Berger–Milon. *Astron. Astrophys. 247*: 247–51.

Osterbrock, D. E. 1958. A study of two comet tails. *Astrophys. J. 128*: 95–105.

Pittichová, J., Sekanina, Z., Birkle, K., Boehnhardt, H., Engels, D., and Keller, P. 1997. An early investigation of the striated tail of comet Hale–Bopp (C/1995 01). *Earth, Moon and Planets 78*: 329–38.

Ryan, O., Birkle, K., Sekanina, Z., Boehnhardt, H., Engels, D., Keller, P., Jäger, M., and Raab, H. 2003. The evolution of striae in the dust tail of comet Hale–Bopp. I. Wide-field imaging, computer processing, and astrometry. Preprint, Jet Propulsion Laboratory, California Institute of Technology: JPL D-17890.

Sekanina, Z. 1973. Existence of icy comet tails at large distances from the Sun. *Astrophys. Lett. 14*: 175–80.

1975. A study of the icy tails of the distant comets. *Icarus 25*: 218–38.

1976. Progress in our understanding of cometary dust tails. In *The Study of Comets*, ed. B. Donn, M. Mumma, W. Jackson, M. A'Hearn, and R. Harrington, pp. 893–939. Washington, DC: NASA SP-393.

Sekanina, Z., and Farrell, J. A. 1980. The striated dust tail of comet West 1976 VI as a particle fragmentation phenomena. *Astron. J. 85*: 1538–54.

Sekanina, Z., and Miller, F. D. 1973. Comet Bennett 1970II. *Science 179*: 565–7.

Sekanina, Z., and Pittichová, J. 1997. Distribution law for particle fragmentation times in a theory for striated tails of dust comets: Application to comet Hale–Bopp (C/1995 01). *Earth, Moon Planet. 78*: 339–46.

Sekanina, Z., Larson, S. M., Emerson, G., Helin, E. F., and Schmidt, R. E. 1987. The sunward spike of Halley's comet. *Astron. Astrophys. 187*: 645–9.

Spinrad, H., Brown, M. E., and Johns, C. M. 1994. Kinematics of the ion tail of comet P/Swift–Tuttle. *Astron. J. 108*: 1462–70.

Sykes, M. V., Lebofsky, L. A., Hunten, D. M., and Low, F. J. 1986. The discovery of dust trails in the orbits of periodic comets. *Science 232*: 1115–17.

Wilson, J. K., Baumgardner, J., and Mendillo, M. 1998. Three tails of comet Hale–Bopp. *Geophys. Res. Lett. 25*: 225–8.

Additional reading

Bessel, F. W. 1836. Beobactungen über die physische Beschaffenheit des Halley'shen Kometen und dadurch veranlasste Bemerkungen. *Astron. Nachr. 13*: 185–232.

Brandt, J. C. 1968. The physics of comet tails. *Ann. Rev. Astron. Astrophys. 6*: 267–86.

Brandt, J. C., Niedner, M. B., Jr., and Rahe, J. 1992. *The International Halley Watch Atlas of Large-Scale Phenomena*. Boulder, CO: LASP-University of Colorado.

Brandt, J.C., Niedner, M. B., Jr., and Rahe, J. 1997. *The International Halley Watch Atlas of Large-Scale Phenomena: Supplement 1997*. Boulder, CO: LASP-University of Colorado.

Donn, B., Rahe, J., and Brandt, J. C. 1986. *Atlas of Comet Halley 1910 II*. Washington, DC: NASA SP-488.

Grün, E., and Jessberger, E. 1990. Dust. In *Physics and Chemistry of Comets*, ed. W. F. Huebner, pp. 113–76. Berlin: Springer-Verlag.

Jaegermann, R. 1903. Prof. Dr. Th. Bredichin's Mechanische Untersuchungen über Cometenformen (St. Petersburg).

Niedner, M. B., Jr. 1986. Dynamics of cometary plasma tails. *Adv. Space Res. 6* (No. 1): 315–27.

Rahe, J., Donn, B., and Wurm, K. 1969. *Atlas of Cometary Forms. Structures Near the Nucleus*. Washington, DC: NASA SP-198.

Chapter 5

A'Hearn, M. F., Feldman, P. D., and Schleicher, D. G. 1983. The discovery of S_2 in comet IRAS–Araki–Alcock 1983d. *Astrophys. J. 274*: L99–103.

A'Hearn, M. F., Hoban, S., Birch, P. V., Bowers, C., and Klinglesmith, D. A., III 1986. Cyanogen jets in comet Halley. *Nature 324*: 649–51.

A'Hearn, M. F., Arpigny, C., Feldman, P. D., Jackson, W. M., Meier, R., Weaver, H. A., Wellnitz, D. D., and Woodney, L. M. 2000. Formation of S_2 in comets. *DPS 32*: 4401.

Arpigny, C. 1977. On the nature of comets. *Proceedings of the Robert A. Welch Foundation Conferences on Chemical Research XXI. Cosmochemistry*. Houston.

Balsiger, H., Altwegg, K., Bühler, F., Geiss, J., Ghielmetti, A. G., Goldstein, B. E., Goldstein, R., Huntress, W. T., Ip, W.-H., Lazarus, A. J., Meier, A., Neugebauer, M., Rettenmund, U., Rosenbauer, H., Schwenn, R., Sharp, R. D., Shelley, E. G., Ungstrup, E., and Young, D. T. 1986. Ion composition and dynamics at comet Halley. *Nature 321*: 330–4.

Bertaux, J. L., Blamont, J. E., and Festou, M. 1973. Interpretation of hydrogen Lyman-alpha observations of Comet Bennett and Encke. *Astron. Astrophys. 25*: 415–30.

Biermann, L. 1968. *On the Emission of Atomic Hydrogen in Comets*. J1LA Report No. 93. Boulder, CO: Joint Institute for Laboratory Astrophysics.

Brockelée-Morvan, D., Lis, D. C., Wink, J. E., Despois, D., Crovieier, J., Bachiller, R., Benford, D. J., Biver, N., Colom, P., Davies, J. K., Gérard, E., Germain, B., Houde, M., Mehringer, D., Moreno, R., Paubert, G., Phillips, T. G., and Rauer, H. 2000. New molecules found in comet C/1995 O1 (Hale–Bopp). Investigating the link between cometary and interstellar material. *Astron. Astrophys. 353*: 1101–14.

Campins, H. and Ryan, E. V. 1989. The identification of crystaline olivine in cometary silicates. *Astrophys. J. 341*: 1059–66.

Combi, M. R., and Delsemme, A. H. 1980. Neutral cometary atmospheres. I. An average random walk model for photodissociation in comets. *Astrophys. J. 237*: 663–40.

Combi, M. R., and Smyth, W. H. 1988a. Monte Carlo particle trajectory models for neutral cometary gases. I. Models and equations. *Astrophys. J. 327*: 1026–43.

1988b. Monte Carlo particle trajectory models for neutral cometary gases. II. The spatial morphology of the Lyman alpha coma. *Astrophys. J. 327*: 1044–59.

Combi, M. R., Bos, B. J., and Smyth, W. H. 1993. The OH distribution in cometary atmospheres: A collisional Monte Carlo model for heavy species. *Astrophys. J. 408*: 668–77.

Combi, M. R., Kabin, K., DeZeeuw, D. L., Gombosi, T. I., and Powell, K. G. 1999. Dust-Gas Interrelations in Comets: Observations and Theory. *Earth, Moon Planet. 79*: 275–306.

Combi, M. R., Reinard, A. A., Bertaux, J.-L., Quemerais, E., and Mäkinen, T. 2000. SOHO/SWAN Observations of the structure and evolution of the hydrogen Lyman-α coma of comet Hale–Bopp (1995 O1). *Icarus 144*: 191–202.

Despois, D., Gerard, E., Crovisier, J., and Kazes, I. 1981. The OH radical in comets – Observations and analysis of the hyperfine microwave transitions at 1667 MHz and 1665 MHz. *Astron. Astrophys. 99*: 320–40.

Eberhardt, P., Kranowsky, D., Schulte, W., Dolder, U., Lämmerzahl, P., Berthelier, J. J., Woweries, J., Stubbemann, U., Hodges, R. R., Hoffman, J. H., and Illiano, J. M. 1987. The CO and N_2 abundance in comet P/Halley. *Astron. Astrophys. 187*: 481–4.

Festou, M. 1981a. The density distribution of neutral compounds in cometary atmospheres. I. Models and equations. *Astron. Astrophys. 95*: 69–79.

1981b. The density distribution of neutral compounds in cometary atmospheres. II Production rates and lifetime of OH radicals in comet Kobayashi–Berger–Milon (1975 IX). *Astron. Astrophys. 96*: 52–7.

Festou, M. 1999. On the existence of distributed sources in comet comae. *Space Science Review 90*: 53–67.

Fulle, M., Colangeli, L., Mennella, V., Rotundi, A., and Bussoletti, E. 1995. The sensitivity of the size distribution to the grain dynamics: simulation of the dust measured by GIOTTO at P/Halley. *Astron. Astrophys. 304*: 622–30.

Gary, G. A., and O'Dell, C. R. 1974. Interpretation of the anti-tail of Kohoutek as a particle flow phenomenon. *Icarus 23*: 519–25.

Giguere, P. T., and Huebner, W. F. 1978. A model of comet comae. I. Gas-phase chemistry in one dimension. *Astrophys. J. 223*: 638–54.

Greenberg, J. M., and Li, A. 1998. From interstellar dust to comets: the extended CO source in comet Halley. *Astron. Astrophys. 332*: 374–84.

Gringauz, K. I., Gombosi, T. I., Remizon, A. P., Apáthy, I., Szemerey, I., Verigin, M. I., Denchikova, L. I., Dyachkov, A. V., Kepple, E., Klimenko, I. N., Richter, A. K., Somogyi, A. J., Szegö, K., Szendrö, S., Tátrallyay, M., Varga, A., and Vladimirova, G. A. 1986. First in situ plasma and neutral gas measurements in comet Halley. *Nature 321*: 282–5.

Haser, L. 1957. Distribution d'intensité dans la tête d'une comète. *Bull. Acad. R. Belg.*, 5e serie 43: 740.

Horányi, M., and Mendis, D. A. 1986. The effects of electrostatic charging on the dust distribution at Halley's Comet. *Astrophys. J. 307*: 800–7.

Horányi, M. 1996. Charged dust dynamics in the solar system. *Ann. Rev. Astron. Astrophys 34*: 383–418.

Huebner, W. F., and Benkhoff, J. 1999a. On the relationship of chemical abundances in the nucleus to those in the coma. *Earth, Moon, Planets 77*: 217–22.

1999b. From coma abundances to nucleus composition. In *Composition and Origin of Cometary Material*, ed. K. Altwegg, P. Ehrenfreund, J. Geiss, and W. F. Huebner, *Space Sci. Review 90*: 117–30.

Huebner, W. F., and Boice, D. C. 1989. Polymers in comet comae. In *AGU Monograph 54*, eds. J. H. Waite, Jr., J. L. Burch, and R. L. Moore, pp. 453–6.

Huebner, W. F. and Giguere, P. T. 1980. A model of comet comae. II. Effects of solar photodissociative ionization. *Astrophys. J. 238*: 753–62.

Huebner, W. F., Boice, D. C., Schmidt, H. U., and Wegmann, R. 1991. Structure of the coma: Chemistry and solar wind interaction. In *Comets in the Post-Halley Era*. ed. R. L. Newburn, Jr., M. Neugebauer, and J. Rahe, pp. 907–36. Dordrecht: Kluwer Academic Publishers.

Keller, H. U. 1973. Hydrogen production rates of comet Bennett (1969i) in the first half of April 1970. *Astron. Astrophys. 27*: 51–7.

Kranowsky, D., Lämmerzahl, P., Herrwerth, I., Woweries, J., Eberhardt, P., Dolder, U., Herrmann, U., Schulte, W., Berthelier, J. J., Illiano, J. M., Hodges, R. R., and Hoffmann, J. H. 1986. In situ gas and ion measurements at comet Halley. *Nature 321*: 326–9.

Laffont, C., Rousselot, P., Clairemidi, J., Moreels, G., and Boice, D. C. 1997. Jets and arcs in the coma of comet Hale–Bopp from August 1996 to April 1997. *Earth, Moon and Planets 78*: 211–17.

Lardière, O., Garro, S., and Merlin, J.-C. 1997. Evolution of dust shells and jets in the inner coma of comet C/1995 (Hale–Bopp). *Earth, Moon and Planets 78*: 205–10.

Liller, W. 1960. The nature of grains in the tails of comets 1956h and 1957d. *Astrophys. J. 132*: 867–82.

Malaise, D. J. 1970. Collisional effects in cometary atmospheres I. Model atmospheres and synthetic spectra. *Astron. Astrophys. 5*: 209–27.

McBride, N., Green, S. F., Levasseur-Regourd, A. C., Goidet-Devel, B., and Renard, J.-B. 1997. The inner dust coma of Comet 26P/Grigg-Skjellerup: multiple jets and nucleus fragments. *Mon. Not. R. Astron. Soc. 289*: 535–53.

McBride, N., Green, S. F., Levasseur-Regourd, A. C., Goidet-Devel, B., and Renard, J.-B. 1997. The inner dust coma of Comet 26P/Grigg-Skjellerup: multiple jets and nucleus fragments. *Mon. Not. R. Astron. Soc. 289*: 535–53.

Mendis, D. A. 1988. Postencounter view of comets. *Ann. Rev. Astron. Astrophys. 26*: 121–49.

O'Dell, C. R., Robinson, R. R., Krishna Swamy, K. S., McCarthy, P. J., and Spinrad, H. 1988. C_2 in comet Halley: Evidence for its being third generation and resolution of the vibrational population discrepancy. *Astrophys. J. 334*: 476–88.

Probstein, R. N. 1968. The dusty gas dynamics of comet heads. In *Problems of Hydrodynamics and Continuum Mechanics*, ed. Society for Industrial Mathematics, Philadelphia, pp. 568–83.

Richter, K., Combi, M. R., Keller, H. U., and Meier, R. R. 2000. Multiple scattering of hydrogen Lyα radiation in the coma of comet Hyakutake (C/1996 B2). *Astrophys. J. 531*: 599–611.

Rodionov, A. V., Jorda, L., Jones, G. H., Crifo, J. F., Colas, F., and Lecacheux, J. 1998. Comet Hyakutake gas arcs: First observational evidence of standing shock waves in a cometary coma. *Icarus 136*: 232–67.

Smyth, W. H., Marconi, M. L., Scherb, F., and Roesler, F. L. 1993. Analysis of the hydrogen Hα observations the coma of comet P/Halley. *Astrophys. J. 413*: 456–763.

Smyth, W. H., Combi, M. R., Roesler, F. L., and Scherb, F. 1995. Observations and analysis of the $O(^1D)$ and NH_2 line profiles for the coma of comet P/Halley. *Astrophys. J. 440*: 349–60.

Thomas, N., Boice, D. C., Huebner, W. F., and Keller, H. U. 1988. Intensity profiles of dust near extended sources on Comet Halley. *Nature 332*: 51–2.

Whipple, F. L., and Huebner, W. F. 1976. Physical processes in comets. *Ann. Rev. Astron. Astrophys. 14*: 143–72.

Woodney, L. M., A'Hearn, M. F., and Meier, R. 2000. Sulfur abundances in comets. *DPS 32*: 4402.

Zanstra, H. 1929. The excitation of line- and band-spectra in comets by sunlight. *Mon. Not. R. Astron. Soc. 89*: 178–97.

Additional reading

Code, A. D., Houck, T. E., and Lillie, C. F. 1972. Ultraviolet observations of comets. In *The Scientific Results from the Orbiting Astronomical Observatory (OAO-2)*, ed. A. D. Code, pp. 109–14. NASA SP-310. Washington, DC: GPO. (Observations of the hydrogen cloud.)

Combi, M. R., and Delsemme, A. H. 1980. Neutral cometary atmospheres. II. The production of CN in comets. *Astrophys. J. 237*: 641–5.

Feldman, P. D. 1991. Ultraviolet spectroscopy of cometary comae. In *Comets in the Post-Halley Era*. ed. R. L. Newburn, Jr., M. Neugebauer, and J. Rahe, pp. 139–48. Dordrecht: Kluwer Academic Publishers.

Ip, W.-H., and Mendis, D. A. 1974. Neutral atmospheres of comets: A distributed source model. *Astrophys. Space Sci. 26*: 153–66.

Keller, H. U., and Lillie, C. F. 1978. Hydrogen and hydroxyl production rates of comet Tago–Sato–Kosaka. *Astron. Astrophys. 62*: 143–7.

Morgenthaler, J. P., Harris, W. M., Scherb, F., Anderson, C. M., Oliversen, R. J., Doane, N. E., Combi, M. R., Marconi, M. L., and Smyth, W. H. 2001. Large-aperture [OI] 6300 Å photometry of comet Hale–Bopp: Implications for the photochemistry of OH. *Astrophys. J. 563*: 451–61.

Newburn, R. L., Jr., Neugebauer, M., and Rahe, J. (eds) 1991. *Comets in the Post-Halley Era.* (2 volumes). Dordrecht: Kluwer Academic Publishers.

Rahe, J., Donn, B., and Wurm, K. 1969. *Atlas of Cometary Forms*. NASA SP-198. Washington, DC: GPO.

Richter, N. B. 1963. *The Nature of Comets*. New York: Dover. (Contains a thorough bibliography of pre-1960 comet literature.)

Vsekhsvyatskij, S. K. 1958, 1966, 1967. Fizicheskie Karakteristiki Komet. Moscow: Nauka. (Three Russian volumes describing physical characteristics of comets.)

Whipple, F. L., and Huebner, W. F. 1976. Physical processes in comets. *Ann. Rev. Astron. Astrophys. 14*: 143–72.

Wilkening, L. L. (ed.). 1982. *Comets*. Tucson: The University of Arizona Press.

Chapter 6

Biermann, L. 1951. Kometenschweife und solar Korpuskularstrahlung. *Z. Astrophys. 29*: 279–86.

Biermann, L., Brosowski, B., and Schmidt, H. U. 1967. The interaction of the solar wind with a comet. *Solar Phys. 1*: 254–84.

Brandt, J. C., and Chapman, R. D. 1992. *Rendezvous in Space: The Science of Comets* New York: W. H. Freeman.

Brandt, J. C., and Snow, M. 2000. Heliospheric latitude variations of properties of cometary plasma tails: A test of the Ulysses comet watch paradigm. *Icarus 148*: 52–64.

Brandt, J. C., Roosen, R. G., and Harrington, R. S. 1972. Interplanetary gas. XVII. An astrometric determination of solar-wind velocities from orientations of ionic comet Tails. *Astrophys. J. 177*: 277–84.

Brandt, J. C., Hawley, J. D., and Niedner, M. B., Jr. 1980. A very rapid turning of the plasma-tail axis of comet Bradfield 1979*l* on 1980 February 6. *Astrophys. J. Lett. 241*: L51–4.

Brandt, J. C., Yi, Y., Petersen, C. C., and Snow, M. 1997. Comet de Vico (122P) and latitude variations of plasma phenomena. *Planet Space Sci. 45*: 813–19.

Brandt, J. C., Caputo, F. M., Hoeksema, J. T., Niedner, M. B., Jr., Yi, Y., and Snow, M. 1999. Disconnection events (DEs) in Halley's comet 1985–1986: The correlation with crossings of the heliospheric current sheet (HCS). *Icarus 137*: 69–83.

Cravens, T. E. 1997. Comet Hyakutake x-ray source: Charge transfer of solar wind heavy ions. *Geophys. Res. Lett. 24*: 105–8.

   2002. X-ray emission from comets. *Science 296*: 1042–5.

Dennerl, K., Englhauser, J., and Trümper, J. 1997. X-ray emissions from comets detected in the Röntgen X-ray all-sky survey. *Science 277*: 1625–30.

Farnham, T. L., and Meech, K. J. 1994. Comparison of the plasma tails of four comets: P/Halley, Okazaki–Levy–Rudenko, Austin, and Levy. *Astrophys. J. Suppl. 91*: 419–60.

Gloeckler, G., Geiss, J., Schwadron, N. A., Fisk, L. A., Zurbuchen, T. H., Ipavich, F. M., von Steiger, R., Balsiger, H., and Wilken, B. 2000. Interception of comet Hyakutake's ion tail at a distance of 500 million kilometers. *Nature 404*: 576–8.

Gombosi, T. I., De Zeeuw, D. L., and Häberli, R. M. 1996. Three-dimensional multiscale MHD model of cometary plasma environments. *J. Geophys. Res. 101*: 15, 233–53.

Gombosi, T. I., Hansen, K. C., De Zeeuw, D. I., Combi, M. R., and Powell, K. G. 1997. MHD simulation of comets: The plasma environment of comet Hale–Bopp. *Earth, Moon and Planets 79*: 179–207.

Grewing, M., Praderie, F., and Reinhard, R., Eds. 1988. *Exploration of Halley's Comet*
    Berlin: Springer-Verlag.

Hansen, K. C., Combi, M. R., Crary, F. J., De Zeeuw, D. L., Gombosi, T. I., and Young,
    D. T. 2001. Global MHD simulations of comet Borrelly's plasma environment:
    Effects of a strong neutral jet. *BAAS 33*: 1075.

Ibadov, S. 1996. Interplanetary dust interaction with comets: Production of X-rays. *Adv.
    Space Res. 17*: (12)93–7.

Jokipii, J. R., and Thomas, B. T. 1981. Effects of drift on the transport of cosmic rays.
    IV. Modulation by a wavy interplanetary current sheet. *Astrophys. J. 243*:
    1115–22.

Jones, G. H., Balogh, A., and Horbury, T. S. 2000. Identification of comet Hyakutake's
    extremely long tail from magnetic signatures. *Nature 404*: 574–6.

Lisse, C. M., and 11 co-authors. 1996. Discovery of X-ray and extreme ultraviolet
    emission from comet C/Hyakutake 1996 B2. *Science 274*: 205–9.

Lisse, C. M., Dennerl, K., Englhauser, J., Trümper, J., Marshall, F. E., Petre, R., Valinia,
    A., Kellett, B. J., and Bingham, R. 1997. X-ray emission from comet Hale–Bopp.
    Earth, Moon, and Planets 77: 283–91.

Lisse, C. M., Christian, D., Dennerl, K., Englhauser, J., Trümper, J., Desch, M., Marshall,
    F. E., Petre, R., and Snowden, S. 1999. X-ray and extreme ultraviolet emission from
    comet P/Encke 1997. *Icarus 141*: 316–30.

Lisse, C. M., Christian, D. J., Dennerl, K., Meech, K. J., Petre, R., Weaver, H. A., and
    Wolk, S. J. 2001. Charge exchange-induced X-ray emission from comet C/1999 S4
    (LINEAR). *Science 292*: 1343–8.

McComas, D. J., Elliott, H. A., Gosling, J. T., Reisenfeld, D. B., Skoug, R. M., Goldstein,
    B. E., Neugebauer, M., and Balogh, A. 2002. Ulysses' second fast-latitude scan:
    Complexity near solar maximum and the reformation of polar coronal holes.
    Geophys. Res. Lett. 29: 9, 10.1029/2001GL014164.

Niedner, M. B., Jr., and Brandt, J. C. 1978. Interplanetary gas. XXIII. Plasma tail
    disconnection events in comets: Evidence for magnetic field reconnection at
    interplanetary sector boundaries. *Astrophys. J. 223*: 655–70.

Niedner, M. B., Jr., and Schwingenshuh, K. 1987. Plasma-tail activity at the time of the
    Vega encounters. *Astron. Astrophys. 187*: 103–8.

Niedner, M. B., Jr., Rothe, E. D., and Brandt, J. C. 1978. Interplanetary gas. XXII.
    Interaction of comet Kohoutek's ion tail with the compression region of a solar-wind
    corotating stream. *Astrophys. J. 221*: 1014–25.

Nordholt, J. E., Reisenfeld, D. B., Wiens, R. C., Crary, F., Delapp, D. M., Elphic, R. C.,
    Funsten, H. O., Gary, S. P., Hanley, J. J., Lawrence, D. J., McComas, D. J., Shappirio,
    M. Steinberg, J. T., Wang, J., and Young, D. T. 2003. Deep Space 1 encounter with
    comet 19P/Borrelly: Ion composition measurements by the PEPE mass spectrometer.
    Geophys. Res. Lett., 30: 18–1.

Northrop, T. G., Lisse, C. M., Mumma, M. J., and Desch, M. D. 1997. A possible source of
    the X-rays from comet Hyakutake. *Icarus 127*: 246–50.

Parker, E. N. 1958. Dynamics of the interplanetary gas and magnetic fields. *Astrophys.
    J. 128*: 664–75.

    1963. *Interplanetary Dynamical Processes*. New York: Interscience Publishers.

Smith, E. J. 2001. The heliospheric current sheet. *J. Geophys. Res. 106*: 15,819–31.

Smith, E. J., and Marsden, R. G. 1998. The Ulysses Mission. *Sci. Am.*, January 1998,
    pp. 74–9.

Smith, E. J., Balogh, A., Forsyth, R. J., and McComas, D. J. 2001. Ulysses in the south
    polar cap at solar maximum: Heliospheric magnetic field. *Geophys. Res. Lett. 28*:
    4159–62.

Snow, M., Brandt, J. C., Yi, Y., Petersen, C. C., and Mikuz, H. 2004. Comet Hyakutake (C/1996 B2): Spectacular disconnection event and the latitudinal structure of the solar wind. *Planet. Space Sci.*, in press.

Sturrock. P. A. 1994. *Plasma Physics*. Cambridge: Cambridge University Press.

von Rosenvinge, T. T., Brandt, J. C., and Farquhar, R. W. 1986. The International Cometary Explorer Mission to comet Giacobini-Zinner. *Science 232*: 353–6.

Wegmann, R. 1995. MHD model calculations for the effect of interplanetary shocks on the plasma tail of a comet. *Astron. Astrophys. 294*: 601–14.

    1998. Comment on "Global MHD simulation of a comet crossing the HCS" by Yi, Walker, Ogino, and Brandt. *J. Geophys. Res. 103*: 6633–5.

    2000. The effect of some solar wind disturbances on the plasma tail of a comet: models and observations. *Astron. Astrophys. 358*: 759–75.

Yi, Y., Walker, R. J., Ogino, T., and Brandt, J. C. 1996. Global magnetohydrodynamic simulation of a comet crossing the heliospheric current sheet. *J. Geophys. Res. 101*: 27 585–601.

    1998. Reply. *J. Geophys. Res. 103*: 6637–9.

Additional reading

Balogh, A., Smith, E. J., Tsurutani, B. T., Southwood, D. J., Forsyth, R. J., and Horbury, T. S. 1995. The heliospheric magnetic field over the south polar region of the sun. *Science 268*: 1007–10.

Brandt, J. C. 1970. *Introduction to the Solar Wind*. San Francisco: W. H. Freeman.

Brandt, J. C. 1982. Observations and dynamics of plasma tails. In *Comets*, ed. L. L. Wilkening, pp. 519–35. Tucson: University of Arizona Press.

Brandt, J. C., and Mendis, D. A. 1979. The interaction of the solar wind with comets. In *Solar System Plasma Physics, Vol. II*, eds. C. F. Kennel, L. J. Lanzerotti, and E. N. Parker, pp. 253–92. Amsterdam: North-Holland.

Hoeksema, J. T. 1991. Large-scale solar and heliospheric magnetic fields. *Adv. Space Res. 11*: (1)15–24.

Hundhausen, A. J. 1972. *Coronal Expansion and Solar Wind*. Berlin: Springer-Verlag.

Ip, W.-H., and Axford, W. I. 1982. Theories of physical processes in the cometary comae and ion tails. In *Comets*, ed. L. L. Wilkening, pp. 588–633. Tucson: University of Arizona Press.

Marsden, R. G., ed. 1995. *The High Latitude Heliosphere*. Dordrecht: Kluwer.

McComas, D. J., Bame, S. J., Barraclough, B. L., Feldman, W. C., Funsten, H. O., Gosling, J. T., Riley, P., Skoug, R., Balogh, A., Forsyth, R., Goldstein, B. E., and Neugebauer, M. 1998. Ulysses return to the slow solar wind. *Geophys. Res. Lett. 25*: 1–4.

Mendis, D. A., and Flammer, K. R. 1984. The multiple modes of interaction of the solar wind with a comet as it approaches the sun. *Earth, Moon, Planet 31*: 301–11.

Niedner, M. B., Jr. 1984. Magnetic reconnection in comets. In *Magnetic Reconnection in Space and Laboratory Plasmas*, pp. 79–89. Geophysical Monograph 30. American Geophysical Union.

Niedner, M. B., Jr., and Brandt, J. C. 1979. Interplanetary Gas. XIV. Are cometary plasma tail disconnections caused by sector boundary crossings or by encounters with high-speed streams? *Astrophys. J. 234*: 723–32.

Phillips, J. L., Bame, S. J., Feldman, W. C., Goldstein, B. E., Gosling, J. T., Hammond, C. M., McComas, D. J., Neugebauer, M., Scime, E. E., and Suess, S. T. 1995. Ulysses solar wind plasma observations at high southerly latitudes. *Science 268*: 1030–3.

Schmidt, H. U., and Wegmann, R. 1982. Plasma flow and magnetic fields in comets. In *Comets*, ed. L. L. Wilkening, pp. 538–60. Tucson: University of Arizona Press.

Smith, E. J. 1000. The Sun, solar wind, and magnetic field. I. In *Proceedings of the International School of Physics "Enrico Fermi," Course CXLII*, eds. B. Coppi, A. Ferrari, and E. Sindoni, pp. 179–204. Amsterdam: IOS Press.

Smith, E. J., Marsden, R. G., and Page, D. E. 1995. Ulysses above the sun's south pole: An introduction. *Science 268*: 1005–7.

Chapter 7

A'Hearn, M. F., Campins, H., Schleicher, D. G., and Millis, R. L. 1989. The nucleus of comet P/Tempel 2. *Astrophys. J. 347*: 1155–66.

A'Hearn, M. F., Millis, R. L., Schleicher, D. G., Osip, D. J., and Birch, P. V. 1995. The ensemble properties of comets: Results from narrowband photometry of 85 comets, 1976–1992. *Icarus 118*: 223–70.

Belton, M. J. S. 1991. Characterization of the rotation of cometary nuclei. In *Comets in the Post-Halley Era, Vol. 2*, ed. R. L. Newburn, Jr., M. Neugebauer, and J. Rahe, pp. 691–721. Dordrecht: Kluwer.

Belton, M. J. S., Julian, W. H., Anderson, A. J., and Mueller, B. E. A. 1991. The spin state and homogeneity of Comet Halley's nucleus. *Icarus 93*: 183–93.

Boehnhardt, H., Babion, J., and West, R. M. 1997. An optimized detection technique for faint moving objects on a star-rich background. *Astron. Astrophys. 320*: 642–51.

Boehnhardt, H., Delahodde, C., Sekiguchi, T., Tozzi, G. P., Amestica, R., Hainaut, O., Spyromilio, J., Tarenghi, M., West, R. M., Schulz, R., and Schwehm, G. 2002. VLT observations of comet 46P/Wirtanen. *Astron. Astrophys. 387*: 1107–13.

Brandt, J. C., and Chapman, R. D. 1992. *Rendezvous in Space: The Science of Comets*. New York: W. H. Freeman and Co.

Campbell, D. B., Harmon, J. K., and Shapiro, I. I., 1989. Radar observations of comet Halley. *Astrophys. J. 338*: 1094–105.

Campins, H., Osip, D. J., Rieke, G. H., and Rieke, M. J. 1995. Estimates of the radius and albedo of comet–asteroid transition object 4015 Wilson–Harrington based on infrared observations. *Planet. Space Sci. 43*: 733–6.

Crovisier, J., Leech, K., Bockelée-Morvan, D., Brooke, T. Y., Hanner, M. S., Altieri, B., Keller, H. U., and Lellouch, E. 1997. The spectrum of comet Hale–Bopp (C/1995 O1) observed with the infrared space observatory at 2.9 astronomical units from the sun. *Science 275*: 1904–7.

Cruikshank, D. P., and Brown, R. H. 1983. The nucleus of comet Schwassmann–Wachmann 1. *Icarus 56*: 377–80.

Emerich, C., Lamarre, J. M., Moroz, V. I., Combes, M., Sanko, N. F., Nikolsky, Y. V., Rocard, F., Gispert, R., Coron, N., Bibring, J. P., Encrenaz, T., and Crovisier, J. 1987. Temperature and size of the nucleus of comet P/Halley deduced from IKS infrared Vega 1 measurements. *Astron. Astrophys. 187*: 839–42.

Enzian, A. 1999. On the prediction of CO outgassing from comets Hale–Bopp and Wirtanen. *Space Sci. Rev. 90*: 131–9.

Fernández, J. A., Tancredi, G., Rickman, H., and Licandro, J. 1999. The population, magnitudes, and sizes of Jupiter family comets. *Astron. Astrophys. 352*: 327–40.

Fernández, Y. R. 1999. Physical Properties of Cometary Nuclei. Thesis: University of Maryland, College Park.

Flammer, K. R., Jackson, B., and Mendis, D. A. 1986. On the brightness variations of comet Halley at large heliocentric distances. *Earth, Moon, and Planets 35*: 203–12.

Hanner, M. S., Aitken, D. K., Knacke, R., McCorkle, S., Roche, P. F., and Tokunaga, A. T. 1985. Infrared spectrophotometry of comet IRAS–Araki–Alcock (1983d): A bare nucleus revealed? *Icarus 62*: 97–109.

Hanner, M. S., Newburn, R. L., Spinrad, H., and Veeder, G. J. 1987. Comet Sugano–Saigusa–Fujikawa (1983 V): A small, puzzling comet. *Astron. J. 94*: 1081–7.

Harmon, J. K., Campbell, D. B., Hine, A. A., Shapiro, I. I., and Marsden, B. G. 1989. Radar observations of comet IRAS–Araki–Alcock 1983d. *Astrophys. J. 338*: 1071–93.

Harmon, J. K., Ostro, S. J., Benner, L. A. M., Rosema, K. D., Jurgens, R. F., Winkler, R., Yeomans, D. K., Choate, D., Cormier, R., Giorgini, J. D., Mitchell, D. L., Chodas, P. W., Rose, R., Kelley, D., Slade, M. A., and Thomas, M. L. 1997. Radar detection of the nucleus and coma of comet Hyakutake (C/1996 B2). *Science 278*: 1921–4.

Harmon, J. K., Campbell, D. B., Ostro, S. J., and Nolan, M. C. 1999. Radar observations of comets. *Planet. Space Sci. 47*: 1409–22.

Huebner, W. F. (ed.) 1990. *Physics and Chemistry of Comets.* Berlin and New York: Springer Verlag.

Hughes, D. W. 1975. Cometary outbursts: A brief survey. *Quart. J. Roy. Astr. Soc. 16*: 410–27.

1991. Possible Mechanisms for Cometary Outbursts. In *Comets in the Post-Halley Era, Vol. 2*, ed. R. L. Newburn, Jr., M. Neugebauer, and J. Rahe, pp. 825–51. Dordrecht: Kluwer.

Jewitt, D., 1990. The persistent coma of comet P/Schwassmann-Wachmann 1. *Astrophys. J. 351*: 277–86.

1997. Cometary rotation: An overview. *Earth, Moon, Planet. 79*: 35–53.

Jewitt, D., and Luu, J. 1989. A CCD portrait of Comet P/Tempel 2. *Astron. J. 97*: 1766–90.

Jewitt, D. C., and Meech, K. J. 1988. Optical properties of cometary nuclei and a preliminary comparison with asteroids. 1988. *Astrophys. J. 328*: 974–86.

Jones, T. J., and Morrison, D. 1974. Recalibration of the photometric/radiometric method of determining asteroid sizes. *Astron. J. 79*: 892–5.

Keller, H. U. 1990. The Nucleus. In *Physics and Chemistry of Comets*, ed. W. F. Huebner, pp. 13–68. Berlin: Springer-Verlag.

Lamy, P. L., and Toth, I. 1995. Direct detection of a cometary nucleus with the Hubble Space Telescope. *Astron. Astrophys. 293*: L43–5.

Lamy, P. L., Toth, I., Jorda, L., Weaver, H. A., and A'Hearn, M. 1998a. The nucleus and inner coma of comet 46P/Wirtanen. *Astron. Astrophys. 335*: L25–9.

Lamy, P. L., Toth, I., and Weaver, H. A. 1998b. Hubble Space Telescope observations of the nucleus and inner coma of comet 19P/1904 Y2 (Borrelly). *Astron. Astrophys. 337*: 945–54.

Lamy, P. L., Toth, I., Weaver, H. A., Delahodde, C., Jorda, L., and A'Hearn, M. F. 2000. The nucleus of 13 short-period comets. *BAAS 32* Paper 36.04.

Larson, S. M. 2003. Inferring the Nucleus Spin State from Groundbased Observations. In *The Nanjing IAU Colloquium Proceedings*, ed. M. F. A'Hearn (to be published).

Lebofsky, L. A., and Spencer, J. R. 1989. Radiometry thermal modeling of asteroids. In *Asteroids II*, ed. R. Binzel, T. Gehrels, and M. S. Matthews, pp. 128–47. Tucson: University of Arizona Press.

Lebofsky, L. A., Sykes, M. V., Tedesco, E. F., Veeder, G. J., Matson, D. L., Brown, R. H., Gradie, J. C., Feierberg, M. A., and Rudy, R. J. 1986. A refined "standard" thermal model for asteroids based on observations of 1 Ceres and 2 Pallas. *Icarus 68*: 239–51.

Lewis, J. S. 1997. *Physics and Chemistry of the Solar System.* San Diego: Academic Press.

Lowry, S. C., Fitzsimmons, A., and Collander-Brown, S. 2003. CCD photometry of distant comets III. Ensemble properties of Jupiter-family comets. *Astron. Astrophys. 397*: 329–43.

Luu, J. X., and Jewitt, D. C. 1992. Near-aphelion CCD photometry of comet P/Schwassmann–Wachmann 2. *Astron. J. 104*: 2243–9.

Lyttleton, R. A. 1953. *The Comets and their Origin*. Cambridge: Cambridge University Press.

Meech, K. J., Jewitt, D., and Ricker, G. R. 1986. Early photometry of comet P/Halley: Development of the coma. *Icarus 66*: 561–74.

Mekler, Y., Prialnik, D., and Podolak, M. 1990. Evaporation from a porous cometary nucleus. *Astrophys. J. 356*: 682–6.

Mekler, Y., and Podolak, M. 1994. Formation of amorphous ice in the protoplanetary nebula. *Planet. Space Sci. 42*: 865–70.

Morrison, D. 1973. Determination of radii of satellites and asteroids from radiometry and photometry. *Icarus 19*: 1–14.

Mumma, M. J., Weaver, H. A., and Larson, H. P. 1987. The ortho-para ratio of water vapor in Comet P/Halley. *Astron. Astrophys. 187*: 419–24.

Peale, S. J., and Lissauer, J. J. 1989. Rotation of Halley's comet. *Icarus 79*: 396–430.

Prialnik, D. 1992. Crystallization, sublimation, and gas release in the interior of a porous comet nucleus. *Astrophys. J. 388*: 196–202.

  1997. A model for the distant activity of comet Hale–Bopp. *Astrophys. J. 478*: L107–10.

  2002. Modeling the comet nucleus interior: Applications to comet C/1995 O1 Hale–Bopp. *Earth, Moon, Planet., 89*: 27–52.

Prialnik, D., and Bar-Nun, A. 1992. Crystallization of amorphous ice as the cause of comet P/Halley's outburst at 14 AU. *Astron. Astrophys. 258*: L9–12.

Prialnik, D., and Podolak, M. 1999. Changes in the structure of comet nuclei due to radioactive heating. *Space Sci. Rev. 90*: 169–78.

Rahe, J., Vanysek, V., and Weissman, P. R. 1994. Properties of cometary nuclei. In *Hazards Due to Comets and Asteroids*, ed. T. Gehrels, pp. 597–634. Tucson: University of Arizona Press.

Rickman, H. 1989. The nucleus of comet Halley: Surface structure, mean density, gas and dust production. *Adv. Space Res. 9*: (3)59–71.

Roemer, E. 1966. The dimensions of cometary nuclei. *Mémoires de la Société Royale des Sciences de Liège, 5th Series, 12*: 23–8.

Samarasinha, N. H., and A'Hearn, M. F. 1991. Observational and dynamical constraints on the rotation of Comet P/Halley. *Icarus 93*: 194–225.

Samarasinha, N. H., and Belton, M. J. S. 1995. Long-term evolution of rotational states and nongravitational effects for Halley-like cometary nuclei. *Icarus 116*: 340–58.

Sekanina, Z. 1981. Rotation and precession of cometary nuclei. *Ann. Rev. Earth Planet. Sci. 9*: 113–45.

  1982. The problem of split comets in review. In *Comets*, ed. L. L. Wilkening, pp. 251–319. Tucson: University of Arizona Press.

Sekanina, Z., and Larson, S. M. 1984. Coma morphology and dust-emission pattern of periodic comet Halley. II. Nucleus spin vector and modeling of major dust features in 1910. *Astron. J. 89*: 1408–25.

  1986. Coma morphology and dust-emission pattern of periodic comet Halley. IV. Spin vector refinement and map of discrete dust sources for 1910. *Astron. J. 92*: 462–82.

Senay, M. C., and Jewitt, D. 1994. Coma formation driven by carbon monoxide release from comet Schwassmann–Wachmann 1. *Nature 371*: 229–31.

Shoemaker, E. M., and Wolfe, R. F. 1982. Cratering time scales for the Galilean satellites. In *Satellites of Jupiter*, ed. D. Morrison, pp. 277–339. Tucson: University of Arizona Press.

Soderblom, L. A., Becker, T. L., Bennett, G., Boice, D. C., Britt, D. T., Brown, R. H., Buratti, B. J., Isbell, C., Giese, B., Hare, T., Hicks, M. D., Howington-Kraus, E., Kirk, R. L., Lee, M., Nelson, R. M., Oberst, J., Owen, T. C., Rayman, M. D., Sandel, B. R., Stern, S. A., Thomas, N., and Yelle, R. V. 2002. Observations of Comet 19P/Borrelly by the miniature integrated camera and spectrometer aboard Deep Space 1. *Science 296*: 1087–91.

Spencer, J. R., Lebofsky, L. A., and Sykes, M. V. 1989. Systematic biases in radiometric diameter determinations. *Icarus 78*: 337–54.

Wallis, M. K. 1980. Radiogenic melting of primordial comet interiors. *Nature 284*: 431–3.

Weaver, H. A., and Lamy, P. L. 1997. Estimating the size of Hale–Bopp's nucleus. *Earth, Moon, Planet. 79*: 17–33.

Weaver, H. A., Feldman, P. D., A'Hearn, M. F., Arpigny, C., Brandt, J. C., Festou, M. C., Haken, M., McPhate, J. B., Stern, S. A., and Tozzi, G. P. 1997. The activity and size of the nucleus of comet Hale–Bopp (C/1995 O1). *Science 275*: 1900–4.

Weaver, H. A., Sekanina, Z., Toth, I., Delahodde, C. E., Hainaut, O. R., Lamy, P. L., Bauer, J. M., A'Hearn, M. F., Arpigny, C., Combi, M. R., Davies, J. K., Feldman, P. D., Festou, M. C., Hook, R., Jorda, L., Keesey, M. S. W., Lisse, C. M., Marsden, B. G., Meech, K. J., Tozzi, G. P., and West, R. 2001. HST and VLT Investigations of the fragments of comet C/1999 S4 (LINEAR). *Science 292*: 1329–33.

Weissman, P. R. 1980. Physical loss of long-period comets. *Astron. Astrophys. 85*: 191–6.

Whipple, F. L. 1982. The rotation of comet nuclei. In *Comets*, ed. L. L. Wilkening, pp. 227–50. Tucson: University of Arizona Press.

Wyckoff, S., Wagner, R. M., Wehinger, P. A., Schleicher, D. G., and Festou, M. C. 1985. Onset of sublimation in comet P/Halley (1982i). *Nature 316*: 241–2.

Zahnle, K., Dones, L., and Levison, H. F. 1998. Cratering rates on the Galilean satellites. *Icarus 136*: 202–22.

Additional reading

Altwegg, K., Ehrenfreund, P., Geiss, J., and Huebner, W. (eds.) 1999. *Composition and Origin of Cometary Materials.* Dordrecht: Kluwer, 412 pages. Reprinted from *Space Science Reviews, Vol. 90*, Nos. 1–2.

Benkhoff, J. 1999. On the flux of water and minor volatiles from the surface of comet Nuclei. *Space Sci. Rev. 90*: 141–8.

Froeschlé, Cl., Klinger, J., and Rickman, H. 1983. Thermal models for the nucleus of comet/P Schwassmann–Wachmann 1. In *Asteroids, Comets, Meteors*, ed. C.-I. Lagerkvist, and H. Rickman, pp. 215–24. Upsalla: Upsalla University Press.

Huebner, W. F., and Benkhoff, J. 1999. From coma abundances to nucleus composition. *Space Sci. Rev. 90*: 117–30.

Keller, H. U., Curdt, W., Kramm, J.-R., and Thomas, N. 1994. *Images of the Nucleus of Comet Halley: Images Obtained by the Halley Multicolour Camera (HMC) on Board the Giotto Spacecraft.* European Space Agency, ESA SP-1127, Vol. 1, 252 pages.

Larson, S. M., and Sekanina, Z. 1984. Coma morphology and dust-emission pattern of periodic comet Halley. I. High-resolution images taken at Mount Wilson in 1910. *Astron. J. 89*: 571–8.

  1985. Coma morphology and dust-emission pattern of periodic comet Halley. III. Additional high-resolution images taken in 1910. *Astron. J. 90*: 823–6.

Prialnik, D. 1997–1999. Modeling gas and dust release from comet Hale–Bopp. *Earth, Moon, Planet. 77*: 223–30.

Prialnik, D., and Bar-Nun, A. 1987. On the evolution and activity of cometary nuclei. *Astrophys. J. 313*: 893–905.

Prialnik, D., and Bar-Nun, A. 1990. Gas release in comet nuclei. *Astrophys. J. 363*: 274–82.

Rickman, H., Fernández, J. A., and Gustafson, B. Å. S. 1990. Formation of stable dust mantles on short-period comet nuclei. *Astron. Astrophys. 237*: 524–35.

Szegö, K., Sagdeev, R. Z., Whipple, F. L., Abergel, A., Bertaux, J.-L., Merényi, E., Szalai, S., and Várhalmi, L. 1995. *Images of the Nucleus of Comet Halley: Obtained by the Television System (TVS) on Board the Vega spacecraft.* European Space Agency, ESA SP-1127, Vol. 2, 255 pages.

Weissman, P. R., and Lowry, S. C. 2003. The size distribution of cometary nuclei. *Science*, in press.

Yeomans, D. K., and Chodas, P. W. 1989. An asymmetric outgassing model for cometary nongravitational accelerations. *Astron. J. 98*: 1083–93.

Chapter 8

Asphaug, E. 1997. Impact origin of the Vesta family. *Meteorites & Planet. Sci. 32*: 965–80.

Asphaug, E. 2000. The Small Planets. *Sci. Am.*, May issue, 46–55.

Asphaug, E., and Melosh, J. 1993. The Stickney impact of Phobos: A dynamical model. *Icarus 101*: 144–64.

Belton, M. J. S., Chapman, C. R., Thomas, P. C., Davies, M. E., Greenberg, R., Klaasen, K., Byrnes, D., D'Amarlo, L., Synnott, S., Johnson, T. V., McEwen, A., Merline, W. J., Davis, D. R., Petit, J.-M., Storrs, A., Veverka, J., and Zellner, B. 1995. Bulk density of asteroid 243 Ida from the orbit of its satellite Dactyl. *Nature 374*: 785–8.

Belton, M. J. S., Chapman, C. R., Klaasen, K. P., Harch, A. P., Thomas, P. C., Veverka, J., McEwen, A. S., and Pappalardo, R. T. 1996. Galileo's encounter with 243 Ida: An overview of the imaging experiment. *Icarus 120*: 1–19.

Binzel, R. P. 2000. Asteroids come of age. *Science 289*: 2065–6.

2001. A new century for asteroids. *Sky & Telescope 102*: 44–51 (July 2001).

Binzel, R. P., and Xu, S. 1993. Chips off of Asteroid 4 Vesta: Evidence for the parent body of basaltic A chondrite meteorites. *Science 260*: 186–91.

Binzel, R. P., Gehrles, T., and Matthews, M. S. 1989. *Asteroids. II.* Tucson: University of Arizona Press.

Bottke, W., Cellino, A., Paolicchi, P., and Binzel, R. P., eds. 2003. *Asteroids III*, Tucson: University of Arizona Press.

Burns, J. A. 2002. Two bodies are better than one. *Science 297*: 942–3.

Bus, S. J., A'Hearn, M. F., Schleicher, D. G., and Bowell, E. 1991. Detection of CN emissions from (2060) Chiron. *Science 251*: 774–7.

Chapman, C. R. 1999. Asteroids. In *The New Solar System*, 4th edn., ed. J. K. Beatty, C. C. Petersen, and A. Chaikin, pp. 337–50. Cambridge, MA: Sky Publishing.

Chapman, C. R., Veverka, J., Thomas, P. C., Klaasen, K., Belton, M. J. S., Harch, A., McEwen, A., Johnson, T. V., Helfenstein, P., Davies, M. E., Merline, W. J., and Dank, T. 1995. Discovery and physical properties of Dactyl, a satellite of asteroid 243 Ida. *Nature 374*: 783–5.

Cheng, A. F., and 11 co-authors. 2001. Laser altimetry of small-scale features on 433 Eros from NEAR–Shoemaker. *Science 292*: 488–91.

Clark, B. E., Lucey, P., Helfenstein, P., Bell, J. F. III, Peterson, C., Ververka, J., McConnochie, T., Robinson, M. S., Bussey, B., Murchie, S. L., Izenberg, N. I., and

Chapman, C. R. 2001. Space weathering on Eros: Constraints from albedo and spectral measurements of Psyche crater. *Meteoritics & Planet. Sci. 36*: 1617–37.

Cunningham, C. J. 1988. *Introduction to Asteroids*. Richmond, VA: Willmann-Bell.

Fernández, Y. R., McFadden, L. A., Lisse, C. M., Helin, E. F., and Chamberlain, A. B. 1997. Analysis of POSS images of comet–asteroid transition object 107P/1949 W1 (Wilson–Harrington). *Icarus 128*: 114–26.

Fujiwara, A., Kamimoto, G., and Tsukamoto, A. 1977. Destruction of basaltic bodies by high-velocity impact. *Icarus 31*: 277–8.

Fujiwara, A., Kamimoto, G., and Tsukamoto, A. 1978. Expected shape distribution of asteroids obtained from laboratory impact experiments. *Nature 272*: 602–3.

Hirayama, K. 1918. Groups of asteroids probably of common origin. *Astron. J. 31*: 185–8.

Kowal, C. T. 1996. *Asteroids: Their Nature and Utilization*, 2d edn. Chichester: Wiley.

Lewis, J. S. 1997. *Physics and Chemistry of the Solar System*. San Diego: Academic Press.

McFadden, L.-A. 1993. The comet-asteroid transition: Recent telescopic observations. In *Asteroids, Comets, Meteors 1993*, ed. A. Milani, M. Di Martino, and A. Cellino, pp. 95–110. IAU Symposium No. 160. Dordrecht: Kluwer.

McSween, H. Y., Jr. 1999. *Meteorites and Their Parent Planets*, 2nd edn. Cambridge: Cambridge University Press.

Meech, K. J., and Belton, M. J. S. 1990. The atmosphere of 2060 Chiron. *Astron. J. 100*: 1323–38.

Melosh, H. J., and Ryan, E. V. 1997. Asteroids: Shattered but not dispersed. *Icarus 129*: 562–4.

Ostro, S. J., and 19 co-authors. 1999. Radar and optical observations of asteroid 1998 KY 26. *Science 285*: 557–9.

Peebles, C. 2000. *Asteroids: A History*. Washington, DC: Smithsonian.

Pieters, C. M. and McFadden, L. A. 1994. Meteorite and asteroid reflectance spectroscopy. *Ann. Rev. Earth Planet. Sci. 22*: 457–97.

Prialnik, D., Brosch, N., and Ianovici, D. 1995. Modelling the activity of 2060 Chiron. *Mon. Not. R. Astron. Soc. 276*: 1148–54.

Richter, I., Brinza, D. E., Cassel, M., Glassmeier, K.-H., Kuhnke, F., Musmann, G., Othmer, C., Schwingenshuh, K., and Tsurutani, B. T. 2001. First direct magnetic field measurements of an asteroidal magnetic field: DS1 at Braille. *Geophys. Res. Lett. 28*: 1913–16.

Ryan, E. V. 2000. Asteroid fragmentation and evolution of asteroids. *Ann. Rev. Earth Planet Sci. 28*: 367–89.

Ryan, E. V., and Melosh, H. J. 1998. Impact fragmentation: From the laboratory to asteroids. *Icarus 133*: 1–24.

Tholen, D. J. 1984. Asteroid Taxonomy from Cluster Analysis of Photometry. Ph.D. thesis, University of Arizona.

Tholen, D. J., and Barucci, M. A. 1989. Asteroid Taxonomy. In *Asteroids II*, ed. R. P. Binzel, T. Gehrels, and M. S. Matthews, pp. 298–315. Tucson: University of Arizona Press.

Thomas, P. C., Binzel, R. P., Gaffey, M. J., Storrs, A. D., Wells, E. N., and Zellner, B. H. 1997. Impact excavation on asteroid 4 Vesta: Hubble Space Telescope results. *Science 277*: 1492–5.

Trombka, J. I., and 23 co-authors. 2000. The elemental composition of asteroid 433 Eros: Results of the NEAR–Shoemaker X-ray spectrometer. *Science 289*: 2101–5.

Veverka, J., and 16 co-authors. 1997. NEAR's flyby of 253 Mathilde: Images of a C asteroid. *Science 278*: 2109–14.

Veverka, J., and 32 co-authors. 2000. NEAR at Eros: Imaging and spectral results. *Science 289*: 2088–97.

Veverka, J., and 38 co-authors. 2001a. The landing of the NEAR–Shoemaker spacecraft on asteroid 433 Eros. *Nature 413*: 390–3.

Veverka, J., and 32 co-authors. 2001b. Imaging of small-scale features on 433 Eros from NEAR: Evidence for a complex regolith. *Science 292*: 484–8.

Wetherill, G. W. 1991. End products of cometary evolution – Cometary origin of earth-crossing bodies of asteroidal appearance. *Comets in the Post-Halley Era. Vol. 1*: 537–56.

Wisdom, J. 1985. Meteorites may follow a chaotic route to Earth. *Nature 315*: 731–3.

Yeomans, D. K., and 12 co-authors. 1997. Estimating the mass of asteroid 253 Mathilde from tracking data during the NEAR flyby. *Science 278*: 2106–9.

Yeomans, D. K., and 15 co-authors. 2000. Radio science results during the NEAR–Shoemaker spacecraft rendezvous with Eros. *Science 289*: 2085–8.

Zanda, B., and Rotaru, M. 2001. *Meteorites.* Cambridge: Cambridge University Press.

Zuber, M. T., and 11 co-authors. 2000. The shape of 433 Eros from the NEAR–Shoemaker laser rangefinder. *Science 289*: 2097–101.

Additional readings

Binzel, R. P., Gehrels, T., and Matthews, M. S. 1989. *Asteroids II.* Tucson: University of Arizona Press.

Chapman, C. R., Paolicchi, P., Zappalà, V., Binzel, R. P., and Bell, J. 1989. Asteroid families: Physical properties and evolution. In *Asteroids II*, ed. R. P. Binzel, T. Gehrels, and M. S. Matthews, pp. 386–415. Tucson: University of Arizona Press.

Valsecchi, G. B., Carusi, A., Kneževič, Z., Kresák, L., and Williams, J. G. 1989. Identification of asteroid dynamical families. In *Asteroids II*, ed. R. P. Binzel, T. Gehrels, and M. S. Matthews, pp. 368–85. Tucson: University of Arizona Press.

Wetherill, G. W. 1991. End products of cometary evolution: Cometary origin of earth-crossing bodies of asteroidal appearance. In *Comets in the Post-Halley Era*, Vol. 1, ed. R. L. Newburn, M. Neugebauer, and J. Rahe, pp. 537–56. Dordrecht: Kluwer.

Chapter 9

Altweg, K., Ehrenfreund, P. Geiss, J., and Huebner eds., W. F. 1999a. Composition and origin of cometary materials. *Space Science Reviews 90*: Nos. 1–2.

Altweg, K., Ehrenfreund, P., Geiss, J., Huebner, W. F., Levassuer-Regourd, A.-R. 1999b. Cometary materials: Progress toward understanding the composition of the outer solar nebula. *Space Sci. Rev. 90*: 373–89.

Bailey, M. E. 2002. Where have all the comets gone? *Science 296*: 2151–3.

Bailey, M. E. and Stagg, C. R. 1988. Cratering constraints on the inner Oort cloud: steady state models. *Mon. Not. R. Astr. Soc. 235*: 1–32.

Binney, J. and Merrifield, M. 1998. *Galactic Astronomy.* Princeton, New Jersey: Princeton University Press.

Cameron, A. G. W. 1988. The origin of the solar system. *Ann. Rev. Astron. Astrophys. 26*: 441–72.

Delsemme, A. H. 1987. Galactic tides affect the Oort cloud: an observational confirmation. *Astron. Astrophys. 187*: 913–18.

1998. Recollections of a cometary scientist. *Planet. Space. Sci. 46*: 111–24.

Delsemme, A. H., and Swings, P. 1952. Hydrates de gaz dans les noyaux cométaires et les grains interstellaires. *Ann. Astrophys. 15*: 1–6.

Duncan, M. J., and Levison, H. F. 1997. A disk of scattered icy objects and the origin of Jupiter-family comets. *Science 276*: 1670–2.

Duncan, M., Quinn, T., and Tremaine, S. 1987. The formation and extent of the solar system comet cloud. *Astron. J. 94*: 1330–8.

Duncan, M., Quinn, T., and Tremaine, S. 1988. The orgin of short period comets. *Astrophys. J. 328*: L69–73.

Edgeworth, K. E. 1949. The origin and evolution of the solar system. *Mon. Not. R. Astron. Soc. 192*: 600–9.

Everhart, E. 1967. Intrinsic distributions of cometary perihelia and magnitudes. *Astron. J. 72*: 1002–11.

Fegley, B., Jr., 1999. Chemical and physical processing of presolar materials in the solar nebula and the implications for preservation of presolar materials in comets. *Space Sci. Rev. 90*: 239–52.

Fernández, J. A. and Ip, W.-H. 1981. Dynamical evolution of a cometary swarm in the outer planetary region. *Icarus 47*: 470–9.

Goldreich, P. and Ward, W. R. 1973. The formation of planetesimals. *Astrophys. J. 183*: 1051–62.

Heisler, J. 1990. Monte Carlo simulations of the Oort comet cloud. *Icarus 88*: 104–21.

Heisler, J., Tremaine, S., and Alcock, C. 1987. The frequency and intensity of comet showers from the Oort cloud. *Icarus 70*: 269–88.

Huebner, W. F. and Benkhoff, J. 1999a. On the relationship of chemical abundances in the nucleus to those in the coma. *Earth, Moon, Planet. 77*: 217–22.

Huebner, W. F. and Benkhoff, J. 1999b. From coma abundances to nucleus composition. In *Composition and Origin of Cometary Material*, ed. K. Altwegg, P. Ehrenfreund, J. Geiss and W. F. Huebner. *Space Sci. Review 90*: 117–30.

Ip. W.-H., and Fernández, J. A. 1997. On dynamical scattering of Kuiper belt objects in 2: 3 resonance with Neptune into short-period comets. *Astron. Astrophys. 324*: 778–84.

Jewitt, D. 1999. Kuiper belt objects. *Annu. Rev. Earth Planet. Sci. 27*: 287–12.

Jewitt, D., and Luu, J. 1993. Discovery of the candidate Kuiper Belt object 1992 QB1. *Nature 362*: 730–2.

Jewitt, D., Luu, J., and Chen, J. 1996. The Mauna Kea–Cerro Tololo (MKCT) Kuiper belt and Centaur survey. *Astron. J. 112*: 1225–38.

Jewitt, D. Luu, J. and Trujillo, C. 1998. Large Kuiper belt objects: The Mauna Kea 8k CCD survey. *Astron. J. 115*: 2125–35.

Kawakita, H., Watanabe, J., Ando, H., Aoki, W., Fuse, T., Honda, S., Izumiura, H., Kajino, T., Kambe, E., Kawanomoto, S., Noguchi, K., Okita, K., Sadakane, K., Sato, B., Takada-Hidai, M., Takeda, Y., Usuda, T., Watanabe, E., and Yoshida, M 2001. The spin temperature of NH_3 in comet C/1999 S4 (LINEAR). *Science 294*: 1089–91.

Levison, H. F., and Duncan, M. J. 1994. The long-term dynamical behavior of short-period comets. *Icarus 108*: 18–36.

Levison, H. F., Dones, L., Duncan, M. J., 2001. The origin of Halley-type comets: Probing the inner Oort cloud. *Astron. J. 121*: 2253–67.

Levison, H. F., Morbidelli, A., Doanes, L., Jedicke, R., Wiegert, P. A., Bottke, W. F., Jr. 2002. The mass disruption of Oort cloud comets. *Science 296*: 2212–15.

Luu, J. and Jewitt, D. 1996. Color diversity among the Centaurs and Kuiper belt objects. *Astron. J. 112*: 2310–18.

Luu, J. X., and Jewitt, D. C. 2002. Kuiper belt objects: relics from the accretion disk of the sun. *Anna. Rev. Astron. Astrophys. 40*: 63–101.

McGlynn, T. A., and Chapman, R. D. 1989. On the non-detection of extrasolar comets. *Astrophys. J. 346*: L105–8.

Mekler, Y. and Podolak, M. 1994. Formation of amorphous ice in the protoplanetary nebula. *Planet. Space Sci. 42*: 865–70.

Mestel, L. and Spitzer, L. 1956. Star formation in magnetic dust clouds *Mon. Not. R. Astron. Soc. 116*: 503–14.

Mumma, M. J., Weissman, P. R., and Stern S. A. 1993. Comets and the origin of the solar system: Reading the Rosetta Stone. In *Protostars and Planets III*, ed. E. H., Lvey, and J. I. Lunine, pp. 1177–1252. Tucson: University of Arizona Press.

Mumma, M. J., Dello Russo, N., DiSanti, M. A., Magee-Sauer, K., Novak, R. E., Brittain, S., Rettig, T., McLean, I. S., Reuter, D. C., and Xu, Li-H. 2001. Organic composition of C/1999 S4 (LINEAR): A comet formed near Jupiter? *Science 292*: 1334–9.

Oort, J. H. 1950. The structure of the cloud of comets surrounding the solar system and a hypothesis concerning its origin. *Bull. Astron. Inst. Neth. 11*: 91–110.

Safronov, V. S. 1969. *Evolution of the Protoplanetary Cloud and Formation of the Earth and Planets.* Moscow: Nauka. Engl. Transl. 1972. Jerusalem: Keter.

Saha, P. and Tremaine, S. 1993. Symplectic integrators for solar system dynamics. *Astron. J. 104*: 1633–40.

Sekanina, Z. 1976. A probability of encounter with interstellar comets and the likelihood of their existence. *Icarus 27*: 123–33.

Shu, F. H., Adams, F. C., and Lizano, S. 1987. Star formation in molecular clouds: Observation and theory: *Annu. Rev. Astron. Astrophys. 25*: 23–82.

Trujillo, C. A., Jewitt, D. C., and Luu, J. X. 2000. Population of the scattered Kuiper Belt. *Astrophys. J. 529*: L103–6.

Weissman, P. R. 1990. Are periodic bombardments real? *Sky & Telescope 79*: 266–70.

Weissman, P. R. 1991. Dynamical history of the Oort cloud. In *Comets in the Post-Halley Era.* ed. R. L. Newburn, Jr., M. Neugebauer, and J. Rahe, pp. 463–86. Dordrecht: Kluwer Academic Publishers.

Yamamoto, T. 1985. Formation environment of cometary nuclei in primordial solar nebula. *Astron. Astrophys. 142*: 31–6.

Additional reading

Alfvén, H., and Arrhenius, G. 1976. *Evolution of the Solar System.* NASA SP-345. Washington, DC: GPO.

Arrhenius, G., and Alfvén, H. 1973. *Asteroid and Comet Exploration.* NASA CR-2991. Washington, DC: GPO.

Bailey, M. E., Clube, S. V. M., and Napier, W. M. 1990. *The Origin of Comets.* New York: Pergamon Press.

Brandt, J. C. (ed.) 1981. *Comets: Readings from Scientific American.* San Francisco: W. H. Freeman and Co.

Cameron, A. G. W. 1973. Accumulation processes in the primitive solar nebula. *Icarus 18*: 407–50.

Cameron, A. G. W., and Pine, M. R. 1973. Numerical models of the primitive solar nebula. *Icarus 18*: 377–406.

Chang, S. 1979. Comets: cosmic connections with carbonaceous meteorites, interstellar molecules and the origins of life. In *Space Missions to Comets*, ed. M. Neugebauer, D. K. Yeomans, J. C. Brandt, and R. W. Hobbs, pp. 59–111. NASA CP-2089. Washington, DC: GPO.

Chebotarev, G. A., Kazimirchaka-Polonskaya, E. L., and Marsden, B. G. (eds.). 1972. *The Motion, Evolution of Orbits and Origins of Comets.* IAU Symposium No. 45. Dordrecht: D. Reidel.

Delsemme, A. H. 1973a. Origin of the short-period comets. *Astron. Astrophys. 29*: 377–81.
1973b. Brightness law of comets. *Astrophys. Lett. 14*: 163–7.
1975. The volatile fraction of the cometary nucleus. *Icarus 24*: 95–110.

Delsemme, A. H. (ed.). 1977a. *Comets, Asteroids, Meteorites: Interrelations, Evolution and Origins.* Toledo: University of Toledo Press.
1977b. The pristine nature of comets. In *Comets, Asteroids, Meteorites: Interrelations, Evolution and Origins*, ed. A. H. Delsemme pp. 3–14. Toledo: University of Toledo Press.

Donn, B., Mumma, M., Jackson, W., A'Hearn, M., and Harrington, R. (eds.) 1976. *The Study of Comets.* NASA SP-393. Washington, DC: GPO.

Hills, J. G. 1981. Comet showers and the steady-state infall of comets from the Oort cloud. *Astron. J. 86*: 1730–40.

Lewis, J. S. 1972a. Metal/silicate fractionation in the solar system. *Earth Planet. Sci. Lett. 15*: 286–90.
1972b. Low temperature condensation from the solar nebula. *Icarus 16*: 241–52.

Luu, J. X. and Jewitt, D. C. 2002. Kuiper belt objects: Relics from the accretion disk of the sun. *Ann. Rev. Astron. Astrophys. 40*: 63–101.

Ovenden, M. W. 1975. Bode's Law – Truth or consequences. *Vistas in Astronomy 18*: 473–96.

Whipple, F. L. 1950a. A comet model. I. The acceleration of comet Encke. *Astrophys. J. 111*: 375–94.
1950b. A comet model. II. Physical relations for comets and meteorites. *Astrophys. J. 113*: 464–74.
1955. A comet model. III. The zodiacal light. *Astrophys. J. 121*: 750–70.

Web pages

Jewitt maintains a web page where you can find links to lists of KBOs, and an up to date and thorough bibliography, as well as other data: www.ifa.hawaii.edu/~jewitt.

Chapter 10

Alvarez, L. W. 1987. Mass extinctions caused by large bolide impacts. *Physics Today*, July, 24–33.

Alvarez, L. W., Alvarez, W., Asaro, F., and Michel, H. V. 1980. Extraterrestrial cause for the Cretaceous–Tertiary extinction. *Science 208*: 1095–108.

Binzel, R. P. 2000. The Torino impact hazard scale. *Planet. Space Sci. 48*: 297–303.

Blake, D. F., and Jenniskens, P. 2001. The ice of life. *Sci. Am.*, August, 45–50.

Ceplecha, Z., Borovička, J., Elford, W. G., Revelle, D. O., Hawkes, R. L., Porubčan, V., and Simek, M. 1998. Meteor phenomena and bodies. *Space Sci. Rev. 84*: 327–471.

Cooper, B. L., Zook, H. A., and Potter, A. E. 1996. Clementine photographs of the inner zodiacal light. In *Physics, Chemistry, and Dynamics of Interplanetary Dust*, ed. B. Å. S. Gustafson and M. S. Hanner, pp. 333–6. *ASP Conference Series 104*.

Delsemme, A. H. 2000. Cometary origin of the biosphere. *Icarus 146*: 313–25.

Delsemme, A. H. 2001. An argument for the cometary origin of the biosphere. *Am. Sci. 89*: 432–42.

Deming, D., and Harrington, J. 2001. Models of the SL9 impacts II. Radiative–hydrodynamic modeling of the plume splashback. *Astrophys. J. 561*: 468–80.

Ehrenfreund, P., and Charnley, S. B. 2000. Organic molecules in the interstellar medium, comets, and meteorites. *Ann. Rev. Astron. Astrophys. 38*: 427–83.

Frank, L. A., and Sigwarth, J. B. 1997. Simultaneous observations of transient decreases of earth's far-ultraviolet dayglow with two cameras. *Geophys. Res. Lett. 24*: 2427–30.

Frank, L. A., Sigwarth, J. B., and Craven, J. D. 1986a. On the influx of small comets into the Earth's upper atmosphere, I, Observations. *Geophys. Res. Lett. 13*: 303–6.

Frank, L. A., Sigwarth, J. B., and Craven, J. D. 1986b. On the influx of small comets into the Earth's upper atmosphere, II, Interpretation. *Geophys. Res. Lett. 13*: 307–10.

Gehrels, T. (ed.) 1994. *Hazards Due to Comets and Asteroids*. Tucson: University of Arizona Press.

Gustafson, B. Å. S., and Hanner, M. S. (eds.) 1996. Physics, chemistry, and dynamics of interplanetary dust. *Astron. Soc. Pacific Conf. Series 104*.

Hoyle, F., and Wickramasinghe, C. 1985. *Living Comets*. Cardiff: University College Cardiff Press.

Hughes, D. W. 2000. A new approach to the calculation of the cratering rate of the Earth over the last 125 ± 20 Myr. *Mon. Not. R. Astron. Soc. 317*: 429–37.

Krinov, E. L. 1963. The Tunguska and Sikhote-Alin Meteorites. In *The Moon, Meteorites, and Comets*, ed. B. M. Middlehurst and G. P. Kuiper, pp. 208–34. Chicago: Univ. of Chicago Press.

Lewis, J. S. 1996. *Rain of Iron and Ice*. Reading, MA: Addison-Wesley.

Lewis, J. S. 1997. *Physics and Chemistry of the Solar System*. San Diego: Academic Press.

Littmann, M. 1999. *The Heavens on Fire*. Cambridge: Cambridge University Press.

Meadows, A. J. 1987. Earth-based observations of Comet Halley dust and gas. *Phil. Trans. R. Soc. Lond. A 323*: 369–79.

Morbidelli, A., Chambers, J., Lunine, J. I., Petit, J. M., Robert, F., Valsecchi, G. B., and Cyr, K. E. 2000. Source regions and timescales for the delivery of water to the Earth, *Meteoritics & Planet. Sci. 35*: 1309–20.

Morrison, D. (ed.) 1992. *The Spaceguard Survey. Report of the NASA International Near-Earth-Object Detection Workshop*. Pasadena: NASA/Jet Propulsion Laboratory.

Noll, K. S., Weaver, H. A., and Feldman, P. D. 1996. *The Collision of Comet Shoemaker–Levy 9 and Jupiter. Space Telescope Science Institute Symposium Series, No. 9*. Cambridge: Cambridge University Press.

Porter, J. G. 1952. *Comets and Meteor Streams*. London: Chapman & Hall.

Sekanina, Z. 1998. Evidence for asteroidal origin of the Tunguska object. *Planet. Space Sci. 46*: 191–204.

Spencer, J. R., and Mitton, J. 1995. *The Great Comet Crash*. Cambridge: Cambridge University Press.

Thomas, P. J., Chyba, C. F., and McKay C. P. (eds.) 1996. *Comets and the Origin and Evolution of Life*. New York: Springer.

Toon, O. B., Zahnle, K., Turco, R. P., and Covey, C. 1994. Environmental Perturbations Caused by Impacts. In *Hazards Due to Comets and Asteroids*. ed. T. Gehrels, pp. 791–826. Tucson: University of Arizona Press.

Verschuur, G. L. 1996. *Impact! The Threat of Comets and Asteroids*. New York: Oxford University Press.

Wickramasinghe, D. T., and Allen, D. A. 1986. Discovery of organic grains in comet Halley. *Nature 323*: 45–6.
Wyatt, S. P., Jr., and Whipple, F. L. 1950. The Poynting–Robertson effect on meteor orbits. *Astrophys. J. 111*: 134–41.

Additional readings

A'Hearn, M. F. 1997. The impacts of D/Shoemaker–Levy 9 and bioastronomy. In *Astronomical and Biochemical Origins and the Search for Life in the Universe*, ed. C. B. Cosmovici, S. Bowyer, and D. Wertheimer, pp. 165–78. IAU Colloquium 161.
McKinley, D. W. R. 1961. *Meteor Science and Engineering*. New York: McGraw-Hill.
Melosh, H. J. 1989. *Impact Cratering*. New York: Oxford University Press.
Rickman, H., and Valtonen, M. J. 1996. *Worlds in Interaction: Small Bodies and Planets of the Solar System*. Dordrecht: Kluwer.

Chapter 11

Ball, A. J., Keller, H. U., and Schulz, R. 1999. Critical questions and future measurements: Collated views of the workshop participants. *Space Sci. Rev. 90*: 363–9.
Belton, M. J. S., and A'Hearn, M. F. 1999. Deep sub-surface exploration of cometary nuclei. *Adv. Space Res. 24*: 1167–73.
Campins, H., Larson, S. M., Weaver, H. A., and Brandt, J. C. 2001. Comets with the LBT. In *Science with the Large Binocular Telescope (LBT)*. ed. T. Herbst, pp. 151–5. Heidelberg: Max-Planck-Institute für Aeronomie.
Saha, P., and Tremaine, S. 1992. Symplectic integrators for solar System dynamics. *Astron. J. 104*: 1633–40.
Schilling, G. 2002. Comet chasers get serious. *Science 297*: 44–5.
Schulz, R., and Schwehm, G. 1999. Coma composition and evolution of Rosetta target comet 46P/Wirtanen. *Space Sci. Rev. 90*: 321–8.
Schwehm, G., and Schulz, R. 1999. Rosetta goes to comet Wirtanen. *Space Sci. Rev. 90*: 313–19.
[*Scientific American*] 2002. Last chance for the last planet. *Sci. Am. 59*: 6.
Stern, S. A. 2002. Journey to the farthest planet. *Sci. Am. 59*: 56–63.

Websites

Because the space missions are ongoing projects, often the best source of information is their websites. While these sites are sometimes not permanent and may not be available indefinitely, a search engine such as GOOGLE should turn up the current version.

http://stardust.jpl.nasa.gov/overview/
http://www.contour2002.org/
http://www.ss.astro.umd.edu/deepimpact/
http://sci.esa.int/content/doc/e7/2279_htm

Chapter 12

Durant, W. 1954. *The Story of Civilization. I. Our Oriental Heritage.* New York: Simon & Schuster.

Goodavage, J. F. 1973. *The Comet Kohoutek.* New York: Pinnacle Books.

Haney, H. G. 1965. Comets: A Chapter in Science and Superstition in Three Golden Ages: The Aristotelian, the Newtonian, and the Thermonuclear. Thesis, University of Alabama. Ann Arbor, MI: University Microfilms.

Niven, L., and Pournelle, J. 1977. *Lucifer's Hammer.* New York: Fawcett-Crest.

Olson, R. J. M. 1979. Giotto's portrait of Halley's Comet. *Sci. Am. 240*: 160–70. (Discusses art and comets.)

   1985. *Fire and Ice: A History of Comets in Art.* New York: Walker and Company; published for Smithsonian Institution, National Air and Space Museum.

Olson, R. J. M., and Pasachoff, J. M. 1987. New Information on comet P/Halley as depicted by Giotto di Bondone and other Western artists. *Astron. Astophys. 187*: 1–11.

Ridpath, I. 1977a. The comet that hit the earth. *Mercury 6*: 2–5.

   1977b. Tunguska the final answer. *New Scientist 75*: 346–7.

Yeomans, D. K. 1991. *Comets: A Chronological History of Observation, Science, Myth, and Folklore.* New York: John Wiley and Sons.

Additional reading

Freitag, R. S. 1984. *Halley's Comet: A Bibliography.* Washington, DC: Library of Congress.

Hellman, C. D. 1944. *The Comet of 1577: Its Place in the History of Astronomy.* New York: AMS Press.

Hoyle, F., and Wickramasinghe, N. C. 1978a. *Lifecloud: The Origin of Life in the Universe.* New York: Harper & Row. (Discusses comets and their connection with the origin of life.)

   1978b. Influenza from space. *New Scientist 79*: 946–8.

Krinov, E. L. 1963. The Tunguska and Sikhote-Alin meteorites. In *The Moon, Meteorites and Comets*, ed. B. M. Middlehurst and G. P. Kuiper, pp. 208–34. Chicago: University of Chicago Press.

General works

Alfvén, H., and Arrhenius, G. 1976. *Evolution of the Solar System.* NASA SP-345. Washington, DC: GPO.

Arrhenius, G., and Alfvén, H. 1973. *Asteroid and Comet Exploration.* NASA CR-2991. Washington, DC: GPO.

Bailey, M. E., Clube, S. V. M., and Napier, W. M. 1990. *The Origin of Comets.* New York: Pergamon Press.

Bobrovnikoff, N. T. 1931. Halley's comet in its apparition of 1909–1911. *Pub. Lick. Observatory XVII*: 309–482.

Boehnhardt, H., Combi, M., Kidger, M. R., and Schulz, R. 2002. *Cometary Science after Hale–Bopp*, Volumes 1 & 2, Dordrecht: Kluwer. These volumes are reprinted from *Earth, Moon, Planet.*, Vol. 89, Nos. 1–4 and Vol. 90, Nos. 1–4.

Brandt, J. C. (ed.) 1981. *Comets: Readings from Scientific American.* San Francisco: W. H. Freeman and Co.

Brandt, J. C., and Chapman, R. D. 1981. *Introduction to Comets*. Cambridge: Cambridge University Press.
  1992. *Rendezvous in Space: The Science of Comets*. New York: W. H. Freeman.
Brandt, J. C., Niedner, M. B., Jr., and Rahe, J. 1992.*The International Halley Watch Atlas of Large-Scale Phenomena*. Boulder, CO: LASP-University of Colorado.
  1997. *The International Halley Watch Atlas of Large-Scale Phenomena: Supplement 1997*. Boulder, CO: LASP-University of Colorado.
Burnham, R. 2000. *Great Comets*. Cambridge: Cambridge University Press.
Chapman, R. D., and Brandt, J. C. 1984. *The Comet Book*. Boston: Jones & Bartlett.
Chebotarev, G. A., Kazimirchaka-Polonskaya, E. I., and Marsden, B. G. (eds.) 1972. *The Motion, Evolution of Orbits and Origins of Comets*. IAU Symposium No. 45. Dordrecht: D. Reidel.
Crovisier, J., and Encrenaz, T. 2000. *Comet Science*. Cambridge: Cambridge University Press.
Delsemme, A. H. (ed.). 1977. *Comets, Asteroids, Meteorites: Interrelations, Evolution and Origins*. Toledo: University of Toledo Press.
Delsemme, A. H. 1998. Recollections of a cometary scientist. *Planet. Space. Sci. 46*: 111–24.
Delsemme, A. 2000. *Our Cosmic Origins*. Cambridge: Cambridge University Press.
Donn, B., Mumma, M., Jackson, W., A'Hearn, M., and Harrington, R. (eds.). 1976. *The Study of Comets*. (Two parts). NASA SP-393. Washington, DC: GPO.
Donn, B., Rahe, J., and Brandt, J. C. 1986. *Atlas of Comet Halley 1910 II*, Washington, DC: NASA SP-488.
Festou, M. C., Rickman, H., and West, R. M. 1993. Comets. I. Concepts and observations. *Astron. Astrophys. Rev. 4*: 363–447.
  1993. Comets. II. Models, evolution, origin, and outlook. *Astron. Astrophys. Rev. 5*: 37–163.
Gary, G. A. (ed.). 1975. *Comet Kohoutek*. NASA SP-355. Washington, DC: GPO.
Glass, B. P. 1982. *Introduction to Planetary Geology*. Cambridge: Cambridge University Press.
Grewing, M., Praderie, F., and Reinhard, R. (eds.) 1988. *Exploration of Halley's Comet*. Berlin: Springer-Verlag.
Huebner, W. F. (ed.) 1990. *Physics and Chemistry of Comets*. Berlin: Springer-Verlag.
Kuiper, G. P., and Roemer, E. (eds.) 1972. *Proceedings of the Tucson Comet Conference*. Tucson: University of Arizona Press.
Lewis, J. S. 1997. *Physics and Chemistry of the Solar System*. San Diego: Academic Press.
Lyttleton, R. A. 1953. *The Comets and their Origin*. Cambridge: Cambridge University Press.
Maran, S. P., and Hobbs, R. W. (eds.) 1974. Special issue on comet Kohoutek. *Icarus 23*: No. 4 (December).
Marsden, B. G. 1974. Comets. *Ann. Rev. Astron. Astrophys. 12*: 1–21.
Mason, J. W. (ed.) 1990. *Comet Halley: Investigations, Results, Interpretations*. (Two volumes). New York: Ellis Horwood.
Mendis, D. A., Houpis, H. L. F., and Marconi, M. L. 1985. The physics of comets. *Fundamentals of Cosmic Physics 10*: 1–380.
Newburn, R. L., Jr., Neugebauer, M., and Rahe, J. (eds) 1991. *Comets in the Post-Halley Era*. (Two volumes). Dordrecht: Kluwer Academic Publishers.
Oliver, C. P. 1930. *Comets*. Baltimore: Williams & Wilkins.
Proctor, M., and Crommelin, A. C. D. 1937. *Comets*. London: The Technical Press.

Rahe, J., Donn, B., and Wurm, K. 1969. *Atlas of Cometary Forms*. NASA SP-198. Washington, DC: GPO.

Richter, N. B. 1963. *The Nature of Comets*. New York: Dover. (Contains a thorough bibliography of pre-1960 comet literature.)

Schaaf, F. 1997. *Comet of the Century*. New York: Copernicus.

Schechner, S. J. 1997. *Comets, Popular Culture, and the Birth of Modern Cosmology*. Princeton: Princeton University Press.

Sekanina, Z., and Fry, L. 1991. *The Comet Halley Archive: Summary Volume*. NASA–Jet Propulsion Laboratory: JPL 400–450 8/91.

Vsekhsvyatskij, S. K. 1958, 1966, 1967. Fizicheskie Karakteristiki Komet. Moscow: Nauka. (Three Russian volumes describing physical characteristics of comets.)

Whipple, F. L., and Huebner, W. F. 1976. Physical processes in comets. *Ann. Rev. Astron. Astrophys. 14*: 143–72.

Wilkening, L. L. (ed.). 1982. *Comets*. Tucson: University of Arizona Press.

Yeomans, D. K. 1991. *Comets: A Chronological History of Observation, Science, Myth, and Folklore*. New York: John Wiley and Sons.

Additional suggested reading

Alter, D. 1956. Comets and people. *Griffith Observer 20*: 74–82.

Ananthakrishnan, S., Bhandari, S. M., and Rao, A. P. 1975. Occultation of radio source PKS 2025-15 by Comet Kohoutek (1973f). *Astrophys. Space Sci. 37*: 275–82.

Arpigny, C. 1965. Spectra of comets and their interpretation. *Ann. Rev. Astron. Astrophys. 3*: 351–76.

Arpigny, C. 1977. On the nature of comets. *Proceedings of the Robert A. Welch Foundation Conferences on Chemical Research XXI, Cosmochemistry*. Houston.

Baxter, J., and Atkins, T. 1976. *The Fire Came By*. Garden City, NY: Doubleday.

Bobrovnikoff, N. T. 1941. Investigations of the brightness of comets. *Popular Astronomy 49*: 467–79.

Bond, G. P. 1862. *Account of the Great Comet of 1858*. Cambridge, MA: Welch, Bigelow, and Co.

Brown, J. C., and Hughes, D. W. 1977. Tunguska's comet and non-thermal ^{14}C production in the atmosphere. *Nature 268*: 512–14.

Brown, P. L. 1974. *Comets, Meteorites and Man*. New York: Taplinger.

Chambers, G. F. 1909. *The Story of Comets*. Oxford: Clarendon Press.

Chang, S. 1979. Comets: cosmic connections with carbonaceous meteorites, interstellar molecules and the origins of life. In *Space Missions to Comets*, ed. M. Neugebauer, D. K. Yeomans, J. C. Brandt, and R. W. Hobbs, pp. 59–111. NASA CP-2089. Washington, DC: GPO.

Chebotarev, G. A., Kazimirchaka-Polonskaya, E. L, and Marsden, B. G. (eds.) 1972. *The Motion, Evolution of Orbits and Origins of Comets*. IAU Symposium No. 45. Dordrecht: D. Reidel.

Cherednichenko, V. I. 1974. On some new sources of formation of molecules observed in cometary atmospheres. *Problems of Cosmic Physics 9*: 155–58. (In Russian.)

Fesenkov, V. G. 1966. *A Study of the Tunguska Meteorite Fall*. Soviet Astronomy – A. J. *10*: 195–213.

French, B. M. 1978. Comet or spacecraft? *Secondlook*, December, pp. 1–3. (On the Tunguska event.)

Gombosi, T. I. (ed.) 1983. *Cometary Exploration* (Three volumes). Budapest: Hungarian Academy of Sciences.

Guillemin, A. 1877. *The World of Comets*. London: Sampson Low, Marston, Searle, & Rivington.

Oppenheimer, M. 1975. Gas phase chemistry in comets. *Astrophys. J. 196*: 251–59.

Ovenden, M. W. 1975. Bode's Law – Truth or consequences. *Vistas in Astronomy 18*: 473–96.

Porter, J. G. 1952. *Comets and Meteor Streams*. London: Chapman & Hall.

Potter, A. E., and Del Duca, B. 1964. Lifetime in space of possible parent molecules of cometary radicals. *Icarus 3*: 103–8.

Rahe, J. 1980. Ultraviolet spectroscopy of comets. In *Second European IUE Conference*, ed. B. Battrick and J. Mort. ESA SP-157. Report of the Comet Science Working Group, Executive Summary. August 1979. NASA TM 80542.

General index

Index of comets